recent advances in phytochemistry

volume 10

Biochemical Interaction
Between Plants and Insects

RECENT ADVANCES IN PHYTOCHEMISTRY

Recent Volumes in the Series

A Continuation Order Plan is available for this series. A continuation order will bring delivery of each new volume immediately upon publication. Volumes are billed only upon actual shipment. For further information please contact the publisher.

recent advances in phytochemistry

volume 10

Biochemical Interaction Between Plants and Insects

Edited by

James W. Wallace
Western Carolina University
Cullowhee, North Carolina

and

Richard L. Mansell
University of South Florida
Tampa, Florida

PLENUM PRESS • NEW YORK AND LONDON

Library of Congress Cataloging in Publication Data

Phytochemical Society of North America.
  Biochemical interaction between plants and insects.

  (Recent advances in phytochemistry; v. 10)
  Includes index.
  1. Botanical chemistry—Congresses. 2. Insect-plant relationships—Congresses.
I. Wallace, James W., 1940-   II. Mansell, Richard L. III. Title. IV. Series.
QK861.R38 vol. 10   582'.13'045   76-13833
ISBN 0-306-34710-5

Proceedings of the Fifteenth Annual Meeting of the Phytochemical
Society of North America held on the campus of the
University of South Florida, Tampa, Florida, August, 1975

© 1976 Plenum Press, New York
A Division of Plenum Publishing Corporation
227 West 17th Street, New York, N.Y. 10011

This volume is dedicated to Professor Kurt Mothes, who pioneered the study of the formation of secondary metabolites and their role in the life of the organisms producing them.

Contributors

S.D. Beck, Department of Entomology, University of Wisconsin, Madison, Wisconsin 53706.

L.P. Brower, Department of Biology, Amherst College, Amherst, Massachusetts 01002.

R. G. Cates, Department of Biology, University of New Mexico, Albuquerque, New Mexico 87131.

Paul Feeny, Department of Entomology and Section of Ecology and Systematics, Cornell University, Ithaca, New York 14850.

C.J. Fix, Department of Chemistry, Pennsylvania State University, University Park, Pennsylvania 16802.

R.C. Gueldner, Boll Weevil Research Laboratory, Agricultural Research Service, USDA, Mississippi State, Mississippi 39762.

P.A. Hedin, Boll Weevil Research Laboratory, Agricultural Research Service, USDA, Mississippi State, Mississippi 39762.

L.B. Hendry, Department of Chemistry, Pennsylvania State University, University Park, Pennsylvania 16802.

D.M. Hindenlang, Department of Chemistry, Pennsylvania State University, University Park, Pennsylvania 16802.

S.H. Korzeniowski, Department of Chemistry, Pennsylvania State University, University Park, Pennsylvania 16802.

J.G. Kostelc, Department of Chemistry, Pennsylvania State University, University Park, Pennsylvania 16802.

D.A. Levin, Department of Botany, University of Texas, Austin, Texas 78712.

C.M. Moffitt, Department of Biology, Amherst College, Amherst, Massachusetts 01002.

Kurt Mothes, Deutsche Akademie der Naturforscher Leopoldina, August- Bebel- Str. 50 a, Halle (Saale), German Democratic Republic.

J.C. Reese, Department of Entomology, University of Wisconsin, Madison, Wisconsin 53706.

D.F. Rhoades, Department of Zoology, University of Washington, Seattle, Washington 98195.

Eloy Rodriquez, Department of Botany, University of British Columbia, Vancouver, B.C., Canada.

C.N. Roeske, Department of Environmental Toxicology, University of California, Davis, California 95616.

J.N. Seiber, Department of Environmental Toxicology, University of California 95616.

A.C. Thompson, Boll Weevil Research Laboratory, Agricultural Research Service, USDA, Mississippi State, Mississippi 39762.

J.K. Wichmann, Department of Chemistry, Pennsylvania State University, University Park, Pennsylvania 16802.

Preface

Botanists and zoologists have recognized for centuries
the specificity of various insects for plants, and entomologists have long been aware that insects defend themselves
from predators by emitting repulsive odors. Only recently
have chemists and biologists established a joint endeavor
for studying the chemical relationships between plants and
insects. The present symposium volume of the Phytochemical
Society of North America's RECENT ADVANCES IN PHYTOCHEMISTRY
consists of eight papers dealing with phytochemical relationships between plants and their insect herbivores. The
fifteenth P.S.N.A. annual symposium and meeting was held
in August, 1975, on the campus of The University of South
Florida, Tampa.

The chemical defenses of apparent and unapparent plants
were contrasted by Feeny. Rodreguiz and Levin illustrated
parallel defense mechanisms of plants and insects and then
Hendry, Kostelc, Hindenlang, Wichmann, Fix and Koreniowski
discussed chemical messengers for both plants and insects.
Subsequently Beck and Reese reviewed plant contributions to
insect nutrition and metabolism. Indepth studies for the
monarch butterfly-milkweed interaction were presented by
Roeske, Seiber, Brower, and Moffitt and for the cotton boll
weevil-cotton plant relationship by Hedin, Thompson, and
Gueldner. In the latter portion of the symposium Rhoades
and Cates presented a general theory concerning the coevolution of insects and plant antiherbivore chemistry.

The symposium was concluded with an overview of the
function of secondary plant chemistry based on over a half
century of research by Kurt Mothes. Mothes, as well as
other authors, expressed the contemporary concern that
secondary products should not be conceived of as waste
products but as essential compounds for the existence of
organisms in an ever changing environment. Some authors
even felt that the term "secondary substances or compounds"
should be dropped from the literature because of the biological importance of these compounds and the misconception that

the term "secondary" implies.

After discussions with some of the symposium authors
it became apparent that a uniform terminology was needed
for the discipline of biochemical interactions between
organisms.

I wish to give a special word of appreciation to
Mrs. Kandy Tucker who contributed much of her time to
the typing of this manuscript.

James W. Wallace, Jr.
February, 1976

Contents

Chapter One

PLANT APPARENCY AND CHEMICAL DEFENSE

PAUL FEENY

*Department of Entomology & Section of Ecology
  & Systematics
Cornell University, Ithaca, New York*

INTRODUCTION

A major objective of insect ecology is to explain observed patterns of interaction between plants and herbivorous insects. We would like to understand both how such patterns are maintained in ecological time and also how they have come about in evolutionary time. A test of how far such understanding has progressed will be our ability to predict how patterns vary from one kind of community to another and how they will change when subjected to natural or human disturbance.

A considerable step forward along the path towards understanding has followed from the realization that different kinds of insects (as well as other groups of organisms) respond differentially to various secondary chemical compounds (natural products) occurring in plants, many of these compounds also acting as toxins to animals, microorganisms and fungi[14,26,76]. Such observations have led to a general theory of coevolution between insects and plants[4,15,17], according to which some of the compounds present in plants were evolved

or elaborated in response to attack by insects. Some of the
associated insects evolved methods of tolerating the new
plant chemicals and were thus able to remain associated with
their particular host species. As plants evolved, some of
their associated insects coevolved with them, often leaving
relatives on the original plant families or genera. One man's
meat became another man's poison and present day insects are
adapted to tolerate only a certain range of chemicals and
therefore only a certain range of plants. Evolving along
with detoxication mechanisms, chemosensory systems responding
differentially to secondary compounds enable insects to
locate and identify the plants to which they are adapted.
Plants and insects thus seem to be engaged in an 'evolutionary arms race': plants must deploy a fraction of their
metabolic budgets on various kinds of defense and insects
must devote a portion of their assimilated energy and
nutrients on various devices for host location and attack[78].

     Plant species, or at least populations, may thus be
regarded as chemically defended "islands ", subject to
colonization by insect populations and species in evolutionary time[7,36]. For an insect to colonize a chemically "new"
food-plant species without drastic reduction in fitness,
chemical adaptation is required at both the toxicological and
behavioral levels[7,15,22]. Evolutionary changes in the range
of host-plants exploited by a particular insect species or
related group of species are therefore most likely to occur
among plant taxa which share similar secondary chemistry[17].
In several observed cases of genetic host race formation,
such as the colonization by the apple maggot fly,
Rhagoletis pomonella, of apple from hawthorn and of cherry
from apple[6,7], the transfer has occurred among plants which
are closely related taxonomically and therefore, probably,
chemically[66].

     Chemical coevolution between plants and phytophagous
insects (or herbivores in general) is sometimes discussed as
if other groups of organisms can be ignored. However,
secondary compounds in plants are known to have both toxicological and behavioral effects on many groups of organisms,
including not only herbivorous insects, mammals, reptiles and
molluscs but also nematodes, viruses, bacteria and fungi;
some plants contain allelopathic compounds which inhibit the
growth of competing plants[55,78]. The chemical defenses of

most plant species are likely to be a consequence of 'coevolution' with a variety of predators, parasites, pathogens
and competitors. Thus while the role of plants in shaping
chemical counter-adaptation by insects is often clear, the
effects of insects on the evolution of plant defenses must
be seen against a broader coevolutionary background.

The theory of chemical coevolution clearly provides
a partial explanation of observed patterns of interaction
between insects and plants. It even permits qualitative
predictions about natural communities; for example the
presence of certain species of insects in a particular
habitat may indicate the presence in that habitat of certain
species, genera or families of plants[25]. As it has been
formulated until recently, however, the theory is strictly
qualitative, setting likely evolutionary bounds on the distribution of insect species. It tells us little about
ecological aspects of interactions between insects and
plants.

During the past decade research on several herbivore-
plant associations has led to the conclusion that the
chemical defenses of plants may have more far-reaching
effects in natural communities than was earlier anticipated.
Such research has already led to some tentative generaliza-
tions about how the chemical defenses of plants against
herbivores and pathogens may vary as a function of plant
persistence and abundance and how, in turn, such variation
might be expected to affect the ecology of herbivores and
their food-plants in different plant communities[8,9,22,28,39,
40,56,57]. Here I would like to elaborate on some of these
ideas (see also Rhoades and Cates, this volume).

THE CONCEPT OF PLANT APPARENCY

*Contrast between ecological effects of tannins and
glucosinolates*. As a result of research during the past
twelve years on patterns of interaction between herbivorous
insects and oak trees, *Quercus robur*, in England and various
crucifers in central New York State, I have become convinced
that there exist major differences between the chemical
defenses of these two groups of plants. Tannins, which appear
to represent the major chemical defense of mature oak
leaves[21], are present in relatively large amounts in the

leaves, up to at least 5% of leaf dry weight[23] and perhaps
higher (E. C. Bate-Smith, personal communication). Their
inhibitory effect on caterpillars of the winter moth,
Operophtera brumata, is "dosage-dependent": the greater is
the concentration of tannins, the greater is the inhibition
of larval growth[19]. The mode of action of tannins, which
reduce the availability of food nitrogen by forming relatively
indigestible complexes with leaf proteins, is such that the
potential for biochemical counteradaptation by insects which
ingest tannins may be limited[20]. Caterpillars of Lepidoptera
species which naturally feed on mature oak leaves are known
to grow slowly unless they are leaf-mining species which
apparently avoid some of the plant's defenses[21].

Glucosinolates, which seem to represent the primary
chemical defense of crucifers[78], seem, by contrast, to be
present in smaller amounts in natural foliage: the allylglucosinolate content of wild *Brassica nigra* leaves in
Tompkins County, N.Y., for example, was found to range from
0.18-0.66% of leaf fresh weight, depending on habitat and
leaf age (P. Feeny and L. Contardo, unpublished results).
At least against some non-adapted insects, these amounts are
sufficient to cause death or at least drastic reduction in
probable fitness[18,47]. Variation in glucosinolate content
does not, however, seem to affect greatly the growth rate of
adapted insects such as the larvae of the cabbage butterfly,
Pieris rapae (Slansky and Feeny, in preparation). Also,
larval growth rates of *P. rapae* and other Lepidoptera species
which naturally feed on crucifers are fast, relative to those
of larvae feeding on mature oak leaves.

Based on these findings, I think that two kinds of
chemical defense can be distinguished. The glucosinolates
seem to be effective, even in small concentrations, against
non-adapted insect species but are evidently susceptible
to rapid detoxication by adapted insects. They seem to act,
therefore, as "qualitative" barriers which, when countered
by specific adaptation, have little effect on the "variable"
(phenotypic) costs of insect growth[22]. The tannins, by
contrast, seem to act as "quantitative" (i.e. dosage-
dependent) barriers, even to insects which naturally feed
on leaves containing them. By increasing the "variable"
costs of growth, tannins have the effect of reducing growth
rates and hence, presumably, fitness of a wide range of
insects which ingest them[22].

   Definition of apparency. A reasonable explanation for
the ecological contrast between the tannins and glucosinolates
follows from a consideration of how vulnerable oaks and
crucifers are to discovery by their various enemies in ecological time. Most crucifers are ephemeral plants, characteristic of early stages of community succession, and are likely
to be relatively "hard to find" by their adapted enemies,
against which they seem to rely primarily on escape in time
and space[56]. Selection in such species seems to have favored
fast growth to early maturity, high reproductive output,
and ability to disperse and colonize new areas rapidly. Such
plants might not be able to "afford" the metabolic cost of
high concentrations of defensive compounds, since this would
be likely to reduce metabolic allocation for growth and
reproduction and, perhaps also, competitive ability against
other plants of both their own and other species. Oaks,
by contrast, are climax forest dominants and are "bound to
be found" by their enemies in ecological time; predation
pressure on undefended plants would probably be severe. In
such plants allocation of the energy and nutrients required
for quantitative defense is more likely to bring commensurate
increases in fitness[22].

   The susceptibility of an individual plant to discovery
by its enemies may be influenced not only by its size, growth
form and persistence, but also by the relative abundance of
its species within the overall community. To denote the
interaction of abundance, persistence and other plant
characteristics which influence likelihood of discovery, I
now prefer to describe "bound to be found" plants by the
more convenient term "apparent", meaning "visible, plainly
seen, conspicuous, palpable, obvious" (Shorter Oxford English
Dictionary, 3rd, edition; Webster's Concise English Dictionary).
Plants which are "hard to find" by their enemies will be
referred to as "unapparent", the antonym of apparent (O.E.D.
and Webster, loc. cit.). The vulnerability of an individual
plant to discovery by its enemies may then be referred to
as its "apparency", meaning "the quality of being apparent;
visibility" (O.E.D. and Webster, loc. cit.). Since animals,
fungi and pathogens may use means other than vision to locate
their host-plants, I shall consider apparency to mean
"susceptibility to discovery" by whatever means enemies may
employ.

Within any one population of a particular plant species, each individual plant is "discovered" by enemies at a certain "discovery rate" during any specified time period. This discovery rate is likely to vary substantially from one individual to another within the population and cannot be predicted precisely in advance. The average discovery rate for all the individuals in the population during the time period is, however, likely to be less variable from one generation to another. This average discovery rate for the population reflects an average per capita "population apparency", which is also likely to be less variable from generation to generation than are the individual apparency values within a population. An analogy may be made with population mortality phenomena: though the age of death of particular individuals is often highly variable and not predictable with precision, the average age of death for all individuals in the population may vary little from generation to generation and corresponds to an average life expectancy for the component individuals. Just as average life expectancy values may vary among component sub-groups within a population, so also average apparency values of different sub-groups within a plant population are likely to differ as a function of clump size, microhabitat and other variables.

Apparencies and the discovery rates which they predict may refer to any specified time period. Of most biological interest are the apparencies and discovery rates of individuals over a single generation period. Different populations within a species may then be compared in terms of the average "generation apparencies" of their individuals. The average generation apparency characteristic of the species as a whole has the additional value of permitting comparison with different species which may differ widely in their persistence and generation time. In this paper, when comparing the apparencies of different plant species in relation to chemical defense, I shall be referring to this "species generation apparency".

The apparency of an individual plant to its enemies is determined both by its genotype, reflected in various adaptations such as growth form and secondary chemistry, and also by various environmental influences which act on the phenotype. It is also dependent upon various character-istics of the microenvironment and of the community as a

whole. Such characteristics include the nature of neighboring
plants, the population density of the plant's own species and
the species, numbers, and host-finding adaptations of all
relevant herbivores and pathogens in the community. Though
it is possible to discuss generation apparencies for the
individual plants of any population, the apparency of any
individual over shorter time periods during its lifespan
may vary considerably depending on age, size, seasonal
changes and other variations in the plant or its environment.
There will also be variation between the component apparencies
of individual plant parts such as roots, shoots, and leaves,
each of which is liable to discovery by differing arrays of
enemy species. The average generation discovery rate of the
individual plant represents the cumulative discovery rates of
the component plant parts by their various enemies over
one generation period.

The discovery rate experienced by an individual plant
must bear some relationship to the consequent reduction of
reproductive output and fitness. In some predator-prey
interactions, such a relationship may be quite clear; once
an undefended caterpillar is discovered by a bird predator,
for example, its immediate death is almost certain. By
contrast, a brightly colored but well defended insect may
be highly apparent to bird predators, but each discovery
is likely to be followed by rejection, with little or no
effect on the insect. The impact of discovery on a plant is
usually much less clear and will depend on many factors
such as the plant's defenses, the timing of attacks relative
to the age of the plant, and the differing degrees of damage,
if any, inflicted by different enemies after discovering
the plant[38].

Though this paper is concerned chiefly with the relationship between plant apparency and defense, it is worth mention-
ing that there are occasions when it is advantageous for
particular parts of many plants to be highly apparent to
other organisms. Two prime examples are apparency of flowers
to pollinating organisms and apparency of fruits to various
agents of dispersal. In these interactions, apparency and
consequent discovery rates are, of course, related to
natality rather than to mortality characteristics of plant
populations.

Relationship between plant apparency and chemical defense. Consideration of the contrast between the chemical defenses of oaks and crucifers led to speculation that more general correlations may exist between the abundance and persistence of a plant species and the nature of its chemical defenses[22]. These ideas can be expanded and restated in the form of the following tentative hypotheses:

1) The proportion of metabolic resources allocated to chemical defense tends to be greater in plant species which are characteristically apparent than in characteristically unapparent species, at least within the same biome. The greater the apparency of a plant species, the greater the concentration of defensive compounds likely to be present in its tissues[9,28].

2) The qualitative nature of defensive compounds present in plants changes along a gradient of species apparency between the following extremes:
   a) Unapparent plants contain compounds which are effective in relatively small concentrations as toxins against non-adapted enemies, for which they may serve also as behavioral deterrents. Such compounds are likely to be susceptible to counter-adaptation, however, and may have little inhibitory effect on the growth and fitness of adapted enemies, which may exploit them as attractants or feeding stimulants.
   b) Apparent plants contain compounds which are not readily susceptible to counteradaptation and which serve to reduce the growth rate and fitness of all enemies, even though relatively high concentrations may be required to be effective. Such chemical defense is generally associated with low nutritive value for herbivores and pathogens, and with relatively tough leaves.

3) The chemical defenses of the component parts (leaves, roots, etc.) of a plant vary both among such parts and also through any generation time as a function of their varying component apparencies to different arrays of enemies, and as a function of varying component importance to plant fitness.

4) Within any one biome, at least, interspecific diversity
of defensive compounds is lower in communities characterized
by apparent plants than in those typified by unapparent
species[28].

The remainder of this paper will be devoted to a
discussion of some of the circumstantial and experimental
evidence which form the background for these hypotheses,
though it is by no means intended to be a comprehensive
review of the subject. Examples will be drawn mainly from
research with insects, since these are the organisms with
which I am most familiar. Since selection exerted by insects
is only partially responsible for the evolution of plant
defenses, however, the hypotheses have been formulated so
as to apply to plant enemies in general.

DEFENSES OF UNAPPARENT PLANTS

Introduction. Appropriate ecological studies of plants
which are rare, but have long generation times, seem to be
few. Southwood (1961, 1972), though his evidence has been
contested by Strong (1974a, 1974b), concluded that the abundance of a tree species in recent evolutionary time may be
correlated positively with the number of insect species which
now attack it. This suggests that rare plants, even if
persistent, may be less apparent also in ecological time.
Janzen (1970) has proposed that the very low relative
abundance of many tree species in various tropical habitats,
where predation pressure is uninterrupted by severe and
unpredictable fluctuations in the physical environment,
results from the efficiency of seed predators and herbivores
in finding and destroying seeds and seedlings in the neighborhood of the parent tree; only very few seeds and seedlings,
dispersed away from this neighborhood, escape such destruction. Whether or not the apparency of a climax forest tree
species can be sufficiently reduced as a result of rarity to
enable it to survive without quantitative defense in its
mature foliage, however, is uncertain. For the tropical
forest species, at least, the answer is likely to be no
(D. H. Janzen, personal communication).

Far better studied have been plants which are unapparent
as a result of being ephemeral and it is to these, primarily
the weedy species characteristic of the early stages of
plant succession, that I shall restrict my discussion. I

shall make the assumption that all such plants contain chemical defenses.

I shall also assume that the behavioral responses, such as feeding stimulation or inhibition, which an herbivorous animal may exhibit to a plant in an ecologically appropriate habitat reflect the suitability of that plant as food for that animal[9]. The assumption is, in other words, that the relationship between "proximate" behavior patterns and "ultimate" ecological food value is adaptive; "herbivores will eat what is good for them". Reasonable as this assumption may appear at first sight, especially for animals which are able to learn by experience, the precision with which such a relationship applies to herbivore species in natural habitats is uncertain.

It will be helpful to distinguish herbivores which are "generalist" feeders from those which are "specialists". It is difficult to make a sharp distinction between these two categories, but for present purposes an animal which naturally attacks plant species within only one family, or group of families which are closely related chemically, will be considered as a specialist. Animals which naturally attack plant species of several families, chemically unrelated, will be considered to be generalists. Among herbivorous insects, though the patterns of sensory responses to nutrients and secondary compounds are complex[59,60], food choice by specialists is often influenced strongly by specific chemical attractants, whereas many generalists seem to eat almost any nutritionally acceptable plant species which is not specifically repellent to them[15,63,64] (D. J. Futuyma, personal communication). Of two species of crucifer-feeding aphid, for example, the crucifer-feeding specialist *Brevicoryne brassicae* exhibits a positive feeding response to sinigrin (allylglucosinolate), whereas this compound inhibits the feeding response by the generalist species *Myzus persicae*[70]. Mammalian herbivores, ourselves included, tend mostly to be generalist feeders[27].

Secondary chemistry of early successional herbs. The diversity of secondary compounds contained in the herbaceous plants characterizing an early successional plant community is high as a result of the typical occurrence of different classes of compound in plant species of different families.

Thus the milkweeds (Asclepiadaceae) contain cardiac glycosides, the crucifers contain glucosinolates, the legumes contain cyanogenic glycosides or alkaloids, the umbellifers contain coumarins and essential oils, *Hypericum* contains the dimeric anthraquinone, hypericin, and so forth[28,29,33]. The diversity is enhanced further by considerable variation of the typical family chemistry between different genera and species within any one family. Within the Cruciferae alone, for example, not only do the different species contain different arrays of individual glucosinolates but, in addition, the genus *Iberis* contains cucurbitacins, the genera *Erysimum* and *Cheiranthus* contain cardiac glycosides, and the genus *Lunaria* contains the unusual macrocyclic 'Lunaria alkaloids'[33]. Such diversity is very well known to phytochemists and hardly needs further elaboration here.

The quantitative content of secondary compounds in the different parts of a plant, and its variation as a function of plant age and habitat, is known for very few plant populations or species. Available results suggest that the levels of compounds in the foliage of herbaceous plants are generally less than 2% of leaf fresh weight and often much less. Among the crucifers, for example, van Emden and Bashford (1969) found that the glucosinolate content, estimated as 'total allylisothiocyanate', decreased from 0.0276% (of dry weight), in young leaves of Brussels sprout plants, to 0.0148% in leaves of middle age. Rees (1969) found that the concentration of hypericin in *Hypericum hirsutum* varied from 0.004% of fresh weight in lower leaves to 0.01% in upper leaves of flowering plants but that concentrations were higher in flower heads, reaching 0.2083% of fresh weight in the sepals.

The proportion of secondary plant chemicals which are potentially toxic to animals, fungi or microorganisms is not known. However, compounds of most classes present in successional herbs are known to be toxic to at least some species in at least one of these taxonomic categories, and it is reasonable to assume that the potential toxic action of secondary chemicals is widespread[26,44,46,78]. The likely defensive function of the compounds in some plant species may also be inferred tentatively from similarity to the ecology of other plant species, the compounds of which have been shown to be toxic.

The biochemical mechanisms used by adapted enemies to detoxify and tolerate potential toxins from their food-plants are poorly known; that some such mechanisms must exist is generally inferred from the ability of such species to utilize plants known to contain compounds which are toxic to other species. For example, glucosinolates or isothiocyanates are demonstrably toxic to certain insects which do not naturally feed on crucifers[18,47] (P. Blau, unpublished results). The larvae of *Pieris rapae* and other insects which are adapted to feed on crucifers must, by implication, be able to tolerate ambient amounts of these compounds. Furthermore, when *P. rapae* larvae are fed on leaves of collards, *Brassica oleracea*, in which the allylgluco-sinolate content has been artificially raised to more than ten times the level in control leaves, no toxic effect can be detected, though some behavioral inhibition of feeding rate is evident on leaves raised to more than three times the normal glucosinolate content (P. Blau, unpublished results). We have also found that growth of *P. rapae* larvae on a wide range of crucifer species can be accounted for satis-factorily in terms of the responses of the larvae to plant nitrogen content, in spite of considerable variation in at least the qualitative pattern of glucosinolates present in the food-plants (Slansky and Feeny, in preparation). For these larvae it seem that the glucosinolates, present at normal concentrations in food-plants, represent no dosage-dependent or "quantiative" barrier to growth. By implication, these compounds seem to represent a "qualitative" barrier to at least some potential enemy species: once overcome by specific adaptation, they may have little effect on growth or fitness.

Generalist insect enemies seem to show wide variation in their degree of adaptation to available food-plants. Some plant species support "normal" growth, others support slow growth, and yet others support no growth at all[62-64]. These responses are likely to reflect, among other things, different degrees of adaptation to defensive chemicals in the food-plants. Larval growth, pupal weight and adult weight of the southern armyworm, *Spodoptera* (=*Prodenia*) *eridania*, are unaffected by incorporation of as much as 2% of the cyanogenic glycoside amygdalin in their diet[54] and they grow normally on a wide variety of plant species which are known to contain potential toxins[63,64]. Though they grow normally on collard leaves, they have less tolerance for leaves

artificially boosted with allylglucosinolate than have larvae
of *P. rapae* (P. Blau, unpublished results). It seems that
some compounds, present in typical amounts in leaves, serve
only as qualitative barriers to such generalist species,
whereas other compounds may have a dosage-dependent effect.
Interpretation of the results of feeding experiments is
complicated by the likely possibilities of both larval
conditioning to particular food-plants[42] and phenotypic
induction of detoxication enzymes (F. E. Hanson, personal
communication).

Most herbivorous mammals are evidently capable of
tolerating a wide variety of toxic compounds, provided
that these are ingested in small amounts when first encountered[27]. Biochemical mechanisms presumably exist in the
liver to detoxify these compounds but the mechanisms are
susceptible to overloading unless sufficient time is allowed
for enzyme induction following first encounter with a toxin[27].
It is likely, therefore, that the small amounts of compounds
present in herbaceous plants have little or no effect on the
growth or fitness of adapted generalist mammals, though
tissues containing large amounts of compounds are likely to
to be avoided. It is unlikely, though, that even a broadly
generalized mammalian herbivore will be adapted to tolerating
all the potential toxins in its environment. We as humans
are notoriously sensitive to even tiny amounts of cyanide;
several examples of chemical toxicity to mammals are discussed by Kingsbury (1964) and Freeland and Janzen (1974).

Escape from adapted enemies in space and time. The
capacity of herbivorous insects to influence the abundance
of herbaceous plants has long been underestimated, partly
because of the difficulty of obtaining convincing experimental
evidence; the successful biological control of several weedy
plant species, after importation of appropriate specialist
insect enemies, has now provided such evidence[79]. A classic
example of the impact of adapted insect enemies is the control
of introduced Klamath weed, *Hypericum perforatum*, in
California. In 1951 this plant was estimated to infest
2 1/3 million acres of range land in that state, covering up
to 80% of the ground area in some places[35]. Following
introduction of the leaf beetles *Chrysolina quadrigemina* and
C. hyperici, begun in 1947, Klamath weed was reduced to less
than 1% of its former abundance by 1959, largely as a result
of the enormous damage inflicted on the plants by the larvae

of these species[35]. Beetle numbers declined rapidly after
their food-plant was exhausted but both the plant and
beetles continued to persist at low densities, the plant
surviving best in shaded areas where the beetles are less
effective[35]. After discussing this example, Wilson (1964)
suggests that "it is highly doubtful if plant ecologists,
examining without knowledge of past events the present
situation of any weed that has been controlled biologically
in any country, would attribute any importance to the effects
of insects on the weed's abundance." *Hypericum* species
contain hypericin, a dimeric quinone, which is known to be
toxic to animals[44] but is evidently tolerated by the
Chrysolina species which specialize on plants of this genus.
One such beetle, *C. brunsvicensis*, possesses chemoreceptors
responding specifically to hypericin, which is exploited as
a feeding stimulant[53].

The specialist insect fauna of the Cruciferae, organized
in part by behavioral responses to the glucosinolates and
their hydrolysis products[52,78], parallels that of *Hypericum*.
During a three-year study of the entire insect fauna on plots
of collards, *Brassica oleracea*, near Ithaca, N.Y., Root (1973)
found that the specialist flea beetle *Phyllotreta cruciferae*
and aphid *Brevicoryne brassicae* consistently reached higher
densities on collard plants grown in pure stands than on
collards planted among nearby diverse meadow vegetation. To
account for this and other findings, Root (1973) proposed a
"resource concentration hypothesis" according to which
"many herbivores, especially those with a narrow host range,
are more likely to find hosts that are concentrated (i.e.
occur in dense or nearly pure stands)." The apparency of
an individual collard plant to some of its adapted enemies
is thus increased by close association with conspecific
individuals. By contrast, however, association with
conspecific individuals reduces the apparency of collard
plants to ovipositing females of *Pieris rapae*[12,57].

Adult flea beetles of *P. cruciferae* and *P. striolata*,
another crucifer-feeding species, locate their food-plants,
in part, by exploiting the mustard oil allylisothiocyanate
as a 'long distance' olfactory attractant[24]. During the
summer of 1975 we found that when this compound, dissolved
in mineral oil, was placed in vials near 2-plant 'islands'
of black mustard, *Brassica nigra*, among diverse meadow
vegetation near Ithaca, colonization of the plants by both

species of flea beetle was accelerated greatly in comparison
with control islands; much higher beetle densities were
reached and most of the experimental plants were rapidly
destroyed as a consequence (P. Feeny, J. Gaasch and L.
Contardo, unpublished results). We conclude that the apparency of natural B. *nigra* plants to these adapted specialist
enemies is influenced by the amounts of compound which the
plants release; plant individuals releasing smaller amounts
of compound can thus be presumed to benefit through reduced
apparency to these enemies.

There is little question, then, that adapted enemies
can have a great impact on the abundance of herbaceous plants
and that reduced apparency is a major factor permitting
survival of at least some such plant species. It may be
that there exists a maximum density which can be reached
by a plant species in any given region in ecological time,
corresponding to a certain level of apparency of the plants
to their adapted herbivores and pathogens present in that
region[37]. Above this level of relative abundance, the
adapted enemies will tend to reduce abundance; below the
critical level, the adapted enemies will be decreasingly
effective as determinants of plant abundance. The maximum
level of abundance would be expected to vary from one plant
species to another, in any given region in ecological time,
depending on a number of factors influencing apparency to
adapted enemies; such factors would include the number of
adapted enemy species in the region, the size and growth
form of the plant (including leaf shape[30]), its secondary
chemistry, and its ability to reproduce and disperse. For
species of sufficiently low apparency, even at high densities,
a critical level of relative abundance may never be reached
and maximum density in suitable habitats may be determined
by other factors, such as competition with other plants.
The finding that introduced species of weeds commonly
reach densities greater than those attained in areas where
they are indigenous may be explained in part by absence of
adapted enemy species in the regions of introduction[79].

Effectiveness of chemical defenses in ecological time.
Since at least some of the defensive compounds of early
successional herbs seem to be susceptible to specific
counteradaptation, their effectiveness as a defense against
any particular enemy species must be related to the degree
of counteradaptation possessed by individuals of that enemy

species. Those enemies, the individual fitness of which
would be reduced to zero by ingesting plant material containing a particular compound, may be described as non-
adapted to that compound. Enemies are unlikely to pose a
major threat, in ecological time, to plants containing
compounds to which they are not adapted, since they are
likely to possess proximate adaptations, including
behavioral responses, which ensure that only suitable plants
are attacked. It is possible, however, that non-adapted
enemies may play some role in eliminating or reducing the
fitness of those individual plants in which, for various
genotypic or phenotypic reasons, the typical defensive or
repellent compounds are absent or present in abnormally
low amounts.

It is the adapted specialist enemies which are likely
to exert the major share of predation-related impact on
herbaceous plants in ecological time[56]. If, as seems likely
from our experiments with *Pieris rapae* larvae on crucifers,
the defensive compounds have little or no effect on the fitness of such enemies, then the defenses can hardly be
considered to be effective against such species. The prime
defense of a plant against such enemies is escape in time
and space, i.e. low apparency. The presence of the compounds
in the plants may even be a liability, since it increases
the apparency of the plants to the adapted specialists, as
has been discussed earlier in connection with the attack of
Chrysolina beetles on *Hypericum* and *Phyllotreta* flea beetles
on crucifers.

What I have been suggesting here is that within broad
limits, rarely exceeded in nature, variation in the concentrations of the defensive compounds in an early successional
herb has little or no effect on the growth and fitness of
enemies which are fully adapted to that plant species. I
must emphasize, though, that the experimental evidence for
this suggestion, discussed earlier, is very limited and the
possibility of quantitative or dosage-dependent effects,
even on adapted specialists, cannot be ignored. There is
some evidence from the experiments of van Emden and Bashford
(1969) and van Emden (1972), for example, that variation in
the glucosinolate concentration may affect fecundity of
the crucifer-feeding aphid *Brevicoryne brassicae* when
feeding on Brussels sprout plants.

Whatever the effectiveness of a plant's chemical defense on those adapted specialist enemies which exploit it as their major or only food-plant, the defenses are likely to be more effective in reducing the growth and fitness of those enemies which, although specialists on the plant family or genus, usually exploit other species. At the present time, it seems reasonable to suppose that although an insect species may possess some general adaptation to tolerate an entire class of chemical compounds, such as the glucosinolates, it may be better adapted to tolerating certain compounds within this class than others. Though we were unable to attribute growth variation of *Pieris rapae* larvae, on different crucifer species, to varying qualitative glucosinolate patterns (Slansky and Feeny, in preparation), our results do not rule out possible effects on fitness which were too subtle to be revealed. Differential host-plant preferences, resulting in part from differing responses to individual glucosinolates and mustard oils, are largely responsible for the maintenance of niche separation between six species of crucifer-feeding flea beetles in central New York[34]. Though not proven, it seems very likely that the differential preferences of these beetles for different crucifer species reflect in part differing degrees of toxicological adaptation to the particular gluco-sinolates present in each species. Furthermore, though a liability in making a particular plant species more apparent to its particular specialist flea beetle species, its glucosinolate defense may be effective in ecological time by reducing the apparency of the plant to those crucifer-feeders which prefer other species.

Whatever the impact on 'family specialist' enemies of chemical variation within the prime class of defensive compounds characterizing a genus or family, there is at least circumstantial evidence to suggest that those species within the genus or family which contain defensive compounds of a different class, either instead of or in addition to compounds of the 'typical' chemical clsss for the family, may be less vulnerable to such enemies. Within the Cruciferae, for example, the genera *Iberis*, *Erysimum* and *Lunaria* all contain glucosinolates but in addition they contain, respectively, cucurbitacins, cardenolides and alkaloids[33]; it may be no coincidence that *Iberis amara* and *Erysimum cheiranthoides* are unusual among crucifer species in being rejected by the crucifer-feeding flea beetle *Phyllotreta cruciferae* and that *E. cheiranthoides* and *Lunaria annua* support unusually

poor larval growth of *Pieris rapae*[24] (Slansky and Feeny, in
preparation).

The effectiveness of plant defenses against inter-family
generalist enemies may be subject to the same considerations,
on a broader scale, as those discussed above for 'family
specialists'. There may be little effect on those generalist
species which are fully adapted to a plant's defenses; against
species which are less well adapted, however, defensive
compounds may be effective in causing a shift of attack to
other plant species, either directly by behavioral repulsion
or indirectly by reducing fitness. Since generalists,
mammals included, may be less adapted to any particular
class of defensive compounds than are the relevant specialists, they may be more susceptible to variation in the
concentrations of such compounds than are the specialists.
Higher concentrations might thus have a commensurate effect
on generalists even if they are ineffective against adapted
specialists. Variation in the glucosinolate content of
Brussels sprout leaves, for example, seems to have a more
pronounced effect on the fecundity of the generalist aphid
species *Myzus persicae* than on that of the specialist
species *Brevicoryne brassicae*[70].

Finally, plant secondary compounds may have a very
different kind of effect on populations of herbivorous
insects in ecological time. Tahvanainen and Root (1972)
showed that the chemical stimuli given off by the non-host
plants tomato, *Lycopersicon esculentum*, and ragweed, *Ambrosia
artemisifolia*, interfered with the ability of *Phyllotreta
cruciferae* flea beetles to find their crucifer host-plants.
The secondary chemistry of one plant species can therefore
be effective in reducing the apparency of plants of another
species to its adapted specialist enemies, thus conferring
what Tahvanainen and Root describe as 'associational resistance' on plants of all species in diverse communities.

Chemical defense and metabolic cost. It is a generally
held assumption among ecologists that the allocation of
metabolic resources to various adaptations manifested by
the individuals of a species is correlated with adaptive
value of those adaptations in terms of fitness[11]. Though
such a 'Principle of Allocation' is often discussed solely
in terms of energy, it is by no means clear that energy
is always or even usually the limiting metabolic ingredient

determining the growth and fitness of organisms in nature.
For many organisms, at least during part of their life
histories, allocation of limiting nutrients, such as
nitrogen, may be far more critical to growth and fitness
than allocation of energy which may be present in super-
abundant supply (Slansky and Feeny, in preparation).   Since
the relative importance of energy and nutrients is not often
known, it seems safest to discuss the principle of allocation
in terms of the relationship between adaptive benefit and
the less specific 'metabolic costs' of adaptations.

According to such a principle, the metabolic costs   to
a plant   of producing defensive chemicals are expected to
bring commensurate rewards in individual fitness[8].   The
proportion of total metabolic resources allocated to defense
within individuals presumably reflects, therefore, the
selective pressures of various kinds of predation on the
plants, relative to all other selective pressures acting on
the plants over the past several generations.   There is
little experimental evidence, as yet, to support the notion
that defensive compounds are "expensive" to plants, but this
is not surprising in view of the difficulties of obtaining
such evidence.

Early successional herbs are likely to be under particularly severe selective pressures favoring fast growth, success
in competition with other plants for nutrients, water and
light, and the production of many propagules as an adaptation
towards effective dispersal and colonization of new areas.
For such species, selection is unlikely to favor a large
metabolic allocation for defensive compounds, especially
since the selective pressures of predation are reduced as a
result of escape from many enemies in space and time.

The adaptive emphasis in such species is   thus   likely
to favor defensive compounds which are especially effective
in small quantities; the metabolic investment to produce
the relatively large amounts of tannins required to be
effective, for example, would probably not bring commensurate
adaptive advantage to early successional herbs, since their
reduced success in competition and dispersal would probably
not offset the advantages of 'quantitative' defense against
predation.   The adaptive emphasis is likely, rather, to favor
small amounts of compounds which are potent toxins to enemies

in spite of the susceptibility of such compounds to effective counteradaptation.

There are other trends towards an increased 'adaptive benefit/metabolic cost' ratio which might be expected in the chemical defenses of unapparent plants. For example, compounds might be favored which do not require utilization of critically limiting nutrients; for plants limited by nitrogen, for instance, a given concentration of alkaloids might be more "expensive" than the same concentration of, say, terpenoids or flavonoids. On the other hand, alkaloids and other nitrogen-containing toxins may be less expensive for legumes, which can exploit atmospheric nitrogen. Those compounds which are effective against several categories of enemy are likely to be favored; the glucosinolates seem to belong to such a category since they (or their aglycones) are effective against bacteria and fungi[77], mammals[44], and insects[18,47]. Adaptive benefit for a given metabolic cost is likely to be increased if compounds combine a defensive function with other adaptive functions, such as allelopathic inhibition of competing plants. Costs may be reduced by producing defensive compounds (phytoalexins), only if or when the plant is actually attacked by its enemies[16]. Finally overall costs may be reduced by apportioning defensive compounds over the life of a plant, and between its component tissues, only in proportion to the relative value (in terms of fitness) of the tissues to the plant and/or in proportion to the differing apparencies of the tissues to enemies. Higher concentrations of defensive compounds in shoots and young leaves than in mature leaves[49,70] may reflect the higher nutritive value of young leaves to enemies and/or the greater impact on plant fitness likely to result from a given amount of damage to young versus old leaves. The markedly greater concentrations of hypericin present in the flowering heads of *Hypericum hirsutum* plants, as compared with other parts of the plant[53], may similarly reflect the differential importance of the tissues to plant fitness[49].

Evolution of chemical defense in early successional herbs. If the predation pressure on a plant population increases over several generations, one conceivable adaptive response which selection might be expected to favor is the production of higher concentrations of defensive compounds already produced by the plant species[8]. Such an evolutionary response by the plants might be adaptive if the increased

predation pressure results mainly from generalist enemies
or perhaps from some 'family specialists' against which the
increased concentrations could have a quantitative impact,
thus forcing the enemies directly or indirectly to switch
to other available food-plants. There are reasons to suspect,
though, that such a response would not bring commensurate
adaptive reward if the increased predation pressure results
primarily from fully adapted specialists. Firstly, since
such enemies are already well-adapted to tolerating the
defensive compound, further counteradaptation to increased
levels would be likely to follow rapidly; the glucosinolate-
boosting experiments discussed earlier indicate that *P. rapae*
larvae can tolerate considerably higher levels of glucosinolates than are typically present in their food plants.
Secondly, the production of increased levels of existing
defensive compounds may render the plants more apparent to
adapted specialists and thus be counterproductive; increased
allylglucosinolate content in *Brassica nigra* plants, for
example, would probably result in increased release of its
volatile aglycone, allylisothiocyanate, which would certainly
increase the apparency of the plants to *Phyllotreta cruciferae*
and *P. striolata* flea beetles as discussed earlier. Further
disadvantages of producing high levels of existing compounds
are the metabolic expense and the possible risk of autotoxicity from high levels of potent compounds in the tissues.

A second category of evolutionary response to increased
predation pressure is the production of different compounds.
Such a response is likely to be particularly effective against
both 'family specialists' and true generalists, since the new
compound may act either directly as a behavioral repellent
or indirectly by reducing the fitness of such enemy individuals as attack the plant. This in turn could bring about
selection in the enemy population for "non-preference" to
the plant species[10] or even enhanced preference for other
suitable food-plant species within the habitat. To what
extent such an evolutionary response can occur 'sympatrically'
within a plant population subjected to increased predation
pressure from 'family specialist' or generalist enemies
would depend on many variables, including the number of
species of alternate food-plants, their defenses and relative
apparencies, and the susceptibility of the new compound
to counteradaptation by the enemy species. The more different
the new compound is chemically, from the previous line of
chemical defense, the less likely is the chance of

counteradaptation; also, the longer such counteradaptation
is postponed the less likely it is to occur, since the
number of individuals attacking the plant will decline.
The production, in addition to the typical crucifer
glucosinolates, of cucurbitacins, cardiac glycosides and
alkaloids in the genera *Iberis*, *Erysimum* and *Lunaria*,
respectively, may result from evolutionary escape from
crucifer-adapted herbivores or pathogens, though it is
hard to see how this could ever be proved. In a sense,
all the plant species of a chemically similar family which
occur in the same habitats are in 'competition' to reduce
their share of the herbivore load resulting from family
specialist enemies. Chemical character displacement away
from the typical family chemistry can also be viewed as an
adaptation to escape the share of the family herbivore
load and "forcing" it onto other species.

Production of new compounds may not be such an effective
defense against "species specialist" enemies which have no
alternate host-plants and which, through continued association
with the host-plant as its defenses evolve, have a higher
chance of "coevolving" appropriate counteradaptations.
Against increased predation pressure from such monophagous
species, a likely evolutionary response may be reduced
apparency; this may occur in ecological time in any event
through reduced abundance of the plant resulting from
attack by the enemy, as seems to have occurred in California
populations of Klamath weed[35]. The plants now survive best
in wooded habitats where the *Chrysolina* beetles are less
effective. Such habitat displacement can clearly occur also
in evolutionary time and it is conceivable, for example,
that the present restriction of the crucifer genus *Dentaria*
to wooded habitats may have resulted from selective pressures
escaping from enemies characteristic of the open habitats
typical of this plant family.

Reduced apparency to adapted specialist enemies might
be achieved by divergence in leaf shape[30] or by loss of
compounds used as behavioral cues by such enemies. *Thlaspi
arvense* is an open-habitat crucifer which, like *Brassica
nigra*, contains only one glucosinolate, allylglucosinolate,
in significant amounts in its leaves (P. Feeny and L.
Contardo, unpublished results). The glucosinolate in *T.
arvense* is broken down on tissue damage, not to the typical
allylisothiocyanate, as produced from *B. nigra* leaves, but

to the atypical allylthiocyanate[31] (P. Feeny and L. Contardo,
unpublished results). Two-plant "islands" of *T. arvense* are
colonized by the flea beetles *Phyllotreta striolata* and *P.
cruciferae* at a much lower rate than equivalent islands of
B. nigra, probably because the isothiocyanate is used as a
specific behavioral attractant by the flea beetles but the
thiocyanate is probably not (P. Feeny, J. Gaasch and L.
Contardo, unpublished results).

In contrast to the predation pressure exerted by
adapted enemies, the effects of non-adapted enemies in
ecological time are, almost by definition, likely to be
trivial. Plants are always subject to the risk, however,
of colonization by new enemy species in evolutionary time,
even though such enemies must undergo adaptive changes at
both toxicological and behavioral levels[7,15,22]. Such
colonization is most likely to occur by species which have
previously been adapted to attacking plants of similar
secondary chemistry, since such species are partially pre-
adapted[17,66]. Though it is hard to see how non-adapted
specialists can exert selective pressures when they do not
actually attack the plants to any significant extent, there
seems to be no question that the possession of chemical
defenses which differ widely from those of other plants
in the same habitats, whatever selective pressures were
responsible for their formation, carries the additional
advantage to a plant species of minimizing the risk of
colonization by new adapted enemies in evolutionary time.

From the above discussion I think it is clear that a
plant is likely to gain several advantages as a result of
containing several lines of chemical defense (i.e. high
intraspecific diversity), and also as a result of being as
different as possible chemically from other plants sharing
the same habitats (thus contributing to high interspecific
chemical diversity). Production of new compounds may permit
escape from adapted enemies through toxic and subsequent
repellent or non-preference effects, and also by reducing
the plant's apparency to such enemies. Also, as a result
of high interspecific chemical diversity, the apparency
of any particular plant species is likely to be reduced
further by associational resistance in ecological time and
the chances of colonization by non-adapted enemies in evolutionary time are likely to be reduced. In other words, the
greater is the diversity of chemical compounds in a community

of early successional herbs, the more effective is any one
compound likely to be both as an ecological and as an
evolutionary barrier to enemies.

These considerations may well be responsible, in large
measure, for the great chemical diversity which characterizes
communities of early succession[28]. Since the plants
characterizing such communities, like many of their enemies,
are capable of relatively fast evolutionary rates compared
with the plants of later stages of succession (V. Grant,
personal communication),the patterns of chemical diversity
may be subject to constant change through time and from one
area to another in response to shifting predation pressures
from animals, fungi and microorganisms. These considerations
may also account in part for the well known phytochemical
finding that though particular secondary compounds are often
characteristic of the plant species of a particular genus
or family, there are usually several exceptional species which
contain atypical classes of compound, either in addition to
or instead of the class typical of the family or genus as a
whole. The effects of plant enemies would be expected to
bring about gradual chemical divergence between the unapparent
plant species of a genus or family, the rate depending in
part on how effective are the original family-wide compounds
as defensive agents[66].

DEFENSES OF APPARENT PLANTS

Mature foliage. The common or pedunculate oak, *Quercus
robur*, in England and continental Europe typifies what I
refer to as an apparent plant species. It is an abundant
member of the prevailing climax vegetation and is apparent
both as a result of its persistence and of its abundance in
ecological time. Its high apparency probably accounts, in
large measure, for the exceptionally large number of insect
species which have colonized the plant in evolutionary time[65].
The lepidopterous insects which attack oak leaves can be
divided into two categories on the basis of their larval
phenology. First there is an early-feeding group which
attacks the relatively nutritious young leaves in the
spring; many insect species in this group, such as the
winter moth, *Operophtera brumata*, are very abundant and
trees are periodically defoliated. Secondly, there is a

late-feeding group of caterpillars which are characterized
by slow growth and by their relative rarity[21].

The slow growth of most lepidopterous caterpillars on
mature oak leaves probably results from an increase in leaf
toughness, accompanied by a reduction in nitrogen and water
content as the leaves mature [21] (J. M. Scriber, unpublished
results). In addition, the tannin content of the leaves
increases as the season progresses[23]; these polyphenolic
compounds inhibit growth of winter moth caterpillars[19]
probably by forming relatively indigestible complexes with
leaf proteins, thus further aggravating the decline of
available nitrogen in the leaves[20]. Tannins are present
in relatively large amounts in summer leaves and their
action on larvae is dosage-dependent; since their action
against proteins, including enzymes, is so generalized,
their susceptibility to counteradaptation by specific
detoxication mechanisms is likely to be low[20]. Tannins
seem to have a wide spectrum of activity and have been
shown to be effective against viruses, bacteria, fungi,
insects and mammals[21]. That tannins are not completely
immune to counteradaptation, however, is suggested by the
variety of sawfly species which feed, as larvae, on the
mature tannin-containing fronds of bracken fern, *Pteridium
aquilinum*; on this plant, insect abundance increases during
the season in spite of decline in protein content[45].

Whatever other functional significance they may have
for oak trees, the properties of declining nitrogen and water
content, combined with tough leaves and the presence of
tannins, all confer upon oak leaves a "quantitative" defense
against herbivores and pathogens; the leaves are plainly
poor food for most potential enemies. Several of the
Lepidoptera species which attack the mature foliage of these
trees have "leaf-mining" larvae which may avoid the tough
leaf cuticle and much of the tannin-containing tissue; the
other species typically grow very slowly, as would be
expected on such a nutritionally poor diet, and many do not
complete larval growth in one season[21]. Slow growth exposes
larvae to increased risk of predation, parasitism, and other
mortality factors, perhaps accounting for the relatively
low population densities of most insect species which attack
mature oak leaves; concentrations of slow-growing larvae,
however cryptic as isolated individuals, would be especially
apparent to predators and parasites, unless their increased

apparency was offset by effective defense. The growth-
inhibitory tannins, however, are less suitable for secondary
exploitation by herbivorous insects than are the potent
toxins characterizing many unapparent plants. Though this
would not prevent *de novo* synthesis of toxins by the insects,
toxic chemical defense and associated warning coloration do
not seem to be common among insects attacking the mature
foliage of climax plants, at least in temperate regions,
whereas such defense is relatively frequent among species
attacking plants of earlier seral stages[58].

The defenses of oak leaves seem to be typical of those
found in the mature leaves of many late successional and
climax plant species. Tannins are found in the leaves of
heather, *Calluna vulgaris*, and other ericaceous shrubs which
form the dominant vegetation in many communities[5]. They occur
in the leaves of many climax tree species in temperate zone
forests[2]. In those tropical habitats which are characterized
by a low diversity of tree species the foliage tends to be
exceptionally well defended by chemical compounds, often
including tannins[40]. Tannins are not typically reported from
early successional herb species[5].

Coniferous trees contain substantial amounts of resins
and many grasses contain silica in their mature leaves.
Conifers and grasses are dominant components of many climax
communities and these materials may be analogous to the tannins of climax deciduous trees and shrubs as quantitative
defenses against herbivores and pathogens[32,51]. The grinding
teeth of ungulate mammals seem to represent a counteradaptation which allows feeding on grasses containing silica.

It might be expected that the quantitative defense of
apparent plants would be enhanced by selective elimination,
from plant tissues, of any nutrients which are required by
animals but not required by plants. Sodium and iron are
such nutrients and both are indeed present in very small
amounts in the leaves of most land-plants[48,80]. The circumstantial evidence suggests that many herbivorous animals may
be limited by the sodium content of their food-plants[1,3].
However, under desert conditions, where ample water for
excretory detoxication is lacking, accumulation of salt[43] may
represent another kind of quantitative defense in an apparent
plant species[1].

If quantitative defense is correlated with plant apparency, one would expect to find that herbivores feeding on the mature foliage of late-successional plant species typically grow more slowly and less efficiently than related herbivore species which feed on early successional plants. This expectation has recently been confirmed by Scriber (1975) who found that larvae of the tiger swallowtail butterfly, *Papilio glaucus*, are considerably less efficient and grow three times more slowly on their natural diet of mature tree leaves than larvae of the black swallowtail, *P. polyxenes*, feeding on their natural diet of umbelliferous herbs. Larvae of the spicebush swallowtail, *P. troilus*, feeding on their natural diet of shrub leaves were intermediate in growth rate and feeding efficiency. These differences are unlikely to reflect differences in feeding specialization (*P. glaucus* is a generalist; *P. polyxenes* is a "family specialist") since a similar comparison between the larvae of several species of saturniid moths, which differ widely in feeding specialization and yet which are all tree-feeders, revealed little variation in growth rate or feeding efficiency[61]. Generally speaking, among the larvae of Lepidoptera which naturally feed on mature leaves, both growth rate and feeding efficiency are inversely correlated with successional status of the food-plant species, regardless of variation in the natural degree of larval feeding specialization[61].

Apparent plant species are probably subject to much lower selective pressure to differentiate chemically from neighboring species than are unapparent plants. Less advantage would be expected from small amounts of compounds which can easily be detoxified. The adaptive emphasis is likely to favor, instead, retention of the most effective compounds for quantitative defense, regardless of whether or not other species in the community share similar defenses. This may account for the sharing of tannins, resins and silica by so many species of climax plants and for the contrast in chemical diversity between early and late successional plant communities[28]. Whereas in early successional plants any particular compound is likely to vary greatly in its behavioral effects on herbivores, being attractive to some species and repellent to others, the widespread distastefulness of plant tannins to animals[21] may reflect the efficacy of these compounds as a component of general quantitative defense against herbivores[9].

An interesting consequence of high apparency is the
likelihood that fruits and seeds will also be apparent. It
is not surprising, therefore, that fruit and seed "masting"
strategies, as a mechanism for predator satiation, tend to
be characteristic of apparent plants, such as climax trees
in communities of relatively low tree species diversity[40].

Immature foliage. For a period of about three weeks
every spring, from the time of bud opening until they have
almost reached full size, the leaves of oak trees (*Quercus
robur*), in England, are especially vulnerable to attack by
insects. The developing leaves have a high nitrogen and
water content and they lack condensed tannins and physical
toughness[21]. The 'phenological window' represented by
these nutritious leaves, is exploited by a large number of
lepidopterous species several of which may reach very high
densities in some years and cause total defoliation of
trees.

The most abundant of these early-feeding species is
the winter moth, *O. brumata*, the wingless females of which
emerge from their pupae in the leaf litter during November
and December and crawl to the nearest large verticle object,
usually an oak trunk; after mating on the trunk, they climb
to the crown of the tree where they lay their eggs among the
mosses and lichens on the twigs and branches. At about the
time of bud-opening the following April, the eggs hatch
and the first instar larvae enter the developing buds. After
feeding for about three weeks on the opening buds and developing leaves, the larvae descend on silk threads to the ground
where the pupae remain until adult emergence the following
winter[75]. The highest mortality of the winter moth, during
this annual life cycle, occurs as a result of failure of
freshly-hatched larvae to find suitable food[73]. The first
instar larvae are unable to penetrate oak buds which have
not yet begun to open[72], nor are they likely to survive on
young leaves which are already a few days old and have
begun to toughen[21]. The relative timing of bud-opening and
egg hatch is thus crucial for larval survival: if peak egg
hatch occurs more than a few days before the time when most
of the buds are just opening, heavy mortality of all except
late-hatching larvae results (Fig. 1A), either through
direct starvation or through dispersal on silk threads,
usually to unfavorable feeding sites[21,72]. If the majority
of buds open before the peak of egg hatch, heavy mortality

is also likely to result since the tiny first instar larvae
are probably unable to survive on the young leaves (Fig. 1B).
When the periods of bud opening and egg hatch substantially
overlap, however, mortality is reduced considerably (Fig. 1C)
and trees may be defoliated (Figs. 2A and 2B).

The relative timing of bud opening and egg hatch in any
one year is determined by climatic factors to which both plant
and insect respond indpendently. The high mortality of winter
moth is a reflection of the lack of perfection in the ability
of the insects to "track" precisely the responses of the

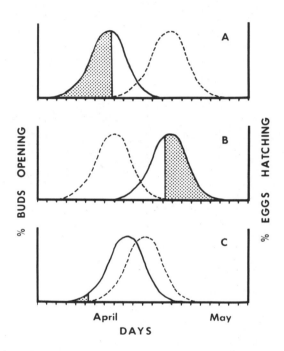

Figure 1. Schematic representation of consequences to
winter moth larvae of variation in the relative timing of
winter moth egg hatch (solid line) and opening of oak buds
(broken line). A: Egg hatch precedes bud-opening: early-
hatched larvae (stippled area) starve or disperse. B: Bud
opening precedes egg hatch; late-hatching larvae starve or
disperse. C: Bud opening broadly overlaps egg hatch; most
larvae find suitable food.

trees to climatic factors. The density independent action of
weather is thus the major factor determining the abundance of
winter moth and other species in any one season, though the
average population densities, over a number of years, are
determined by the density dependent regulating action of
parasitoids and predators[73,75].

Oak trees respond to defoliation by producing new leaves
(Fig. 2C) and the full photosynthetic area is usually restored
by mid-July, not, however, without having a significant effect
on the trees' metabolic resources. Varley and Gradwell (1962b)
have estimated that in the absence of defoliating insects the
trees would add, on average, 40% to their annual timber production. In spite of the impact of defoliation, young oak
leaves do not seem to be chemically protected. This could
result from competitive pressures to establish full photosynthetic area as rapidly as possible or, perhaps, from the
risk of autotoxicity which might follow from the presence of
tannins in rapidly developing leaves[50]. Whatever the reasons
for this lack of chemical defense, however, there seems little
question that immature oak leaves would suffer much more
drastic damage if it were not for the unpredictable timing
of bud opening each season and the inability of the herbivores
to track this precisely. In other words, the immature foliage
of oak trees derives protection by being unapparent.

Individual oak trees in a particular area may differ
by many days in their relative timing of bud-opening, some
trees always being 'early' trees and others always being
'late', even though the average bud-opening date for all trees
in the area may, depending on weather, vary from one season
to another[72]. As a consequence, some trees may escape substantial attack by insects even when the majority of the trees
around them are totally defoliated (Fig. 2A); the insects,
having dropped to the ground from defoliated trees (Fig. 2B),
are evidently unable to locate and crawl up trees which have
not been stripped of foliage. Although atypically early or
late opening trees may thus further reduce their apparency
to insect enemies, there are presumably limits to how far
such deviation from the average is advantageous. Early trees,
for example, could suffer enhanced risk of frost damage and
late-opening trees may suffer a competitive loss of photosynthetic time. Varley and Gradwell (1962b) found that
individual trees which opened their buds late, though less
subject to caterpillar damage, nevertheless grew more slowly

Figure 2. (a) Oak trees defoliated by caterpillars
(mostly winter moth) in Devon, England (6/65); some trees
(arrow) have escaped defoliation as a result of their late
bud-opening dates. (b) Boulder from the foreground of Fig. 2
(a) indicating density of caterpillars which have dispersed
from defoliated trees above. (c) New leaves on the branch of
an oak tree in 6/65, following earlier defoliation (near
Oxford, England); note remains of defoliated leaves. (d) Winter
moth pair in copulation on a tree trunk at night, (12/65);
wingless female is uppermost.

than an early opening tree with a much higher caterpillar
population density.

The low apparency of young oak leaves could explain
why the females of the winter moth, and several other early-
feeding species which attack young oak leaves, are wingless
(Fig. 2D). By walking from the pupal case to the nearest
tree trunk, the female is likely to find and subsequently
lay her eggs on the tree on which she fed as a larva; this
will not only be a tree of the appropriate species, but
also a tree to which she herself was appropriately synchronized
as an emerging first instar caterpillar. Conceivably, therefore, her eggs would have a greater chance of being synchronized with bud-opening the following spring than if they had
been laid on oak trees at random.

Presumably, the leaves of all apparent plants must pass
through a period of especial vulnerability to enemies since
the formation of structural materials associated with toughness
the associated dilution of nutrient content and the formation
of relatively large amounts of quantitative defensive compounds
cannot be achieved without some delay. It is very likely,
therefore, that the dependence of young oak leaves on reduced
apparency to herbivores may be typical of the young foliage
of apparent plants. Escape of such foliage may occur in space
as well as time. The migrations of many grazing mammal species
in East Africa, for example, seem to be related to discovery
of the nutritious, but ephemeral, young foliage of grasses
which appears in response to somewhat unpredictable patterns
of local rainfall.

The young foliage of apparent plants seems to resemble
in several ways the foliage of early successional herbaceous
plants; it is a nutritious source of food for herbivores,
it is not defended by any of the components of quantitative
defense and it escapes catastrophic defoliation by being
unapparent. It would not be very surprising, therefore, to
find that the young foliage of many apparent plants is
defended, also, by small amounts of the same kinds of potent
toxins which occur as qualitative or "evolutionary" barriers
in the foliage of unapparent herbs. Although no such defenses
have as yet been demonstrated in young oak leaves, the young
fronds of bracken fern, *P. aquilinum*, contain thiaminase and
the cyanogenic glucoside prunasin, both of which decline in
concentration as the fronds mature and as levels of tannin

and silicate increase[45]. A similar decline in cyanide and
rise in tannin levels takes place in the maturing leaves
of the chaparral shrub *Heteromeles arbutifolia*[13].

CONCLUSIONS

The basic hypotheses. I hope that I have been able
to demonstrate that the basic hypotheses relating plant
apparency to chemical defense are reasonable. Before the
hypotheses can be considered comfortably robust, however,
they will have to be tested rigorously with a variety of
herbivore-plant interactions in a variety of different
communities and biomes. They may have to be modified
significantly to take account of additional phenomena such
as the relative importance in defense of chemical versus
physical defenses (thorns, spines, etc.), the relative
impact on plants of specialist enemies, versus generalists,
the relative importance of the different taxa of enemies
(mammals, insects, fungi, bacteria, viruses, etc.) and
the relative importance of chemical compounds as proximate
behavioral stimuli (e.g. attractants and repellents) and as
ultimate toxins or growth inhibitors. I am optimistic that
the basic hypotheses will stand the test of time, in spite
of refinements which will undoubtedly have to be made to
them. Even in their present form, I think that they improve
our understanding of at least some patterns of interaction
between herbivores and plants and permit tentative prediction
of likely trends in ecological phenomena such as life histories, phenology, abundance, diversity and succession. As
such, I think that they enhance the potential of basic coevolutionary theory as an important aid in the explanation and
prediction of patterns of interaction between herbivores and
plants.

Influence of plant chemistry on insect life histories.
When preparing this paper, I hoped originally to be able to
discuss the likely effects of insects on patterns of chemicals
in plants. Such is the prevailing ignorance of the relative
importance as selective pressures of different taxa of herbivores and pathogens on the evolution of chemical defenses by
plants that I do not feel sufficiently confident to identify
any defensive plant secondary compound as being solely the
result of selective pressures exerted by insects. This is
not to deny that insects have a strong effect on the

diversity and abundance of chemicals in plants; the powerful
effects that insects may exert are well illustrated by the
effect of *Chrysolina* beetles on Klamath weed, of *Phyllotreta*
beetles on crucifers and of winter moth larvae on oak trees,
to name but a few examples. I think, rather, that the very
trends and patterns which have led to hypotheses relating
plant apparency to chemical defense reflect the likely fact
that the selective pressures exerted by insects are frequently
in the same direction as those exerted by other groups of
herbivores and by pathogens. The taxonomic spectra of
toxicological activity of very few secondary compounds are
known in any detail. Where they are known, even in outline,
toxic activity has frequently been shown against a wide
variety of taxa; examples include both the glucosinolates
and the tannins, which were discussed earlier.

   Whatever the selective contribution of insects to
patterns of chemicals in plants, there is no question that
patterns of chemicals in plants may have profound effects on
the life histories of insects. In early successional communities much of the foliage is highly nutritious but protected
by a diverse array of toxins which represent evolutionary
barriers to insects. The adaptive emphasis among the insects
must therefore favor specific toxicological counteradaptation
combined with elaborate behavioral mechanisms to permit
location and identification of appropriate plants within the
chemical maze; once discovered these plants generally support
rapid growth. The ephemeral plants thus impose an ephemeral
'r-selected' life history on many of their associated insects.

   The mature foliage of apparent plants, by contrast,
imposes slow growth on many associated insects, thus exposing
them to prolonged vulnerability to predation. The adaptive
emphasis among such insects is likely to favor enhanced
crypsis or other defenses against predators and parasitoids.
The 'K-selected' plants thus impose 'K-selected' life histories
on many of their associated insects. Here, however, carrying
capacity may be determined not so much by availability of food
or ability to find it as by apparency to predators and parasitoids (though this is itself a direct consequence of the plant
defenses). Thus, in central New York, the tree-feeding, slow-
growing larvae of the tiger swallowtail, *Papilio glaucus*,
display several anti-predator adaptations not found in herb-
feeding, fast-growing larvae of its relative, the black
swallowtail, *P. polyxenes*. These adaptations include mimicry

of tree snakes, which overwintering warblers might encounter
in central or South America (R. B. Root, personal communication), resting in a leaf roll (which may also reduce the
risk of desiccation), and chewing off the petioles of
damaged leaves after feeding, thus allowing damaged leaves
to fall to the ground and probably reducing larval apparency to
predators and parasitoids (Feeny and Scriber, unpublished
results).

    Agricultural implications. Most of the vegetables
cultivated by man are derived from early successional herbs
which have evolved as naturally unapparent plants. Agriculture consists in part, then, of making unapparent plants
apparent by planting them in monocultures. In addition to
making the plants easier to harvest, however, such practices
also increase the apparency of the plants to other herbivores
and to pathogens. Since the crops have evolved defenses
inappropriate for survival as apparent plants it is hardly
surprising that large amounts of artificial chemical defenses
must frequently be applied to such monocultures in order to
prevent widespread devastation by herbivores and pathogens.
The discontinuation of the older practices of crop rotation
and interplanting of different species aggravated the situation by further increasing apparency to herbivores and
pathogens.

    It seems that progress towards the current goal of
maintaining or increasing agricultural productivity, while
reducing pesticide applications to avoid harmful effects to
the environment and to human health, might benefit from a
study of the naturally evolved chemical defenses of plants.
Clearly the reintroduction of crop rotation and diversiculture
would reduce plant apparency and might be economically
feasible under some circumstances. A second lesson to
be learned from naturally unapparent plants is that high
intraspecific and interspecific chemical diversity are
likely to have several beneficial consequences, including
inhibition of the evolution of resistance by pests[66] and
reduction of plant apparency. Interspecific diversity
might be imitated in an agricultural monoculture by
interplanting different genotypes of the same crop, each
with a different line of chemical resistance.

    It might also be possible to make agricultural use of
some aspects of the quantitative defenses found in plant

communities of naturally low diversity. The breeding of
plant varieties for low nutrient content, increased toughness
or high content of silica or tannins would clearly be
counterproductive unless restricted to parts of a crop
plant which are not used for human consumption or animal
feed. However, if the sodium content of crop plants, for
example, could be sufficiently reduced to inhibit the
growth of pest populations, salt could be restored to the
crop during processing or at the table. Since the nutritive
requirements of insects do not differ greatly from those of
mammals, the greater the nutritional imbalance that can be
bred into a crop variety, the greater its effectiveness is
likely to be as a quantitative barrier to insect growth.
Nutritional deficiencies can be corrected for, after
harvesting, either by addition of deficient ingredients or
by combining complementary foods in the diet.

These suggestions, whatever their economic feasibility,
must be considered cautiously in view of the currently
speculative nature of some of the underlying hypotheses.
It seems clear, though, that the trend towards increasing
interaction and collaboration between phytochemists and
ecologists will not only increase our understanding of
herbivore-plant interactions and of the distribution and
abundance of secondary plant compounds, but that it may
also lead to some practical applications of agricultural
importance.

Acknowledgments. I would like to thank Karen Arms,
Patricia Blau, William Blau, John Christy, Lorraine
Contardo, Daniel Janzen, Robert Lederhouse, Peter Price,
Mark Rausher, Richard Root and Mark Scriber for valuable
discussion and advice. The research leading to this paper
has been supported by research grants GB 43846 and BMS-
7409868 from the National Science Foundation and Hatch
Grant NYC-139413.

REFERENCES

1.   Arms, K., P. Feeny, and R. C. Lederhouse. 1974. *Science*
        185:372.
2.   Bate-Smith, E. C., and C. R. Metcalfe. 1957. *J. Linn.*
        Soc. London 55:669.

3. Botkin, D. B., P. A. Jordan, A. S. Dominski, H. S.
     Lowendorf, and G. E. Hutchinson. 1973. *Proc. Nat.
     Acad. Sci.* (U.S.A.) *70*:2745

4. Brower, L. P., and J. Van Z. Brower. 1964. *Zoologica
     49*:137.

5. Brown, B. R., C. W. Love, and W. R. C. Handley. 1962.
     Rep. on Forest Res. (1962) p. 90, H.M.S.O., London.

6. Bush, G. L. 1974. *In* "Genetic Mechanisms of Speciation
     in Insects" (M.J.D. White, ed.), pp. 3-23. Australian
     and New Zealand Book Co., Sydney.

7. Bush, G. L. 1975. *In* "Evolutionary Strategies of
     Parasitic Insects and Mites" (P. W. Price, ed.), pp.
     187-206. Plenum Publishing Corporation, New York.

8. Cates, R. G. 1975. *Ecology 56*:391.

9. Cates, R. G., and G. H. Orians. 1975. *Ecology 56*:410.

10. Chew, F. S. 1974. "Strategies of Foodplant Exploitation
     in a Complex of Oligophagous Butterflies (Lepidoptera)".
     Ph.D.thesis, Yale University.

11. Cody, M. L. 1966. *Evolution 20*:174.

12. Cromartie, W. J., Jr. 1974. "The Effect of Stand Size
     and Vegetational Background on the Colonization of
     Plants by Herbivorous Insects." Ph.D. thesis,
     Cornell University, Ithaca.

13. Dement, W. A., and H. A. Mooney. 1974. *Oecologia 15*:65.

14. Dethier, V. G. 1947. "Chemical Insect Attractants and
     Repellents." Blakiston Co., Philadelphia.

15. Dethier, V. G. 1954. *Evolution 8*:33.

16. Deverall, B. J. 1972. *In* "Phytochemical Ecology" (J.
     B. Harborne, ed.), pp. 217-233. Academic Press,
     London and New York.

17. Ehrlich, P. R., and P. H. Raven. 1964. *Evolution 18*:586.

18. Erickson, J. M., and P. Feeny. 1974. *Ecology 55*:103.

19. Feeny, P. P. 1968. *J. Insect Physiol. 14*:805.

20. Feeny, P. P. 1969. *Phytochemistry 8*:2119.

21. Feeny, P. 1970. *Ecology 51*:565.

22. Feeny, P. 1975. *In* "Coevolution of Animals and Plants"
     (L. E. Gilbert and P. H. Raven, eds.), pp. 3-19. Univ.
     of Texas Press, Austin and London.

23. Feeny, P. P., and H. Bostock. 1968. *Phytochemistry 7*:871.

24. Feeny, P., K. L. Paauwe, and N. J. Demong. 1970. *Ann.
     Ent. Soc. Amer. 63*:832.

25. Forbes, W. T. M. 1958. *Proc. 10th Int. Congr. Ent. 1*:313.

26. Fraenkel, G. S. 1959. *Science 129*:1466.

27. Freeland, W. J., and D. H. Janzen. 1974. *Amer. Nat. 108*:
     269.

28. Futuyma, D. J. 1976. *Amer. Nat.* (in press).
29. Gibbs, R. D. 1974. "Chemotaxonomy of Flowering Plants" McGill-Queen's Univ. Press, Montreal and London.
30. Gilbert, L. E. 1975. *In* "Coevolution of Animals and Plants" (L. E. Gilbert and P. H. Raven, eds.), pp. 210-240. Univ. Texas Press, Austin and London.
31. Gmelin, R., and A. I. Virtanen. 1959. *Acta. Chem. Scand. 13:*1474.
32. Harris, P. 1960. *Can. J. Zool. 38:*121.
33. Hegnauer, R. 1962-1973. "Chemotaxonomie der Pflanzen", vols. I-VI. Berkhäuser Verlag, Basel and Stuttgart.
34. Hicks, K. L., and J. O. Tahvanainen. 1974. *Amer. Midl. Natur. 91:*406.
35. Huffaker, C. B., and C. E. Kennett. 1959. *J. Range Management 12:*69.
36. Janzen, D. H. 1968. *Amer. Nat. 102:*592.
37. Janzen, D. H. 1970. *Amer. Nat. 104:*501.
38. Janzen, D. H. 1973a. *Amer. Nat. 107:*786.
39. Janzen, D. H. 1973b. *Pure and Applied Chem. 34:*529.
40. Janzen, D. H. 1974. *Biotropica 6:*69.
41. Jermy, T. 1966. *Ent. exp. et appl. 9:*1.
42. Jermy, T., F. E. Hanson and V. G. Dethier. 1968. *Ent. exp. et appl. 11:*211.
43. Kenagy, G. J. 1972. *Science 178:*1094.
44. Kingsbury, J. M. 1964. "Poisonous Plants of the United States and Canada." Prentice-Hall, Inc., Englewood Cliffs, New Jersey.
45. Lawton, J. H. 1975. *In* "The Biology of Bracken" (F. H. Perring, ed.), Academic Press, London (in press).
46. Levin, D. A. 1971. *Amer. Natur. 105:*157.
47. Lichtenstein, E. P., D. G. Morgan, and C. H. Mueller. 1964. *J. Agric. Food Chem. 12:*158.
48. Likens, G. E., and F. H. Bormann. 1970. "Chemical Analyses of Plant Tissues from the Hubbard Brook Ecosystem in New Hampshire." *Yale Univ. School of Forestry Bull. No. 79.*
49. McKey, D. 1974. *Amer. Nat. 108:*305.
50. Orians, G. H., and D. H. Janzen. 1974. *Amer. Nat. 108:*581.
51. Pathak, M. D. 1969. *Ent. exp. et appl. 12:*789.
52. Read, D. P., P. P. Feeny, and R. B. Root. 1970. *Canad. Ent. 102:*1567.
53. Rees, C. J. C. 1969. *Ent. exp. et appl. 12:*565.
54. Rehr, S. S., P. P. Feeny, and D. H. Janzen. 1973. *J. Anim. Ecol. 42:*405.

55. Rice, E. L. 1974. "Allelopathy." Academic Press, New York.

56. Root, R. B. 1973. *Ecol. Monogr. 43:*95.

57. Root, R. B. 1975. *In* "Ecosystem Analysis and Prediction" (S. A. Levin, ed.), pp. 83-97. Siam Institute for Mathematics and Society, Philadelphia.

58. Rothschild, M. 1972. *In* "Insect/Plant Relationships" (H. F. van Emden, ed.), pp. 59-83. Blackwell Scientific Publications, Oxford.

59. Schoonhoven, L. M. 1969. *Ent. exp. et appl. 12:*555.

60. Schoonhoven, L. M. 1972. *In* "Insect/Plant Relationships" (H. F. van Emden, ed.), pp. 87-99. Blackwell Scientific Publications, Oxford.

61. Scriber, J. M. 1975. "Comparative Nutritional Ecology of Herbivorous Insects: Generalized and Specialized Feeding Strategies in the Papilionidae and Saturniidae (Lepidoptera)." Ph.D. thesis, Cornell University, Ithaca.

62. Shapiro, A. M. 1968. *Ann. Ent. Soc. Amer. 61:*1221.

63. Soo Hoo, C. F., and G. Fraenkel. 1966a. *J. Insect Physiol. 12:*693.

64. Soo Hoo, C. F., and G. Fraenkel. 1966b. *J. Insect Physiol. 12:*711.

65. Southwood, T. R. E. 1961. *J. Anim. Ecol. 30:*1.

66. Southwood, T. R. E. 1972. *In* "Insect/Plant Relationships" (H. F. van Emden, ed.), pp. 3-30. Blackwell Scientific Publications, Oxford.

67. Strong, D. A. 1974a. *Proc. Nat. Acad. Sci.* (U.S.A.) *71:*2766.

68. Strong, D. A. 1974b. *Science 185:*1064.

69. Tahvanainen, J. O., and R. B. Root. 1972. *Oecologia 10:*321.

70. van Emden, H. F. 1972. *In* "Phytochemical Ecology" (J. B. Harborne, ed.), pp. 25-43. Academic Press, London and New York.

71. van Emden, H. F., and M. A. Bashford. 1969. *Ent. exp. et appl. 12:*351.

72. Varley, G. C., and G. R. Gradwell. 1958. *Proc. 10th Int. Congr. Ent.* (1956)*4:*133.

73. Varley, G. C., and G. R. Gradwell. 1962a. *Proc. 18th Ann. Sess. Ceylon Assoc. Adv. Sci.* p. 142.

74. Varley, G. C., and G. R. Gradwell. 1962b. *Proc. 11th Int. Congr. Ent.* (1960)*2:*211.

75. Varley, G. C., G. R. Gradwell, and M. P. Hassell. 1973.
     "Insect Population Ecology." Blackwell Scientific
     Publications, Oxford.
76. Vershaffelt, E. 1911. *Proc. Acad. Sci. Amsterdam 13:*
     536.
77. Virtanen, A. I. 1965. *Phytochemistry 4:*207.
78. Whittaker, R. H., and P. P. Feeny. 1971. *Science 171:*
     757.
79. Wilson, F. 1964. *Ann. Rev. Ent. 9:*225.
80. Woodwell, G. M. 1974. *Amer. J. Bot. 61:*749.

Chapter Two

INSECT-PLANT INTERACTIONS: NUTRITION AND METABOLISM

STANLEY D. BECK and JOHN C. REESE

*Department of Entomology, University of Wisconsin
Madison, Wisconsin*

INTRODUCTION

From a phytochemical standpoint, plants are producers
of chemicals, and insects are consumers. The biology of
the consumer role played by insects is a very complex
and intriguing subject, of which we will focus primarily
on only two aspects -- nutrition and metabolism. But
even having so delimited the subject of discourse, it is
immediately obvious that other aspects of insect-plant
interaction must be considered, at least peripherally,
for nutrition and metabolism are not isolated processes.
They occur in conjunction with the behavioral and chemosensory facets of the biology of the phytophagous insects.
It is an obvious truism that no insect is capable of
utilizing every plant species, and conversely that no
plant species is susceptible to attack by every species
of plant-feeding insect. From both practical and
theoretical viewpoints, the most important questions
pertain to the identity of factors determining host

specificity among the insects and susceptibility to insect depredation among the plants.

Three postulates are basic to our analysis of the nutritional and metabolic aspects of insect-plant relationships. They are: (1) that phytophagous insects have coevolved with their host plants, with the burden of adaptation being on the insects; (2) that different species of plant-feeding insects do not differ greatly in their specific biochemical nutritional requirements; and (3) that allelochemics (so-called secondary plant biochemicals) control the host specificity and the evolutionary processes of phytophagous insects. In the latter postulate, we are not asserting that physical, ecological, and phenological factors play no part; we are dealing here only with chemical factors.

NUTRITION AND HOST SPECIFICITY

Early 20th century biologists, confronting the great diversity of insect food habits, postulated that both sensory and nutritional factors played important roles in determining host plant specificity. Different plants were thought to vary greatly in their food value for different insects, so that host plants specificity was based in large part on the insect's unique nutritional requirements. Some insects were considered to be nearly autotrophic; e.g., a prominent nutritionist reported that *Drosophila* sp. could be reared on an artificial diet consisting of sugar, tartaric acid, potassium phosphate, and magnesium sulfate in an agar base[114]. He later reported that the addition of ammonium salts or amino acids was beneficial, and speculated that perhaps bacterial contaminants synthesized some protein[115]. Pearl (1926) used a medium nearly identical to that of Loeb, but found that the addition of yeast was beneficial. An early comprehensive review of insect nutrition and metabolism[188] clearly reflected the view that insects were able to subsist on a wide variety of plants and other substances because of their peculiar and relatively simple nutritional requirements.

With the rise of modern nutritional concepts, the isolation and identification of essential vitamins, amino acids, etc., the field of insect nutrition advanced very rapidly. An insect requirement for a dietary source of the sterol nucleus was described by Hobson[76], and has been shown to be true of all insect species investigated, with the exception of those harboring symbiotic microorganisms[35,36,83,84]. Except for the lack of being able to synthesize sterols, it is now quite apparent that insects do not differ markedly from other animal forms in their fundamental nutritional requirements.

The research of the past three decades has also firmly established the concept that the multitude of species making up the class Insecta do not differ greatly in their fundamental qualitative requirements for biochemical nutrients that must be obtained by ingestion and assimilation. Some differences and specializations occur, of course, and have been reported in respect to sterol identity, ascorbic acid requirement, and requirements for choline, carnitine, and other of the minor B vitamins; specific amino acid ratio requirements have also been found to differ among the species. But these specialized differences do not seem to be any greater than the differences found among the nutritional requirements of different species of birds, or mammals, or any other major group. Based on what is now known of the biochemistry of insect nutritional requirements, it does not seem at all likely that the insect's nutritional requirements play more than a minor role in determining host plant specificity.

The nutritional quality of plant tissues has been shown to influence its utilization by insects in some instances. Although different species of green plants may not differ much in their qualitative content of nutrients required by insects, they may differ in a quantitative sense. The importance of dietary proportions of required nutrients may be greater than their absolute quantities [85-87,117,196]. The pea aphid, *Acyrthosiphon pisum*, was shown to be more successful in colonizing pea plants having a high content of free amino acids than those containing lower concentrations[4,5]. The fecundity of the Colorado potato beetle, *Leptinotarsa decemlineata*

was found to be dependent in part on the phospholipid
content of its solanceous host plant[61], and larvae of
the species grew faster on young foliage than on old,
apparently because of a higher amino acid content in
the younger foliage[34]. The biochemical composition of
phloem sap, particularly in respect to amino acids, has
been shown to be among the several factors influencing
alatae determination in *Myzus persicae* and other
aphids[65,113,123-125,198]. Other examples might also be
cited, but they are reports of similarly minor influences
of nutritional quality of plant tissues on the developmental biology of phytophagous insects.

The examples of nutritional influences cited above
are all seen to occur within the normal host range of
the insects concerned; host plant specificity is not so
determined, only small degrees of differences in host
utilization are involved. This important point may be
underlined by citing the work of Parrott, *et al.* (1969)
on the boll weevil, *Anthonomus grandis.* A comparison
was made of the amino acids (both free and protein) found
in the buds of a series of hosts and nonhosts of the boll
weevil. The amino acid profiles were qualitatively
similar, and all contained the amino acids required by
the weevil. The investigators concluded that there were
insufficient qualitative and quantitative differences to
explain the host or nonhost status of the plant species
sampled.

Modern studies have shown that phytophagous insects
do not vary widely in the biochemical nutrients required,
and that those needed chemicals are present in almost any
green plant. Nevertheless, it has been known for a great
many years that plant species vary in their adequacy as
hosts for even the most polyphagous insects. Grasshoppers
(*Melanoplus* spp.) displayed quite different mortality,
rates of nymphal development, adult longevity, and fecundity with different host plants [22,77,169]. The effects of
different food plants on the development of the pale
western cutworm, *Agrotis orthogonia*[155] and of the
variegated cutworm *Peridroma margaritosa*[170] were also
studied. Both species of cutworm displayed great differences in rates of larval development, survival, adult
emergence, and adult longevity, depending on the host plant
on which they were fed. Survival to the adult staged ranged

from 0 to 85%; larval periods varied from 22 to 44 days, and pupal weights from 232 to 480 mg. These studies demonstrated that the apparent food value of different acceptable host plants was far more variable than could be accounted for on the basis of qualitative or quantitative differences that might be measurable in the nutrients contained in the plant tissues. Obviously, host plant utilization depends on more than the plant's content of the specific nutrient biochemicals required by the insect.

ALLELOCHEMIC INTERACTIONS

The striking similarities in the nutritional require-ments of a broad range of insect species led Fraenkel (1959) to postulate that the basic nutritional requirements of all plant-feeding insects are virtually identical, so that any phytophagous insect could thrive on the tissues of any green plant, if the insect could be induced to eat enough of it. Food habits and host plant specificity were postulated to be determined by the effects of "secondary plant chemicals" that attracted or repelled the insects and influenced their locomotor, ovipositional, and feeding behavior. (See Chapman (1974) for a recent detailed review of the feeding behavior of phytophagous insects.) Fraenkel's theory was updated in 1969 to include the possible effects of secondary chemicals on developmental and physiological processes as well as behavioral patterns[49]. The theory has had a great impact, in that it provided a theoretical basis for broader based studies of insect—plant interactions. It has also encouraged a shift of emphasis from purely biochemical determinations of requirements for various amino acids, vitamins, etc., to more relevant aspects of the host specificity problem. To distinguish it clearly from the purely nutritional, the broader consideration can properly be referred to as "insect dietetics"[9]. Under the rubric of dietetics, it is recognized that the feeding insect must ingest a substrate that not only meets its nutritional requirements, but is also capable of being assimilated and converted into the energy and structural substances required for normal activity and development. It is in the dietetic sense that we will deal with aspects of insect-plant interactions in this paper.

According to the Fraenkel theory, "secondary plant chemicals" are defensive substances tending to render the plant repellent, toxic, or otherwise chemically unsuitable for utilization by herbivores (including insects) and microorganisms. These defense chemicals were dubbed "secondary" because they play no known role in the essential metabolic processes involved in the plant's physiology, and are thus distinguished from the "primary" plant chemicals whose physiological functions have been elucidated. The secondary plant chemicals include a vast array of glycosides, alkaloids, steroids, saponins, tannins, essential oils, phenolics, and other organic compounds whose metabolic functions are currently obscure. There may be considerable question as to whether all such compounds are purely defensive in function and devoid of any other metabolic significance[62,145], but that question lies outside the realm of the subject under discussion.

Plant defense substances and their roles in the chemical interactions between species has been critically analyzed and reorganized on a broad basis[199,200]. Whittaker (1970) proposed the term "allelochemics" to replace "secondary plant substances", and defined an allelochemic as being a non-nutritional chemical produced by an organism of one species, and which affects the growth, health, behavior, or population biology of another species. Allelopathy, production of phytoalexins, attractants, repellents, deterrents, inhibitors, and toxicants are examples of allelochemic interactions. Two classes of allelochemic effects are of particular pertinence in insect-plant interactions: (1) allomones, which are chemical factors giving an adaptive advantage to the producing organism (host plant); and (2) kairomones, which are chemical factors giving an adaptive advantage to the receiving organism (insect)[200].

From a purely chemical standpoint, the coevolutionary give-and-take between insects and plants can be viewed, perhaps simplistically, as the production of allomones by the plant, and their subsequent neutralization or eventual exploitation as kairomones by the insect population. Having arisen by chance mutation or genetic recombination, an allomone might tend to confer a selective advantage, enabling the plant to enter a new adaptive zone. Ensuing evolutionary radiation might

result in what started as a chance genetic characteristic
becoming a characteristic typical of an entire plant
population, species, or even family. On the other hand,
a genetic mutation or recombination appearing in a
population of insects might enable some individuals to
utilize the previously protected plant group, with the
mutant insects responding positively (feeding, oviposition)
to the prohibitive allomone, and thereby converting the
allomone into a kairomone. The selective advantage
thereby conferred on the mutant insects would allow them
to enter a new adaptive zone, where diversification in
the absence of competition from other herbivores (including
the parent population) could occur[10,43].

According to this theory, the token stimuli by which
a phytophagous insect identifies suitable host plants are
kairomones (attractants, stimulants) that were originally
evolved as allomones (repellents, deterrents), the obvious
exceptions for pollinators. An efficient plant allomone
must not only deter the insect behaviorally, but must also
exert an adverse metabolic effect, thereby severely
punishing errancy and enforcing the selective advantage of
the allomone as a deterring token stimulus. Insect
exploitation of allomones as host-identifying kairomones
must, therefore, include metabolic adaptations by which
the plant chemical factor is degraded or utilized by the
insect. In assessing the role of allelochemics in the
evolution of insect-plant interactions, behavioral
allomones and kairomones must also be investigated from
a metabolic standpoint. The extensive compilation of
behavioral allelochemics by Hedin *et al.* (1974) includes
many plant chemicals that exert adverse metabolic as well
as behavioral responses.

The complex effects of allelochemics on the feeding
behavior, dietetics, and ecology of a phytophagous insect
are well illustrated by a recent study of the effects of
sinigrin on the black swallowtail butterfly, *Papilio
polyxenes*[44]. Sinigrin is a mustard oil glycoside, the
active portion of which is the aglucone, allylisothiocyanate (Fig. 1). A number of crucifer-feeding insect species
respond to sinigrin as the kairomone identifying acceptable
hosts for oviposition and/or feeding[46,88,128,185,190,191,197].
Butterflies of the genus *Papilio* do not attack crucifers,

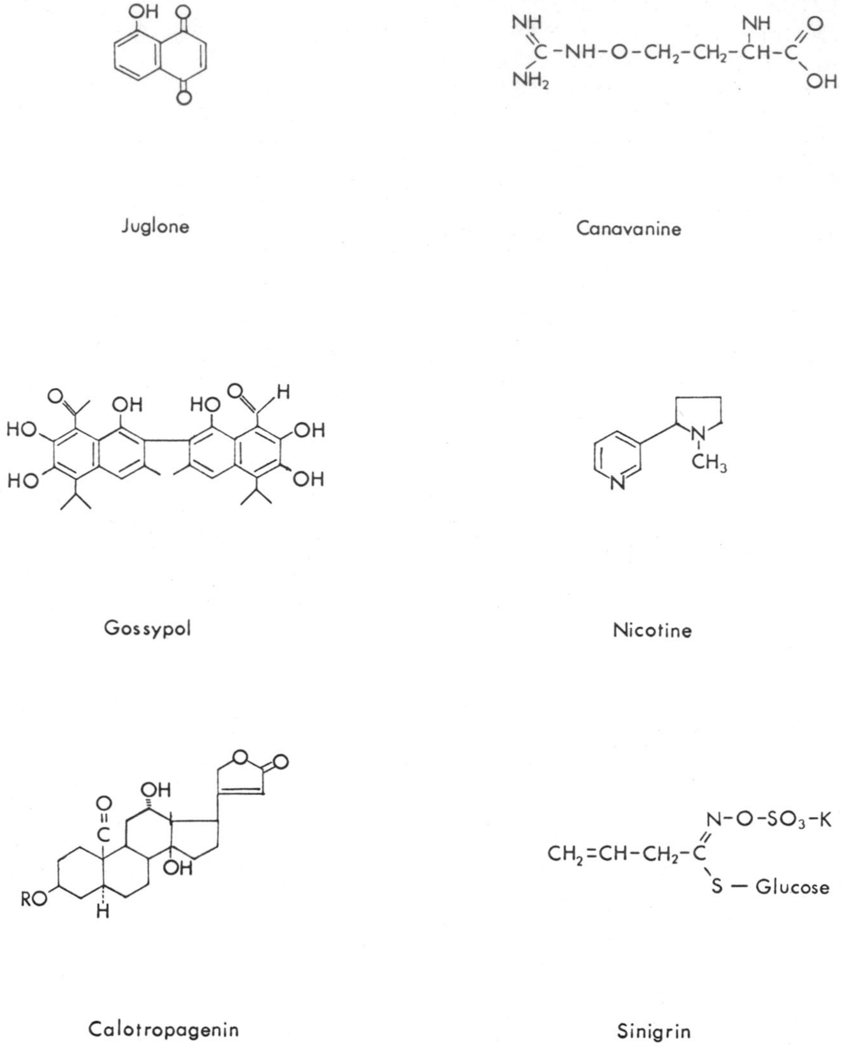

Juglone

Canavanine

Gossypol

Nicotine

Calotropagenin

Sinigrin

Figure 1.   Some plant produced allelochemics of
significance in the dietetics of phytophagous insects.

but utilize host plants of the family Umbelliferae. A
number of essential oils from umbelliferous plants serve
as token stimuli, attracting and stimulating larval
feeding[39]. Erickson and Feeny (1974) reared *Papilio*
larvae on celery leaves (Umbelliferae, *Apium graveolens*)
that had been perfused with known amounts of sinigrin.
Larval feeding was not deterred by the presence of sinigrin
in the celery tissue, but larval growth was greatly
inhibited, with complete mortality occurring at sinigrin
concentrations typically found in cruciferous plant
tissues. The sinigrin had an adverse effect on the
assimilation of food from the gut lumen and on the
efficiency of conversion of ingested nutrients to larval
tissues. The investigators concluded that the utilization
of cruciferous hosts by the black swallowtail butterfly
could occur only after two evolutionary adaptations.
Adaptation to crucifers would involve not only behavioral
evolution (oviposition and feeding acceptance) but also
a metabolic adaptation to reduce the toxic effects of
allylisothiocyanate. Such a breakthrough to cruciferous
host plants would be even further complicated by the
necessity of an adult behavioral mutation enabling
oviposition on crucifers, because Wiklund (1975) showed
that adult and larval host plant preferences are controlled
by separate gene complexes.

     There are many examples of plant-produced allelochemics
that function as allomones to some organisms but as
kairomones to others. Juglone (5-hydroxy-1,4-
naphthoquinone) (Fig. 1), for example, is an allomone
in allelopathic interactions of walnut, hickory, and
perhaps other trees. Juglone has also been shown to be
a feeding deterrent to the smaller European elm bark
beetle, *Scolytus multistriatus*, but not to the closely
related hickory bark beetle, *S. quadrispinosis*[54,55].
Sinigrin was shown to stimulate the feeding of the cabbage
aphid, *Brevicoryne brassicae*, but to deter feeding by
the pea aphid, *Acyrthosiphon pisum*[128]. Similarly,
phlorizin, a phenolic compound found in leaves of apple
(*Malus*), deters probing attempts by the non-apple-feeding
aphids, *Myzus persicae* and *Amphorophora agathonica*, but
not probing by the apple aphid, *Aphis pomi*[126]. Phlorizin
has been reported to be a feeding stimulant for the apple-
feeding aphids *Aphis pomi* and *Rhopalosiphum insertum*[104].

At the metabolic level, insects have converted some allomones into kairomones in some interesting ways. The poisonous cardenolides (principally calactin and calotropin, Fig. 1) contained in many milkweed tissues (*Asclepias* spp.) are allomones that protect these plants from most herbivores. The monarch butterflies, *Danaus plexippus* and *D. chrysippus*, and the larger milkweed bug, *Oncopeltus fasciatus*, are brightly colored aposematic species that sequester cardenolides from their milkweed hosts and utilize them as their allomones for protection against predation[40,140]. Carlisle *et al*. (1965) found that reproductive activity of the desert locust, *Schistocerca gregaria*, was stimulated by the terpenoids produced by the aromatic shrub, *Commiphora myrrhae*. The terpenoids included α-pinene, β-pinene, limonene, and eugenol. A single contact with any of these terpenoids was sufficient to evoke vitellogenesis and reproductive activity. There is also evidence that insects may utilize host plant constituents as pheromones (sex attractants), sequestering these compounds from the host tissues and releasing them later as pheromones[73]. Riddiford (1967) demonstrated that the volatile compound trans-2-hexanal from oak leaves was a required stimulus for the release of sex pheromone by female polyphemus moths, *Antheraea polyphemus*.

ALLELOCHEMICS AND NUTRITIONAL PHYSIOLOGY

As was discussed in an earlier section, plant-feeding insects do not thrive equally well on all plant tissues, even when there are no behavioral barriers to their feeding. The differences in growth, survival, etc., are much greater than can be accounted for on the basis of possible differences in nutrient content of the different plant tissues. It seems likely, therefore, that allelochemics characteristic of various plant species exert strong influences on the developmental success of feeding insects. A number of investigators have studied the effects of plant tissues and known allelochemics on the nutritional physiology of insects, from the standpoints of rates of ingestion, digestion, assimilation, and conversion to insect tissues. Such investigations involve the use of nutritional indices calculated from measurements of the amounts of food ingested, amounts of fecal output, and the weight gained by the feeding insects. For greatest

accuracy, all measurements should be on a dry-weight basis.
Three indices are of greatest interest:

(1)   Approximate Digestibility (AD)

$$AD = \frac{\text{wt of food ingested} - \text{wt of feces}}{\text{wt of food ingested}} \times 100$$

(2)   Efficiency of Conversion of Digested Food (ECD)

$$ECD = \frac{\text{wt gained}}{\text{wt of food ingested} - \text{wt of feces}} \times 100$$

(3)   Effieciency of Conversion of Ingested Food (ECI)

$$ECI = \frac{\text{wt gained}}{\text{wt of food ingested}} \times 100$$

The nutritional index AD provides an estimate of the
proportion (%) of ingested materials that is assimilated.
It is only an approximation of actual uptake of nutrients
across the gut wall, because fecal weights are not corrected
for urine content and insect-produced debris such as
peritrophic membrane and exuviae. AD, therefore, under-
estimates actual assimilation to a small degree. ECD
provides an estimate of the efficiency of the biological
system to convert assimilated substances into biomass.
Nutrients expended for energetic metabolism are not accounted
for by this nutritional index, as there is no measure of
either CO_2 production or the synthesis of detoxification
metabolites. Therefore, ECD values will decrease as the
proportion of assimilated substances metabolized for energy
increases. ECI estimates the overall efficiency by which
ingested substrate is converted into insect mass. ECI is
influenced by both AD and ECD, in that a low AD accompanied
by a high ECD may result in an ECI that is about the same
as that produced by a high AD and low ECD. For details
of the experimental techniques and the precautions attending
the determination of these and other nutritional indices,
the reader is referred to the detailed review by Waldbauer
(1968) and to Reese and Beck (in press).

Determination of nutritional indices displayed by
phytophagous insects on host and nonhost plants has revealed
very marked differences. Using the southern armyworm,
Prodenia eridania, Soo Hoo and Fraenkel (1966a, b) compared
the feeding behavior, growth, survival, and nutritional
indices of the insect on plant tissues representing 32
families. This polyphagous insect refused to feed on the
plants of only 6 families (Polypodiaceae, Taxaceae,
Pinaceae, Salicaceae, some Gramineae, and some Umbelliferae).
All others tested were fed upon either immediately or
within a 24-hour period of exposure. Of the wide range of
plants on which the larvae would feed, host suitability
was quite varied, ranging from normal rapid larval growth
to complete mortality prior to attaining the pupal stage.
Nutritional indices were determined for larvae feeding on
fresh plant tissues representing 12 different plant families.
AD values ranged from 36% on a poor host to 73% on good host
tissues; ECD values were from 16 to 57%; and ECI values were
from 10 to 38%. Waldbauer (1964) determined nutritional
indices of the tobacco hornworm, *Manduca sexta*, larvae on
its normal solanaceous host plants and on nonhosts. In the
latter case, the larvae were induced to feed on the nonhost
plant tissues by surgical removal of maxillary sense organs
that enable the insect to distinguish host from nonhosts.
The ECI values tended to be slightly higher when the larvae
were fed solanaceous hosts than when induced to feed on
nonhosts (Compositae and Scrophulariaceae).

The use of nutritional indices as a basis for comparing
host plant utilization by a phytophagous insect being reared
on different host plants has some serious limitations[195].
Variations in water and fiber contents may result in
variations in the indices[109,171,172,194]. A relatively low
AD might be the result of high fiber content of the diet,
but might also be caused by the presence of wound-induced
antiproteinases in the plant tissue[57,147] which would
reduce the digestibility of the ingesta. A low ECD might
result from antimetabolites present in the dietary material,
but could also be produced by unfavorable amino acid ratios
that would prevent the synthesis of normal structural
proteins. Plant tissues containing acute toxins preclude
the determination of meaningful nutritional indices,
because the larvae would sicken and/or die during the
experimental feeding period. For these, among other reasons,
it is apparent that nutritional indices are useful in a

preliminary assessment of host plant utilization, but are not
sufficient to identify the specific factors influencing the
efficiency of that utilization. The effects of chemical
plant factors, including allomones and nutritional factors,
on the efficiency of dietary utilization can best be
studied by means of nutritional indices if the dietary
substrate is standardized so that the effects of known
chemical factors can be assessed without equivocation. The
study by Erickson and Feeny (1974) of the effects of sinigrin
on development of *Papilio polyxenes* conformed to this
standard. They used the same type of plant tissue in each
assay, but added known amounts of sinigrin by perfusion.
They were then able to demonstrate that sinigrin did not
affect the rate or amounts of ingestion, but exerted an
adverse effect on AD. Of the material that was assimilated,
however, conversion to biomass was unimpaired by sinigrin;
that is ECD was not affected.

In our laboratory, a continuing study is being conducted
on the effects of allelochemics on the development and
dietetics of a polyphagous lepidopteran, the black cutworm
(*Agrotis ipsilon*). The insect is reared on a modification
of the artificial diet developed by Reese *et al.* (1972),
to which allelochemics are added at known concentrations.
The effects of the allelochemics on larval feeding behavior,
survival, developmental rates, and nutritional indices are
then measured. A detailed description of the methods
employed and some of the experimental results are currently
in the process of being published[137] (Reese and Beck, in
press). A number of plant chemicals with possible allomone
functions (Fig. 2) were tested at a wide range of concentrations ($3.75 \times 10^{-7}M$ to $3.75 \times 10^{-2}M$) for effects on
larval survival, growth rate, pupation rate, and pupal
weight. Any compound that resulted in a statistically
significant correlation between its dietary concentration
and one of the above biological parameters was then used
(at $3.75 \times 10^{-2}M$) in nutritional index determinations.
Nutritional indices obtained with the active compounds are
shown in Table I. Hydroquinone, dopamine, chlorogenic
acid, and ferulic acid were inactive.

The experimental results (Table I) show a number of
points of interest. ρ-Benzoquinone reduced the amount of
ingestion and the approximate digestibility (AD). The
reduced efficiency of assimilation was compensated for by

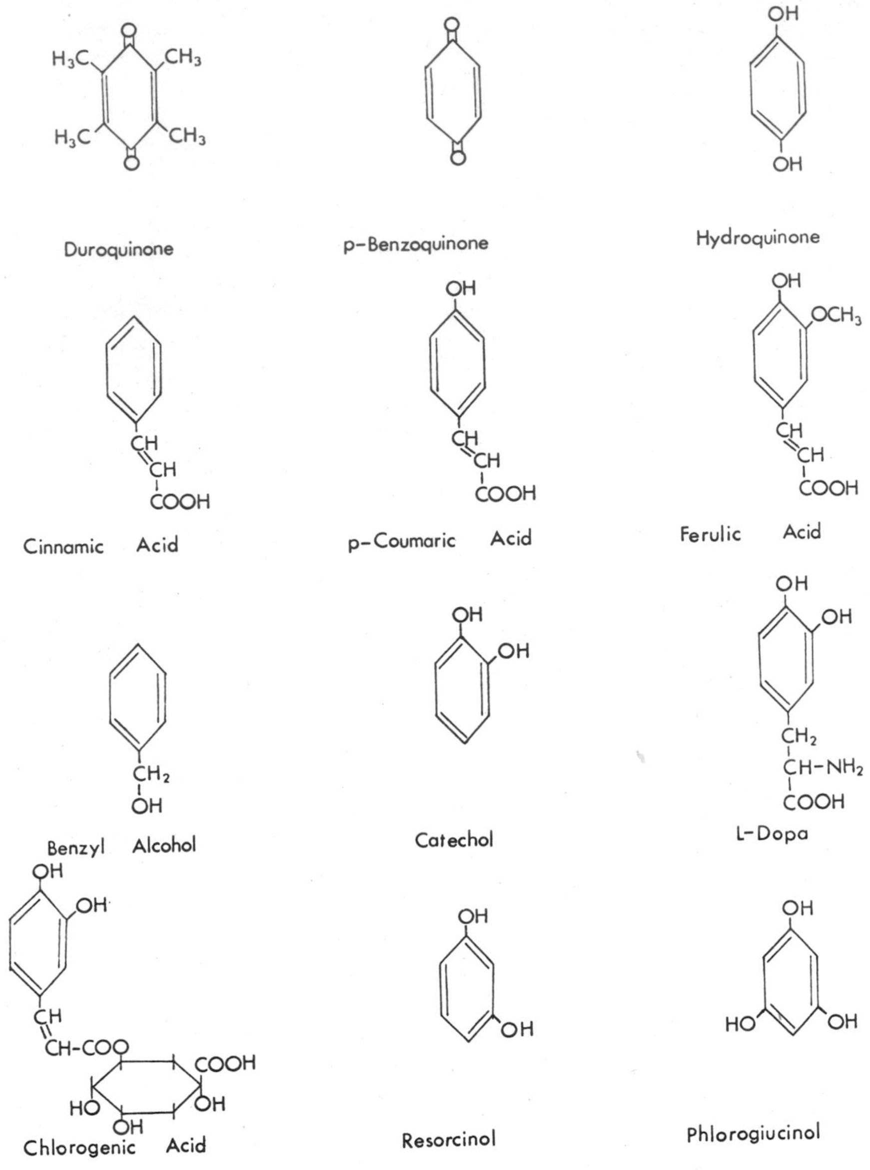

Figure 2.   Allelochemics used in developmental and nutritional index experiments with the black cutworm, *Agrotis ipsilon*.

Table I. Effects of some allelochemics (3.75 X 10^{-2}M) on larval weight gain, ingestion and nutritional indices in the black cutworm, *Agrotis ipsilon*. All values expressed as percent of control and all determinations were made on a dry weight basis (% = $\frac{\text{experimental value}}{\text{control value}}$ X 100).

Allelochemic	Amount Ingested	Wt. gain	AD	ECD	ECI
ρ-Benzoquinone	71.3**	71.7**	74.7**	139.8**	102.5
Duroquinone[a]	26.6**	5.5**	210.3**	10.4**	18.1**
Catechol	95.9	75.3**	97.1	82.2**	79.4**
L-Dopa	111.2	64.3**	77.0**	74.2**	56.3**
Resorcinol	43.8**	51.4**	108.7	114.9**	100.4
Phloroglucinol	80.7	61.3**	88.0	85.1*	73.1**
Cinnamic Acid	90.7	88.2	99.7	98.3	98.5
Benzyl Alcohol	123.9	115.7	108.3	82.4*	88.0

a A synthetic compound not reported in plants.
* Difference from the control significant at the 0.05 level of probability.
** Difference from the control significant at the 0.01 level of probability.

a greatly enhanced ECD, resulting in an overall ECI not
significantly different from that of the control larvae.
The effect of catechol, on the other hand, was through its
inhibition of ECD, resulting in a low ECI. L-Dopa exerted
no effect on ingestion, but inhibited both assimilation and
conversion (AD, ECD). The nutritional indices gave no
indication of the effect of resorcinol, and the data suggest
only that the reduced growth was the result of a reduced
rate of ingestion. Phlorglucinol inhibited growth mainly
through a reduction of ECD and consequently a reduced ECI.
Cinnamic acid inhibited pupation in the preliminary tests,
but showed no effects on the nutritional indices. Benzyl
alcohol also inhibited pupation in the preliminary experiments, but inhibited ECD only slightly. Duroquinone was
included in our experiments because of its structural
relationship to ρ-benzoquinone, but it has not been reported
to occur in plant tissues, as far as we know. Duroquinone
greatly increased AD, but strongly inhibited all other
parameters measured.

When chemicals are tested as single variables under
defined experimental conditions, nutritional index data
are useful in the identification of the physiological
processes influenced. Such data do not elucidate specific
biochemical modes of action; nor do they eliminate the
possibility of additional or non-nutritional effects. They
show that growth can be inhibited by plant allelochemics
in different ways, and they clearly show that the nutritional
physiology of the insects can be influenced by non-
nutritional substances. The nutritional index is a little-
used tool, but it may turn out to be very important in the
elucidation of insect-plant interactions.

We are currently just beginning research on the
possibility of synergistic effects among allelochemics.
If, for example, an allelochemic such as ρ-benzoquinone
inhibits AD but not ECD; whereas, catechol has no significant
effect on AD but inhibits ECD, the two together may exert
much greater effects than would be expected on an additive
basis. This phase may be quite important, because many of
these compounds are formed in response to plant tissue injury
(Fig. 3). Although the injury complex may differ somewhat
from species to species, it seems likely that the insect
must cope with a species-typical or cultivar-typical profile
of such substances. The additive and synergistic effects of

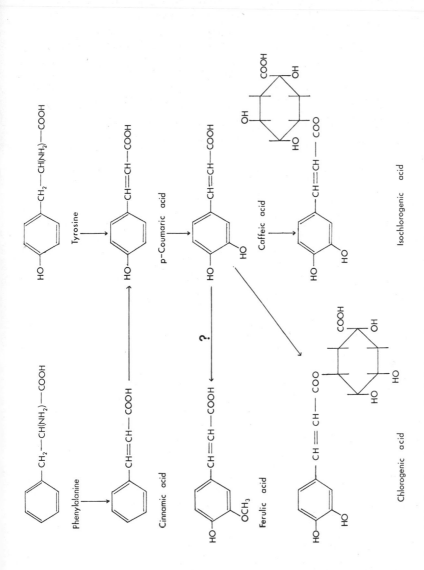

Figure 3. Allelochemics produced by plant tissues in response to injury or infection (modified from Galston and Davies, 1970).

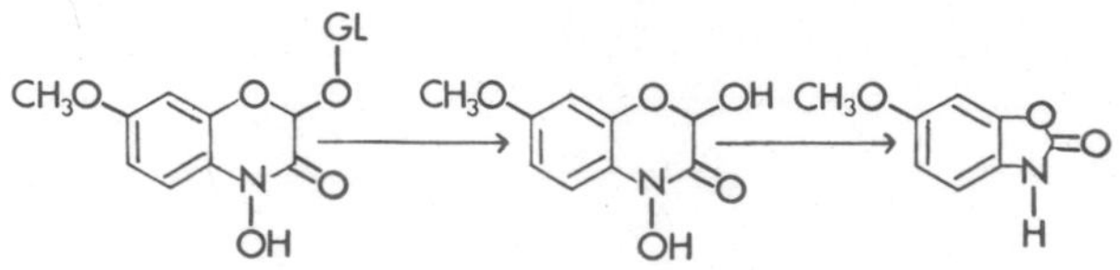

Figure 4.   Benzoxazolinone pathway following tissue injury in maize and other Graminae.   (Adapted from Klun, 1974).

the components of the injury complex may well spell the difference between a susceptible and a resistant host plant. The importance of allelochemics produced in response to injury or infection is also indicated by the benzoxazolinone series of compounds occurring in maize, wheat, and other grain plants.   The benzoxazolinone series (Fig. 4) have been shown to be of importance in both insect resistance[6,7,8,12,168,186] and fungal resistance[192,193].   Although the components of the series may differ in both concentration and biological activity, the feeding insect must deal with the complex as it is formed in the plant tissue being ingested.

ALLOMONES IN INSECT-PLANT INTERACTIONS

Amino acid analogs.   A little-explored aspect of the dietetics of phytophagous insects is the possible importance of antimetabolites in plant tissues.   Particularly, amino acid analogs might greatly influence the apparent nutritional value of plant tissues by exerting adverse effects on amino acid assimilation, metabolism, and the synthesis of proteins. Amino acid analogs have been isolated from the seeds and tissues of a number of plant species, many of which exert toxic effects on animal organisms (see Bell (1972) for a recent review of toxic amino acids in legumes).

The role of amino acid antimetabolites in insect-plant interactions has been investigated in only a few instances.

L-Canavanine, an analog of arginine (Fig. 1), occurs in the seeds of some legumes and has a toxic or growth-inhibiting effect on insects feeding on the seeds[16]. When incorporated into artificial diets of the tobacco hornworm, *Manduca sexta*, canavanine caused larval mortality and developmental deformities[38]. Canavanine has also been demonstrated to exert adverse effects on the development of silkworm larvae, *Bombyx mori*[92], and to act as a competitive inhibitor of arginine metabolism when present in the diet of the boll weevil, *Anthonomus grandis*[189].

Phenylpropane derivatives. The plant chemicals discussed above in the section on nutritional indices fall into the phenylpropane series of allelochemics as classified by Whittaker and Feeny (1971). These compounds include relatively simple phenols, quinones, tannins, and lignins that occur widely throughout the plant kingdom. The phenylpropane series of compounds are apparently synthesized via the shikimic acid-aromatic amino acid pathway[20]. Chemicals of this group that have been reported to exert allelochemic influence in the nutritional and metabolic aspects of insect host plant specificity and/or host plant resistance have been compiled (Table II). As was pointed out previously, many if not most of these chemicals function as behavioral as well as developmental and metabolic allomones (see Hedin, *et al.* (1974) for a compilation of phytochemicals affecting behavior).

Most of the phenylpropane-derived allelochemics appear to be general plant defense substances, rather than specific adaptations evolved in response to depredation by a specific herbivore or pathogen. Some that are of general distribution are thought to be produced in response to tissue injury caused by pathogen invasion[15] and the synthetic pathway is shown in Fig. 3. That such an injury reaction may also occur in response to insect feeding is suggested by the results of Hori (1973), who reported that the phenylalanine content of insect-injured sugar beet tissue was over twice that of uninjured tissue; a number of other amino acid concentrations also increased in response to injury.

The allelochemics underlying plant resistance to pathogens are, in many instances, the same substances implicated in resistance to insects. For example, L-dopa has been implicated in disease resistance of bananas[15] and

Table II. Representative plant allelochemics that have been shown to affect insect growth, development, reproduction or metabolism.

COMPOUND	INSECT	REFERENCE
Catechol	*Schizaphis graminum*	187
Resorcinol	" "	"
Hydroquinone	" "	"
Pyrogallol	" "	"
Phlorglucinol	" "	"
Salicylic Acid	" "	"
Gentisic Acid	" "	"
Protocatechuic Acid	" "	"
Gallic Acid	" "	"
Vanillic Acid	" "	"
Syringic Acid	" "	"
o-Coumaric Acid	" "	"
m-Coumaric Acid	" "	"
p-Coumaric Acid	" "	"
cis-trans Caffeic Acid	" "	"
cis-Caffeic Acid	" "	"
Sinapic Acid	" "	"
6-Methoxybenzoxazo-linone (MBOA)	*Ostrinia nubilalis*	13
	" "	7
	" "	106
	" "	12
Benzoxazolinone	" "	"
2,4-Dihydroxy-7-methoxy-1,4-benoxazin-3-one (DIMBOA)	" "	107
α-Pinene	*Schistocerca gregaria*	26
β-Pinene	" "	"
Limonene	" "	"
Eugenol	" "	"
Giberellic Acid	*Spodoptera littoralis*	148
	Heliothis vivescens	63
Sesamin	*Bombyx mori*	96
Kobusin	" "	"
Chlorogenic Acid	*Schizaphis graminum*	187
	Macrosiphus euphorbiae	31
Chlorogenic Acid	*Bombyx mori*	100

Table II - Continued

COMPOUND	INSECT	REFERENCE
Myristicin	*Bombyx mori*	92
Gossypol	*Heliothis zea*	161
	H. virescens	
	Pectinophora gossypiella	162
	Heliothis vivescens	
	Heliothis zea	
	Anthonomus grandis	119
	Heliothis zea	116
	H. vivescens	
	Spodoptera exigua	21
	Heliothis zea	
	Trichoplusia ni	
	Estigmene acrea	
Hydrocoumarin	*Schizaphis graminum*	187
Scopoletin	" "	"
d-Catechin	" "	"
Naringenin	" "	"
Esculin	*Acheta domesticus*	129
Hesperidin Methyl-chalcone	" "	"
Hesperidin	" "	"
	Pectinophora gossypiella	160
	Heliothis zea	
	H. virescens	
Morin	" "	"
Quercitrin	*Pectinophora gossypiella*	"
	Heliothis	
	H. virescens	
Isoquercetin	" "	"
Quercetin	*Schizaphis graminum*	187
	Anthonomus grandis	119
	Pectinophora gossypiella	160
	Heliothis zea	
	H. virescens	
	Heliothis zea	116
	H. virescens	
Rutin	*Acheta domesticus*	129
	Pectinophora gossypiella	160
	Heliothis zea	
	H. virescens	

Table II - Continued

COMPOUND	INSECT	REFERENCE
Rutin	*Heliothis zea*	116
	H. virescens	
Coumarin	*Macrosiphum euphorbiae*	31
Myristicin	*Bombyx mori*	92
Ponasterone A	" "	163
Ecdysterone	" "	"
Ecdysterone	" "	95
Inokosterone	" "	163
	" "	95
β-Sitosterol	*Heliothis zea*	63
	Spodoptera littoralis	148
Tigogenin	*Melanoplus bivittatus*	64
Digitonin	" "	"
Diosgenin	" "	"
Hecogenin	" "	"
Medicagenic Acid	*Tribolium castaneum*	159
Berberine	*Bombyx mori*	91
Nicotine	*Manduca sexta*	133
	" "	204
	Bombyx mori	
Nornicotine dipicrate	*Melanoplus bivittatus*	64
Solanine	" "	"
Tomatine	" "	"
Santonine	" "	"
Lupinine	" "	"
Pilocereine	8 *Drosophila* spp.	103
L-Dopa	*Prodenia eridania*	139
L-Canavanine	*Bombyx mori*	91
	Anthonomous grandis	189
	Manduca sexta	38
	Bombyx mori	93
Indican	*Melanoplus bivittatus*	64
Indole-3-acetonitrile	*Ostrinia nubilalis*	167
	Galleria mellonella	
Cytokinin (N[6]benzyladenine)	*Chaetosiphon fragaefolii*	150
Sinigrin	*Papilio polyxenes*	44

in insect resistance of *Mucina* seeds[139]. Similarly,
catechol is apparently a factor in disease resistance in
onions[15] and has been found to be toxic to the greenbug,
Schizaphis graminum[187] and to inhibit the growth and
digestive physiology of the black cutworm (Table IO. The
complexity of allelochemic interactions is further illustrated in the case of catechol by the finding that it is a
feeding stimulant for the smaller European elm bark beetle,
Scolytus multistriatus[122,130]. Catechol functions, therefore, as an allomone against some pathogens and insects,
but as a kairomone in the case of the bark beetle.
Chlorogenic acid is another example, in that it is toxic
to the greenbug[187] but is a required dietary component
for the silkworm[100].

Although many phenylpropanoid compounds have been shown
to affect insect behavior, relatively few have been investigated for their influences on insect growth and metabolism
(Table II). The metabolic pathways by which these phytochemicals are degraded or metabolized by the ingesting
insect are unknown, but Krieger *et al.* (1971) suggested
the possible involvement of microsomal oxidases in the
metabolic disposal of such allomones.

Quercitin and its disaccharide, rutin, were also
tested with the boll weevil by Maxwell *et al.* (1967).
Dietary quercitin did not influence either feeding or
oviposition, but had a slight stimulating effect on larval
growth rates.

Terpenoids. The dimeric sesquiterpenoid gossypol
(Fig. 1), a constituent of both foliage and seeds of the
cotton plant, *Gossypium hirsutum*, is of importance in
insect utilization of cotton varieties[119-121,131,132]
Maxwell and coworkers (1967) reported that gossypol stimulated feeding and oviposition of the boll weevil, but
relatively high concentrations (2% or more) had a slight
inhibitory effect on growth and development. The cotton
bollworm, *Heliothis zea*, and the budworm, *H. virescens*,
grew at lower than normal rates when reared on artificial
diets containing gossypol[161,162]. These investigators found
that gossypol caused a reduction in the nutritional index
AD in the case of *H. zea*, but not *H. virescens*; ingestion
rates declined with increasing concentrations of dietary
gossypol with both species.

For other terpenoids influencing insect dietetics, see Table
II.

Of the terpenoids contained in plant tissues, much
recent interest has been focused on those substances showing
juvenile hormone (JH) activity. An unexpected juvenilizing
effect on the plant bug *Pyrrhocoris apterus* was traced to
a factor contained in paper pulp[165,166]. The JH activity
was traced to two sesquiterpenoid esters: methyl todomatuate
(juvabione)[22] and a compound differing from juvabione only
by the presence of a double bond at C4 of the side chain
(dehydrojuvabione)[28] (see Fig. 5). The juvabiones occur in
the bark and wood of many conifers, but at the highest
concentrations in balsam fir, *Abies balsamea*[22]. JH activity

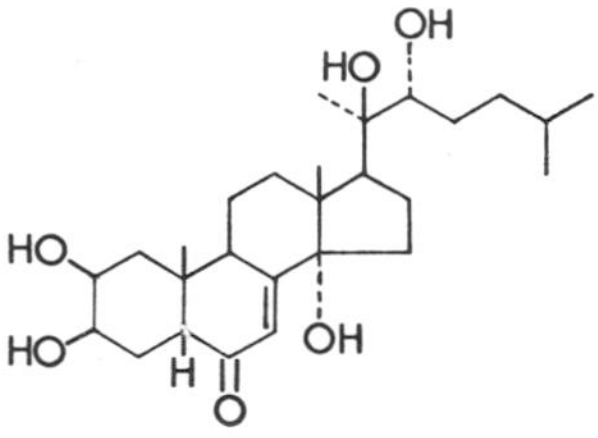

Ponasterone A

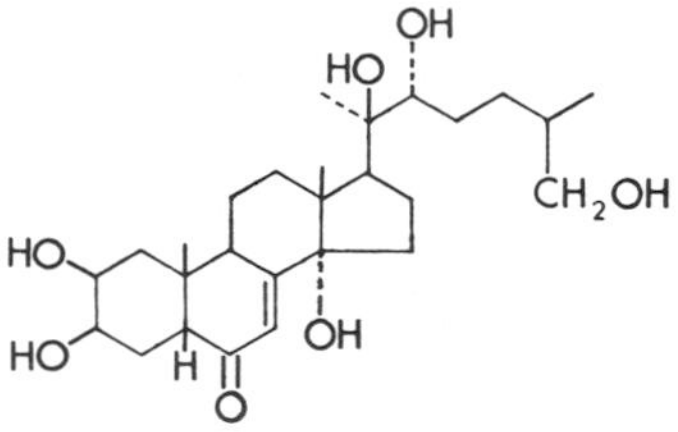

Inokosterone

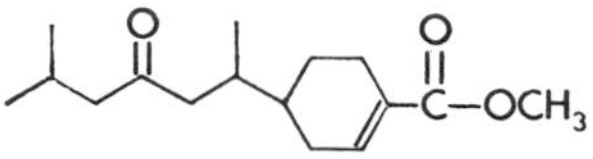

Juvabione

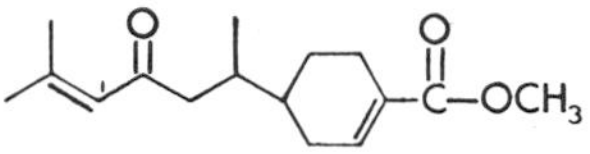

Dehydrojuvabione

Figure 5.   Examples of insect hormone analogs that
have been identified in plant tissues.

has been reported in extracts of bark and wood of red cedar, spruce, hemlock, and pine as well as balsam[118]. A survey of 48 species of common midwestern plants showed that JH activity was detectable in the tissue extracts of only five (*Grindelia squarrosa*, *Boltonia asteroides*, *Trifolium repens*, *Echinacea pallida*, and *Nepeta cataria*)[182].

Farnesol occurs in many plant tissues, and its pyrophosphate is an important intermediate in the mevalonate pathway of terpenoid and steroid synthesis. Farnesol was found to have JH activity by Schmialek (1963), and a large array of farnesol derivatives and analogs have been shown subsequently to exert varying degrees of JH activity[55,74]. Naturally occurring insect-produced juvenile hormones are sesquiterpenes that are farnesyl esters (i.e., methyl 10, 11-epoxy-7-ethyl-3, 11-dimethyl-2, 6-tridecadienoate)[146]. The mevalonic acid-farnesol pathway may be involved in JH biosynthesis, but this point has not been clearly established[55].

However interesting the occurrence of JH-active compounds in plants may be, there is no currently available evidence to suggest that these substances play any role in the host plant relationships of phytophagous insects.

Steroids. With the exception of those that possess specialized internal symbiotes, insects require a dietary source of sterols. Insects have the metabolic mechanisms required to effect side chain and substituent changes in ingested ·sterols, but are incapable of synthesizing the sterol nucleus[18]. Although insects possess the mevalonic acid metabolic pathway, it appears to terminate at squalene, and the insect is unable to synthesize sterols from either squalene or lanosterol[18,35,143].

Cholesterol is the principal structural sterol in most insects, with low levels of 7-dehydrocholesterol also being present. These sterols are essential structural components of cell membranes and nervous tissues. In addition, the ecdysones (molting hormones) are steroids synthesized from cholesterol via 7-dehydrocholesterol (Fig. 6). Four ecdysones have been identified from insects of different species and at different stages of development: α-ecdysone, 20-hydroxyecdysone, 20,26-dihydroxyecdysone, and (from some lepidopterous embryos) 26-hydroxyecdysone[176]. With

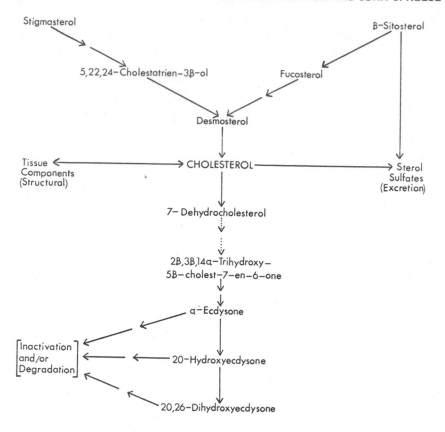

Figure 6.   Summarized sterol metabolism pathways of phytophagous insects (from Robbins, *et al.*, 1971).

cholesterol and its derivatives being involved in both structural and developmental physiology, sterol metabolism has very great significance in insect biology, and has been the subject of much intensive research.

A number of different phytosterols occur in plant tissues, of which cholesterol is a minor constituent[62,70]. The cholesterol content of plant tissues is far below the dietary amounts required by phytophagous insects. Plant-feeding insects have solved this problem by means of either of two adaptations, or in some cases perhaps by a combination

of the two. First, some phytophagous forms have insured
sterol availability by means of symbiotic microorganisms
that are harbored internally or with which the insect lives
in constant association. The great majority of phytophagous
insects have made the second adaptation, which is to evolve
metabolic pathways by which phytosterols are converted into
cholesterol. Both of these adaptations have an influence
on the insects' host plant relationships, and are, therefore,
of interest to the subject under discussion.

Homopterous insects -- aphids, leafhoppers, scales, are
phloem feeders. Phloem sap has very little sterol-content,
and no more than traces of cholesterol[47]. These insects
possess specialized organs (mycetomes) that contain intracellular bacteria-like symbiotes[24,29,41,58,75]. Studies
of the nutritional requirements of several aphid species,
reared on chemically defined diets, have shown that the
aphids do not require a dietary source of sterol with
Myzus persicae[37]; with *Acyrthosiphon pisum*[1,2]; and with
Neomyzus circumflexus[42].

Most of the cited investigators suggested that the
intracellular symbiotes synthesized the sterols required by
the aphids. In a study of lipid metabolism of symbiotes
isolated from the pea aphid, *A. pisum*, demonstrated cholesterol synthesis from ^{3}H-mecalonic acid and ^{14}C-acetate[83].
Griffiths and Beck (1975) demonstrated that the symbiotes
of the pea aphid synthesized cholesterol *in situ* in the
mycetocytes and partially elucidated the secretory
sequence by which the sterols were transferred from
symbiote to host aphid. Pea aphids that had been rendered
aposymbiotic by means of chlortetracycline grew poorly and
produced only a few progeny, none of which survived the
first stadium. This apparent deficiency could be only
partially corrected by the addition of cholesterol to the
diet, suggesting that the symbiotes contribute more than
just cholesterol synthesis to the physiology of their
hosts[59].

Beetles of the family Scolytidae (bark beetles) contain
symbiotic fungi that are thought to provide their insect
hosts with sterols[33]. Chu and coworkers reported that
aposymbiotic beetles, *Xyleborus ferrugineus*, were able to
utilize dietary cholesterol and, surprisingly, lanosterol
for reproduction, but the progeny failed to reach the

pupal stage in the absence of a Δ7 sterol (ergosterol, 7-dehydrocholesterol). These investigators interpreted their results as suggesting that the symbiotic fungi synthesized the ergosterol normally utilized by the beetle larvae.

Phytophagous insects that have no sterol-synthesizing symbiotes are dependent on plant tissues as their source of sterols. The common phytosterols are metabolized to cholesterol via demosterol (Fig. 6)[177]. Phytosterols are assimilated by the midgut epithelium, and the conversion to cholesterol occurs almost exclusively in these cells[175]. Excess sterols or metabolically degraded sterols are excreted as sterol sulfates[89]. Confinement of assimilation, conversion, degradation, and excretion to midgut cells is significant to the present discussion, because it suggests that plant-derived sterols and steroids are not translocated in an unchanged form; the only sterols entering the insect's general circulation are those that have passed through the metabolic and regulatory mechanisms of the midgut epithelium. The probable importance of this feature will become more apparent when we consider the effects of plant-produced ecdysone analogs -- the phytoecdysones.

α-Ecdysone is synthesized from cholesterol. The biochemical intermediates in this synthesis are largely unknown, except for 7-dehydrocholesterol and, probably, 2β, 3β, 14α-trihydroxy-5β-cholest-7-en-6-one (Fig. 6). This synthesis occurs in the prothoracic gland but may not be only in that organ[32,102]. Ecdysone may then be metabolized to form the more active hormone, 20-hydroxyecdysone, or the 26-hydroxy and 20, 26-dihydroxy analogs. These conversions do not appear to be the province of the prothoracic glands, but are effected by fat body, Malpighian tubes, and gut tissues[51,101,127].

The synthesis and release of ecdysones into the hemolymph of the insect must be under very close physiological regulation, if the step-wise pattern of insect developmental processes are to be effectively controlled and coordinated. An essential component of the regulartory system must be a mechanism for the rapid inactivation of the hormones. Ecdysones that have been administered to experimental insects are rapidly degraded or inactivated, with the "biological half-life" being from 2 to 6 hours [11,158]. Inactivation of ecdysones is apparently accomplished by degradation to polar metabolites and by excretion in the form of sulfates and glucosides[184,205]. Degradation was found to be carried out by gut tissues, fat body, and Malpighian tubes[78,98,99,111,158,184].

Plants synthesize sterols that are nutritionally utilizable by phytophagous insects, but, surprisingly, they also produce insect ecdysones and a large number of ecdysone analogs. These analogs may show a high level of biological activity when administered by injection to experimental insects. Of the four known insect-produced ecdysones, both α-ecdysone and 20-hydroxyecdysone have been isolated in large amounts from the ferns *Pteridium aquilinum* and *Polypodium vulgare*[72,97]. A number of phytoecdysones were identified from species of *Achyranthes* (Amaranthaceae), including α-ecdysone, 20-hydroxyecdysone, ponasterone A inokosterone[108,178-181]. *Podocarpus* species (Australian pines) have been shown to contain 20-hydroxyecdysone[51]. Ecdysone and 20-hydroxyecdysone have been shown to occur in a large variety of Pteridophytes, Gymnosperms, and Angiosperms, often in concentrations higher than those found in insects[82]. Nearly 30 different phytoecdysones have now been demonstrated, most of which have relatively low insect hormone activity, but some (cyasterone, polypodine A) have hormonal potency equal to or greater than that of 20-hydroxy-ecdysone[69,74,82,202].

Plant biosynthesis of phytoecdysones appears to be approximately similar to the synthetic pathways described in insects; that is, *Podocarpus elata* seedlings synthesized radiolabelled 20-hydroxyecdysone from applied ^{14}C-cholesterol[71,149]. However, other phytosterols may also serve as precursors for phytoecdysone synthesis[74]. A very interesting similarity between plant and insect ecdysone synthesis was disclosed when 25-deoxyecdysone, a synthetic analog, was injected into 3rd instar fly larvae (*Calliphora stygia*); the larvae converted the analog into 20-hydroxyecdysone and also into two analogs -- ponasterone A and inokosterone (Fig. 5)[183]. These two analogs are not normally found in insects, but have been reported as phytoecdysones; 25-deoxyecdysone may be a normal precursor in plant tissues, but not in insects[143]. It is of extreme interest that ponasterone A and, to a smaller degree, inokosterone are also the only phytoecdysones that have been found to influence insect development when ingested (discussed below).

Regulation of hormone titer and the degradative metabolism of ecdysone and 20-hydroxyecdysone have been studied in some detail. Metabolism of those phytoecdysones that are analogs of insect ecdysones has not been determined, but it is generally assumed to be similar to that of the insect-produced hormones, involving excretion as polar metabolites, sulfates, glucuronides, and glycosides. There is reason to suggest that phytoecdysones that are ingested are degraded by the midgut epithelium and do not reach the other tissues of the insect. Desert locusts, *Schistocerca gregaria*, that were fed bracken fern (high in both ecdysone and 20-hydroxyecdysone) did not show abnormalities in growth or ecdysis[27]. These investigators suggested that the phytoecdysones were dehydroxylated by the midgut and excreted. Others have also reported that neither ecdysone nor 20-hydroxyecdysone influence insect growth processes when they are ingested[74,105,144]. Of the phytoecdysones that have been tested, only ponasterone A in the diet of houseflies, *Musca domestica*[144], cecropia larvae, *Hyalophora cecropia*[202] and the silkworm, *Bombyx mori*[163], has produced pronounced growth abnormalities. Shigematsu *et al.* (1974) also observed that ingested inokosterone resulted in minor aberrations in silkworm growth. Alcoholic solutions of a number of phytoecdysones were found to penetrate the pupal cuticle of the silkworm when the insects were dipped in the solutions for several seconds[66,67], showing that such analogs have biological activity when the route of entry bypasses the midgut.

There is some question as to the function of phyto-ecdysones in the biology of the plant. It has been suggested that the phytoecdysones have an allelochemic function in defense against insect herbivores, and that plants such as *Podocarpus* spp. and *Polypodium* spp. that contain relatively high concentrations of phytoecdysones are little attacked by insects[164]. This interpretation is surely simplistic, as phytoecdysones occur in the tissues of a large number of plants, none of which is devoid of insect herbivores. Whether the phytoecdysones have a defensive or allelochemic function is a question that is yet to be answered definitively. The only instances in which the ingestion of phytoecdysones has led to develop-mental aberrations were under experimental conditions in which the insects were fed artificial or unnatural diets. There are no published reports known to us in which

phytoecdysones have been demonstrated to play any role in
an insect's host plant relationships, either in host plant
specificity or in varietal resistance within the insect's
normal host plant range.

The possible allelochemic function of phytoecdysones
should not be dismissed, however, and the question of their
functions should be vigorously researched. We have included
them in this analysis of allelochemic effects because of
their metabolic significance, rather than because of any
proven role in insect-plant interactions.

Alkaloids. Of the great multitude of different alkaloids
synthesized by diverse plant species, only a relatively few
have been investigated for their roles in insect-plant
interactions (Table II). Most of those that have been so
studied are the major alkaloids of solanaceous plants. In
most cases, the alkaloids have been described as either
repellents, feeding inhibitors, or acute toxins, so that the
insects either fail to feed or die very shortly after
feeding. These effects tend to place these alkaloids outside
the purview of our discussion of nutritional and metabolic
aspects. The paucity of studies pertinent to our subject
and the wide distribution of biologically active alkaloids
suggest that this is a subject that needs a great deal more
research.

The effects of a number of alkaloids on feeding and
growth of the Colorado potato beetle, *Leptinotarsa decemlineata*, have been described. Solanine, chaconine, leptine,
leptinine I and II, demissine, and tomatine were found to
inhibit feeding[173]. Physostigmine, colchicine, veratrine,
aconitin, delphinine, canthardine, quassin, and picrotoxin
have been described as highly toxic to potato beetle larvae[25].
Schreiber (1958) suggested that the steroidal alkaloids
(Fig. 7) affected the feeding insect by blocking steroid
metabolism and perhaps the assimilation of phytosterols.
This mode of action was also suggested by Harley and
Thorsteinson (1967), investigating the effects of a number
of plant chemicals on the feeding behavior and growth of
the two-striped grasshopper, *Melanoplus bivittatus*. The
toxicity of *Solanum* and *Veratrum* alkaloids to housefly larvae
did not prove to be the result of an antagonism to cholesterol[17]. Nevertheless, a possible relationship between

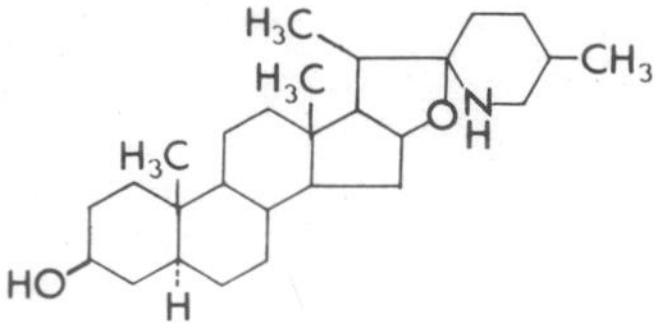

Tomatidine

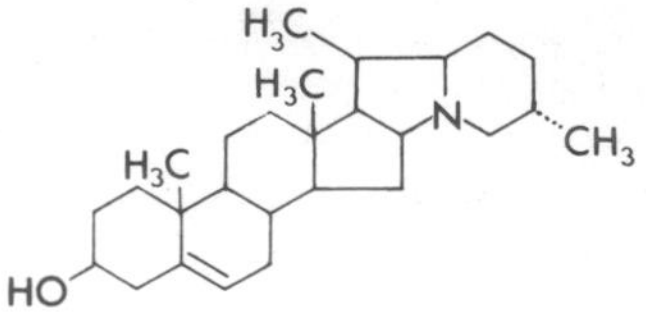

Solanidine

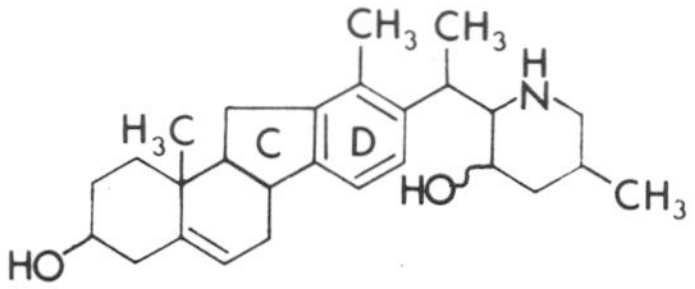

Veratramine

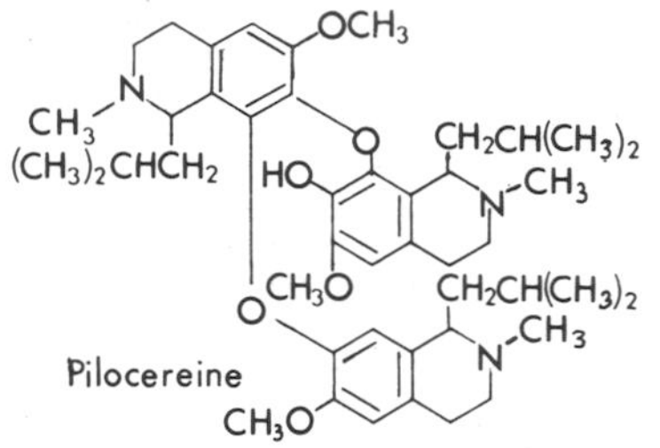

Pilocereine

Figure 7. Steroidal alkaloids.

steroid alkaloids and sterol metabolism should be subjected to further investigation.

The tobacco hornworm, *Manduca sexta*, utilizes a number of solanaceous hosts, including tobacco. The adaptation of hornworm larvae to the ingestion of nicotine has been studied and it was found that nicotine was apparently assimilated, but did not exert a toxic effect at the hemolymph concentrations observed[156,157]. Parr and Thurston (1972) found that tobacco hornworm larvae grew poorly on synthetic diets containing more than 1.5% pure nicotine, but lower dietary levels had no adverse effects on larval development. In view of the fact that green tobacco leaves contain only 0.1-0.5% nicotine, larval tolerance to this alkaloid seems very great.

Two complex alkaloids from the senita cactus, *Lophocereus schotti* (pilocereine and lophocereine), were tested against a series of desert-inhabiting *Drosophila* species, and only *D. pachea* larvae and adults could tolerate pilocereine (Fig. 7). *D. pachea* is the only species of *Drosophila* that utilizes the tissues of the senita cactus[103].

The hypothesis that plant alkaloids play an important chemosensory role in host plant specificity of many phytophagous insects has a sizable body of supporting evidence[19,45,49,110]. In respect to alkaloids as defensive substances at the metabolic level, the evidence is far from persuasive. In his recent analysis of metabolism and function of alkaloids, Robinson (1974) concluded, "Although alkaloids have been considered as protective substances, there is very little hard evidence that such is the case."

One aspect of alkaloid metabolism in plants appears to have been completely ignored by entomological investigators; that is the diurnal fluctuations of alkaloid contents in plant tissues (see Robinson, 1974). These fluctuations should be taken into account not only when tissue samples are to be assayed, but also when considering the dietetics of the phytophagous insects.

Saponins. These plant chemicals are glycosides in which the aglucone is a steroidal sapogenin, such as diosgenin (Fig. 1). They function in the plant in seed germination and also in the control of root growth[70]. The saponins are quite toxic to some fungi and bacteria, and have been implicated as defensive allomones against insects in a few cases.

Unidentified saponins contained in soybean seeds were shown to inhibit larval development of the bruchid beetle *Callosobruchus chinensis*[3]. A developmental inhibition of whitegrub larvae, *Melolontha vulgaris*, was attributed to the complex of saponins in the roots of alfalfa cultivars, *Medicago sativa*[80].

The metabolic effects of saponins on phytophagous insects have been postulated to be the result of inhibited sterol assimilation[64,80]. Saponins have also been shown to inhibit the proteinase α-chymotrypsin[90]. Saponins have been implicated as inhibitors of insect feeding (repellents and deterrents) in a few cases[56,79,174].

CONCLUDING REMARKS

     This brief study of the nutritional and metabolic aspect
of insect-plant interactions is by no means complete. We
hope that it is a reasonable cross-section of the current
"state of the art." There are some gaping lacunae in current
biological knowledge of the subject. We presently know next
to nothing about the effects of non-nutrient phytochemicals
on the assimilation and conversion processes in the gut and
body tissues of the insect. Nor do we have even a reasonable
understanding of the effects of such phytochemicals on amino
acid utilization in protein synthesis and other synthetic
pathways in the developmental and reproductive physiology of
the insect. A beginning toward elucidation of some of these
questions can be made through the use of the nutritional
index technique, but it is a severely limited technique.
Nutritional indices can do no more than serve as indicators
identifying some of the aspects that merit detailed biochemical and physiological study. Another useful approach
to some of these problems is that of energy budget determinations, using techniques such as those applied by Schroeder
(1971, 1972) and Randolph *et al*. (1975). Such studies
would involve measurements of energy flow of insects
reared on defined diets to which known allelochemic factors
(singly or as plant-typical complexes) had been added.

     In the past, most of us have been obsessed with the
idea that an insect-plant relationship can be explained
on the basis of the presence or the quantity of one
identifiable chemical factor, or at least not more than two
or three such factors. It seems to us unlikely that such
a simple approach will often succeed. The plant contains
a typical complex of non-nutrient allelochemics, and it is
this complex profile of substances with which the insect
must cope. An example might be the case of a particular
agricultural plant in which there are insect-resistant and
insect-susceptible cultivars. These cultivars may differ
in their profiles of wound-induced allelochemics in only
quantitative ways, in which case there may not be an
identifiable "resistance factor" that would account for
the differences between cultivars.

     Although hallowed through traditional use, the term
"secondary plant chemical" should be abandoned. Recent
studies have shown it to be a misnomer at best and to

discourage inspired research at the worst. The term implies that many phytochemicals are waste products or evolutionary vestiges that may have no function other than incidental discouragement of pathogens and herbivores, and although their marginal defensive role might be of interest to a few entomologists and plant pathologists, there is little reason for any attention from plant pathologists and phyto-chemists. It must now be recognized that the steroids, alkaloids, and phenylpropane derivatives play important functions in the life of a plant. The rapid turnover rate of alkaloids[145], the possible involvement of steroids in plant hormone functions[70], and the large number of polyphenols, flavonoids, etc., in wound reactions[15] all point to dynamic roles being carried out by these so-called secondary substances. The concepts of allelochemic interactions as introduced by Whittaker (1970) is a large step in the right direction, in that it recognizes the dynamic nature of the interaction without relegating whole metabolic systems to a second-class status. In this discussion, our primary interest has been in the insect-plant interaction, and we have no reason to believe that this dynamic interaction is, in any way, of secondary importance to either the insect or the plant.

Acknowledgment. Research supported by the College of Agricultural and Life Sciences, University of Wisconsin, and by a research grant (BMS-74-24001) from the National Science Foundation.

REFERENCES

1. Akey, D. H. and S. D. Beck, 1971. Continuous rearing of the pea aphid, *Acyrthosiphon pisum*, on a holidic diet. *Ann. Ent. Soc. Amer. 64*:946-949.
2. Akey, D. H. and S. D. Beck, 1972. Nutrition of the pea aphid, *Acyrthosiphon pisum*. Requirements for trace metals, sulphur, and cholesterol. *J. Insect Physiol. 18*:1901-1914.
3. Applebaum, S. W., B. Gestetner, and Y. Birk, 1965. Physiological aspects of host specificity in the Bruchidae. IV. Developmental incompatibility of soybeans for *Callosobruchus*. *Jour. Insect Physiol. 11*:611-616.

4. Auclair, J. L., 1963. Aphid feeding and nutrition.
    *Ann. Rev. Ent. 8:*439–490.
5. Auclair, J. L., J. B. Maltais, and J. J. Cartier, 1957.
    Factors in resistance of peas to the pea aphid,
    Acyrthosiphon pisum (Harr) (Homoptera:Aphididae).
    II. Amino acids. *Canad. Ent. 89:*457–464.
6. Beck, S. D., 1957. The European corn borer *Pyrausta
    nubilalis*, Hubn., and its principal host plant.
    VI. Host plant resistance to larval establishment.
    *Jour. Insect Physiol. 1:*158–177.
7. Beck, S. D., 1960. The European corn borer, *Pyrausta
    nubilalis* (Hbn.) and its principal host plant.
    VII. Larval feeding behavior and host plant
    resistance. *Ann. Ent. Soc. Amer. 53:*206–212.
8. Beck, S. D., 1965. Resistance of plants to insects.
    Ann. Ent. Soc. Amer. 53:207–232.
9. Beck, S. D., 1972. Nutrition, adaptation and environ-
    ment. *in* Insect and Mite Nutrition, J. G.
    Rodriguez, Ed., North Holland Publ. Co., Amsterdam.
10. Beck, S. D., 1974. Theoretical aspects of host plant
    specificity in insects. *in* Proc. Summer Inst.
    on Biol. Control of Plant Insects and Diseases,
    F. G. Maxwell and F. A. Harris, Eds., p. 290–311.
    Univ. Press of Mississippi, Jackson.
11. Beck, S. D. and J. L. Shane, 1969. Effects of ecdysones
    on diapause in the European corn borer, *Ostrinia
    nubilalis. J. Insect Physiol. 15:*721–730.
12. Beck, S. D. and E. E. Smissman, 1961. The European
    corn borer, *Pyrausta nubilalis*, and its principal
    host plant. IX. Biological activity of chemical
    analogs of corn resistance factor A (6 methoxyben-
    zoxazolinone). *Ann. Ent. Soc. Amer. 54:*53–61.
13. Beck, S. D. and J. F. Stauffer, 1957. The European
    corn borer *Pyrausta nubilalis* (Hbn.), and its
    principal host plant. III. Toxic factors
    influencing larval establishment. *Ann. Ent. Soc.
    Amer. 50:*166–170.
14. Bell, E. A., 1972. Toxic amino acids in the Leguminosae.
    in Phytochemical Ecology, J. B. Harborne, ed., p.
    163–177. Acad. Press.
15. Bell, E. A., 1974. Biochemical bases of resistance of
    plants to pathogens. *in* Proceedings of the Summer
    Institute on Biological Control of Plant Insects and
    Diseases, F. G. Maxwell and F. A. Harris, eds., Univ
    Press, Miss. Jackson, p. 453–462.

16. Bell, E. A. and D. H. Janzen, 1971. Medical and ecological considerations of L-dopa and 5-HTP in seeds. *Nature 229*:136–137.

17. Bergmann, E. D., Z. H. Levinson and R. Mechoulam, 1958. The toxicity of *Veratrum* and *Solanum* alkaloids to housefly larvae. *Jour. Insect Physiol. 2*:162–177.

18. Bloch, K., R. G. Langdon, A. J. Clark and G. Fraenkel, 1956. Impaired steroid biogenesis in insect larvae. *Biochim. Biophys. Acta 21*:176.

19. Bongers, W., 1970. Aspects of host–plant relationship of the Colorado beetle. *Meded. Landbouwhogeschool Wageningen, Nederland 70-10*:1–77.

20. Bonner, J. and J. E. Varner, 1965. *Plant Biochemistry*. Academic Press, New York and London. 1054 pp.

21. Bottger, G. T. and R. Patana, 1966. Growth, development and survival of certain Lepidoptera fed gossypol in the diet. *J. Econ. Ent. 59*:1166–1169.

22. Bowers, W. S., H. M. Fales, M. J. Thompson, and E. C. Uebel, 1966. Juvenile hormone: Identification of an active compound from balsam fir. *Science 154*: 1020–1021.

23. Brett, C. H., 1947. Interrelated effects of food, temperature, and humidity on the development of the lesser migratory grasshopper, *Melanoplus mexicanus mexicanus* (Saussure). *Oklahoma Agr. Exp. Sta. Tech. Bull. T-26*.

24. Buchner, P., 1965. *Endosymbiosis of Animals with Plant Microorganisms*. Wiley, New York.

25. Buhr, H., R. Toball, and K. Schreiber, 1958. Die Wurkung von einigen Pflanzlichen Sonderstoffen, insbesondere von Alkaloiden, auf die Entiwicklung der Larven des kartoffelkäfers (*Leptinotarsa decemlineata*, Say). *Ent. Exp. & Appl. 1*:209–224.

26. Carlisle, D. B., P. E. Ellis and E. Betts, 1965. The influence of aromatic shrubs on sexual maturation in the desert locust, *Schistocerca gregaria*. *Jour. Insect Physiol. 11*:1541–1558.

27. Carlisle, D. B. and P. E. Ellis, 1968. Bracken and locust ecdysones: Their effects on molting in the desert locust. *Science 159*:1472–1474.

28. Cerny, V., L. Dolejs, L. Labler, F. Sorm and K. Slama, 1967. Dehydrojuvabione -- a new compound with juvenile hormone activity from balsam fir. *Tetrahedron Letters*, March, 1967(12):1053–1057.

29.   Chang, K. P. and A. J. Musgrave, 1972. Multiple sym-
      biosis in a leafhopper, *Helochara communis* Fitch
      (Cicadellidae:Homoptera): Envelopes, nucleoids
      and inclusions of the symbiotes. *J. Cell. Sci. 11:*
      275-293.
30.   Chapman, R. F., 1974. The chemical inhibition of
      feeding by phytophagous insects: A review. *Bull.
      Ent. Res. 64:*339-363.
31.   Chawla, S. S., J. M. Perron and M. Cloutier, 1974.
      Effects of different growth factors on the potato
      aphid, *Macrosiphum exphorbiae*, fed on an artifical
      diet. *Canad. Ent. 106:*273-280.
32.   Chino, H., 1974. Biosynthesis of α-ecdysone by protho-
      racic glands *in vitro*. *Science 183:*529-530.
33.   Chu, H. M., D. M. Norris and L. T. Kok, 1970. Pupation
      requirement of the beetle, *Xyleborus ferrugineus:*
      Sterols other than cholesterol. *J. Insect. Physiol.
      16:*1379-1387.
34.   Cibula, A. B., R. H. Davidson, F. W. Fisk and J. B.
      Lapidus, 1967. Relationship of free amino acids
      of some solanaceous plants to growth and development
      of *Leptinotarsa decemlineata* (Coleoptera:Chrysomeli-
      dae). *Ann. Ent. Soc. Amer. 60:*626-631.
35.   Clark, A. J. and K. Bloch, 1959. The absence of sterol
      synthesis in insects. *J. Biol. Chem. 234:*2578-2582.
36.   Clayton, R. B., 1960. The role of intestinal symbionts
      in the sterol metabolism of *Blattella germanica*.
      *Jour. Biol. Chem. 235:*3421-3425.
37.   Dadd, R. H. and T. E. Mittler, 1966. Permanent culture
      of an aphid on a totally synthetic diet. *Experientia
      22:*832.
38.   Dahlman, D. L. and G. A. Rosenthal, 1975. Non-protein
      amino acid-insect interactions. 1. Growth effects
      and symptomology of L-canavanine consumption by
      tobacco hornworm, *Manduca sexta* (L.). *Comp.
      Biochem. Physiol. 51A:*33-36.
39.   Dethier, V. G., 1941. Chemical factors determining
      the choice of food plants by papilio larvae.
      *Amer. Nat. 75:*61-73.
40.   Duffey, S. S. and G. G. E. Scudder, 1972. Cardiac
      glycosides in North American Asclepiadaceae, a basis
      for unpalatability in brightly coloured Hemiptera
      and Coleoptera. *J. Insect Physiol. 18:*63-78.

41. Ehrhardt, P., 1966. Die Wirkung von Lysozyminjektionen
     auf Aphiden und deren Symbioten. *Z. vergl. Physiol.*
     53:130-141.
42. Ehrhardt, P., 1968. Der Vitaminbedarf einer
     siebröhrensaugenden Aphide, *Neomyzus circumflexus*
     Buckt. (Homoptera, Insecta). *Z. Vergl. Physiol.*
     60:416-426.
43. Ehrlich, P. R. and P. H. Raven, 1964. Butterflies and
     plants: A study in coevolution. *Evol. 18*:586-608.
44. Erickson, J. M. and P. Feeny, 1974. Sinigrin: A
     chemical barrier to the black swallowtail butterfly.
     Ecology 55:103-111.
45. Feeny, P. P., 1968. Effects of oak leaf tannins on
     larval growth of the winter moth *Operophtera brumata*.
     J. Insect Physiol. 14:805-817.
46. Feeny, P., K. L. Paauwe, and N. J. Demong, 1970. Flea
     beetles and mustard oils: Host plant specificity of
     Phyllotreta cruciferae and *P. striolata* adults
     (Coleoptera:Chrysomelidae). *Ann. Ent. Soc. Amer.*
     63:832-841.
47. Forrest, J. M. S. and B. A. Knights, 1972. Presence of
     phytosterols in the food of the aphid, *Myzus persicae.*
     J. Insect Physiol. 18:723-728.
48. Fraenkel, G. S., 1959. The Raison d'être of secondary
     plant substances. *Sci. 129*:1466-1470.
49. Fraenkel, G. S., 1969. Evaluation of our thoughts on
     secondary plant substances. *Ent. Exp. and Appl.*
     12:473-486.
50. Galbraith, M. N. and D. H. S. Horn, 1969. Insect
     moulting hormones: Crustecdysone (20-hydroxyecdysone)
     from *Podocarpus elatus*. *Austral. J. Chem. 22*:1045-
     1057.
51. Galbraith, M. N., D. H. S. Horn, and E. J. Middleton.
     1973. Ecdysone biosynthesis in the blowfly
     Calliphora stygia. *Chem. Commun.* (*3*):179-180.
52. Galston, A. W. and P. J. Davies, 1970. *Control*
     Mechanisms in Plant Development. Prentice-Hall, Inc.,
     Englewood Cliffs, N. J.   184 pp.
53. Gilbert, B. L., J. E. Baker, and D. M. Norris, 1967.
     Juglone (5-hydroxy-1,4,naphthoquinone) from *Carya*
     ovata, a deterrent to feeding by *Scolytus multistria-*
     tus. *J. Insect Physiol. 13*:1453-1459.
54. Gilbert, B. L. and D. M. Norris, 1968. A chemical basis
     for bark beetle (*Scolytus*) distinction between host
     and non-host trees. *J. Insect Physiol. 14*:1063-1068.

55. Gilbert, L. I. and D. S. King, 1973. Physiology of growth and development: Endocrine aspects. *Physiology of Insecta* 1:249-370.

56. Granich, M. S., B. P. Halpern and T. Eisner, 1974. Gymnemic acids: Secondary plant substances of dual defensive action? *J. Insect Physiol. 20:* 435-439.

57. Green, T. R. and C. A. Ryan, 1972. Wound-induced proteinase inhibitor in plant leaves: A possible defense mechanism against insects. Science 175: 776-777.

58. Griffiths, G. W. and S. D. Beck, 1973. Intracellular symbiotes of the pea aphid, *Acyrthosiphon pisum*. *J. Insect Physiol. 19:*75-84.

59. Griffiths, G. W. and S. D. Beck, 1974. Effects of antibiotics on intracellular symbiotes in the pea aphid, *Acyrthosiphon pisum*. *Cell Tiss. Res. 148:* 287-300.

60. Griffiths, G. W. and S. D. Beck, 1975. Ultrastructure of pea aphid mycetocytes: Evidence for symbiote secretion. *Cell Tiss. Res. 159:*351-367.

61. Grison, P., 1958. L'influence de la plant-hôte sur la fecondite de l'insecte phytophage. *Ent. Exp. & App. 1:*73-93.

62. Grunwald, C., 1975. Plant sterols. *Ann. Rev. Pl. Physiol. 26:*209.

63. Guerra, A., 1970. Effects of biologically active substances in the diet on development and reproduction of *Heliothis* spp. J. Econ. Ent. 63:1518-1521.

64. Harley, K. L. S. and A. J. Thorsteinson, 1967. The influence of plant chemicals on the feeding behavior, development, and survival of the two-striped grasshopper, *Melanoplus bivittatus* (Say), Acrididae: Orthoptera. *Canad. J. Zool. 45:*305.

65. Harrewijn, P., 1973. Functional significance of indole alkylamines linked to nutritional factors in wing development of the aphid, *Myzus persicae*. *Ent. Exp. & Appl. 16:*499-513.

66. Hasegawa, K. and A. M. Ata, 1971. Studies on the effect of ecdysone -- analogues on the development of the silkworm, *Bombyx mori* L. (Lepidoptera:Bombycidae). I. Penetration of phytoecdysones in larval cuticle. *Appl. Ent. Zool. 6:*147-155.

67. Hasegawa, K. and A. M. Ata, 1972.   Penetration of
    phytoecdysones through the pupal cuticle of the
    silkworm, *Bombyx mori*.   *J. Insect Physiol. 18*:959-
    971.
68. Hedin, P. A., F. G. Maxwell, and J. N. Jenkins, 1974.
    Insect plant attractants, feeding stimulants,
    repellents, deterrents, and other related factors
    affecting insect behavior.   *in* Proceedings of the
    Summer Institute on Biological Control of Plant
    Insects and Diseases.   F. G. Maxwell and F. A.
    Harris, eds., Univ. Press, Mississippi, Jackson,
    p. 494-527.
69. Heftmann, E., 1970.   Insect molting hormones in plants.
    Rec. Adv. Phytochem. 3:211-277.
70. Heftmann, E., 1975.   Functions of steroids in plants.
    Phytochem. 14:891-901.
71. Heftmann, E., H. H. Savas and R. D. Bennett, 1968.   Bio-
    synthesis of ecdysterone from cholesterol by a plant.
    Naturwiss 55:37.
72. Heinrich, G. and H. Hoffmeister, 1967.   Ecdyson als
    Begleitsubstanz des Ecdysterons in *Polypodium
    vulgare* L.   *Experientia 23*:995.
73. Hendry, L. B., J. K. Wichmann, D. M. Hinderlang, R. O.
    Mumma, and M. E. Anderson, 1975.   Evidence for origin
    of insect sex pheromones: Presence in food plants.
    Science 188:59-62.
74. Herout, V.   1970.   Some relations between plants, insects
    and their isoprenoids.   *Rec. Adv. Phytochem. 2*:143.
75. Hinde, R., 1971.   The fine structure of the mycetome
    symbiotes of the aphids *Brevicoryne brassicae*, *Myzus
    persicae*, and *Macrosiphum rosae*.   *J. Insect Physiol.
    17*:2035-2050.
76. Hobson, R. P., 1935.   On a fat-soluble growth factor
    required by blowfly larvae.   II.   Identity of the
    growth factor with cholesterol.   *Biochem. J. 29*:
    2023-2026.
77. Hodge, C., 1933.   Growth and nutrition of *Melanoplus
    differentials* Thomas Orthoptera, acrididoe) I.   Growth
    on a satisfactory mixed diet and on diets of single
    food plants.   *Physiol. Zool. 6*:306-328.
78. Hoffman, J. A., J. Koolman, P. Karlson and P. Joly, 1974.
    Molting hormone titer and metabolic fate of injected
    ecdysone during the fifth instar and in adults of
    Locusta migratoria.   *Gen. Comp. Endocrinol. 22*:90-
    97.

79. Horber, E., 1965. Isolation of components from the roots
     of alfalfa (*Medicago sativa* L.) toxic to white grubs
     (*Melolontha vulgaris* F.). Proc. XII. Int. Congr. Ent.
     pp. 540–541.
80. Horber, E., 1972. Alfalfa saponins significant in resis-
     tance to insects. *in* Insect and Mite Nutrition, J. B
     Rodriguez, Ed., North Holland Publ. Co., Amsterdam.
     pp. 611–627.
81. Hori, K., 1973. Studies on the feeding habits of *Lygus
     disponsi* Linnavuori (Hemiptera: Miridae) and the
     injury to its host plant. IV. Amino acids and sugars
     in injured tissues of sugar beet leaf. *Appl. Ent.
     Zool. 8:*138–142.
82. Horn, D. H. S., 1971. The ecdysones. *in* Naturally
     Occurring Insecticides, M. Jacobson, D. G. Crosby,
     Eds., Dekker, New York. pp. 333–459.
83. Houk, E. J., G. W. Griffiths, and S. D. Beck, 1975.
     Lipid metabolism in the symbiotes of the pea aphid,
     Acyrthosiphon pisum. Comp. Biochem. Physiol. (in
     press).
84. House, H. L., 1962. Insect nutrition. *Ann. Rev.
     Biochem. 31:*653–672.
85. House, H. L., 1969. Effects of different proportions
     of nutrients on insects. *Ent. Exp. & Appl. 12:*651–
     669.
86. House, H. L., 1970. Choice of food by larvae of the
     fly, *Agria affinis*, related to dietary proportions
     of nutrients. *J. Insect Physiol. 16:*2041–2050.
87. House, H. L., 1971. Relations between dietary propor-
     tions of nutrients, growth rate, and choice of food
     in the fly larva *Agria affinis. J. Insect Physiol.
     17:*1225–1238.
88. Hovanitz, W., V. C. S. Chang and G. Honch., 1963. The
     effectiveness of different isothiocyanates on
     attracting larvae of *Pieris rapae. J. Res. Lepid.
     3:*159–173.
89. Hutchins, R. F. N. and J. N. Kaplanis, 1969. Sterol
     sulfates in an insect. *Steroids 13:*605–614.
90. Ishaaya, I. and Y. Birk, 1965. Soybean saponins. IV.
     The effect of proteins on the inhibitory activity of
     soybean saponins on certain enzymes. *J. Food Sci.
     30:*118–120.

91. Isogai, A., C. Chang, S. Murakoshi, A. Suzuki, 1973. Screening search for biologically active substances to insects in crude drug plants. *J. Agr. Chem. Soc. Japan* 47:443-447.

92. Isogai, A., S. Murakoshi, A. Suzuki and S. Tamura, 1973b. Isolation from "Astragali Radix" of L-canavanine as an inhibitory substance to metamorphosis of silkworm, *Bombyx mori*. *J. Agr. Chem.* 47:449-453.

93. Isogai, A., S. Murakoshi, A. Suzuki and S. Tamura, 1973c. Growth inhibitory effects of phenylpropanoids in nutmeg on silkworm larvae. *J. Agric. Chem. Soc. Jap.* 47:275-279.

94. Isogai, A., S. Murakoshi, A. Suzuki and S. Tamura, 1973d. Isolation from "Astragali Radix" of L-canavanine as an inhibitory substance to metamorphosis of silkworm, *Bombyx mori* L. *J. Agric. Chem.* 7:449-453.

95. Ito, T., Y. Horie and K. Watanabe, 1970. Effect of phytoecdysones on the length of the fifth instar and the quality of cocoons in the silkworm, *Bombyx mori*. *Annot. Zool. Japon.* 43:175-181.

96. Kamikado, T., C. F. Chang, S. Murakoshi, A. Sakurai, and S. Tamura, 1975. Isolation and structure elucidation of growth inhibitors on silkworm larvae from *Magnolia kobus*. *Agric. Biol. Chem.* 39:833-836.

97. Kaplanis, J. N., M. J. Thompson, W. E. Robbins, and B. M. Bryce, 1967. Insect hormones: Alpha ecdysone and 20-hydroxyecdysone in bracken fern. *Science* 157: 1436-1437.

98. Karlson, P. and C. Bode, 1969. Die inaktivierung des ecdysons bei der schmeissfliege *Calliphora erythrocephala* Meigen. *J. Insect Physiol.* 15:111-118.

99. Karlson, P. and J. Koolman, 1973. On the metabolic rate of ecdysone and 3-dehydroecdysone in *Calliphora vicina*. *Insect Biochem.* 3:409-417.

100. Kato, M. and H. Yamada, 1966. Silkworm requires 3,4-dihydroxybenzene structure of chlorogenic acid as a growth factor. *Life Sciences* 5:717-722.

101. King, D. S., 1972. Ecdysone metabolism in insects. *Amer. Zool.* 12:343-345.

102. King, D. S., W. E. Bollenbacher, D. W. Borst, W. V. Vedeckis, J. D. O'Connor, P. I. Ittycheriah and L. I. Gilbert, 1974. The secretion of α-ecdysone by the prothoracic glands of *Manduca sexta in vitro*. *Proc. Nat. Acad. Sci.* 71:793.

103. Kircher, H. W., W. B. Heed, J. S. Russell, and J.
      Grove, 1967. Senita cactus alkaloids: Their signi-
      ficance to Sonoran Desert *Drosophila* ecolody. *J.
      Insect Physiol.* *13*:1869-1874.

104. Klingauf, F., 1971. Die wirkung des clucosids phlorizin
      auf das wirtswahlverhalten von *Rhopalosiphum insertum*
      (Walk.) und *Aphis pomi* DeGeer (Homoptera:Aphididae).
      Z. ang. Ent. 63:41-55.

105. Klun, J. A., 1974. Biochemical basis of resistance of
      plants to pathogens and insects: Insect hormone
      mimics and selected examples of other biologically
      active chemicals derived from plants. *in* Proceedings
      of the Summer Institute of Biological Control of
      Plant Insects and Diseases, F. G. Maxwell and F. A.
      Harris, eds., Univ. Press, Jackson, Mississippi.
      pp. 463-484.

106. Klun, J. A. and T. A. Brindley, 1966. Role of 6-
      methoxybenzoxazolinone in inbred resistance of host
      plant (maize) to first-brood larvae of European
      corn borer. *Jour. Econ. Ent.* *59:*711-718.

107. Klun, J. A., C. L. Tipton, and T. A. Brindley, 1967.
      2,4-Dihydroxy-7-methoxy-1,4-benzoxazin-3-one (DIMBOA)
      an active agent in the resistance of maize to the
      European corn borer. *J. Econ. Ent.* *60:*1529-1533.

108. Kobayashi, M., T. Takemoto, S. Ogawa, and N. Nishimoto,
      1967. The moulting hormone activity of ecdysterone
      and inokosterone isolated from *Acyranthis radix.*
      J. Insect Physiol. *13*:1395-1399.

109. Kogan, M. and D. Cope, 1974. Feeding and nutrition of
      insects associated with soybeans. 3. Food intake,
      utilization, and growth in the soybean looper,
      Pseudoplusia includens. *Ann. Ent. Soc. Amer.* *74:*
      66-72.

110. Kogan, M. and R. D. Goeden, 1971. Feeding and host-
      selection behavior of *Lema trilineata daturaphila*
      larvae (Coleoptera:Chrysomelidae). *Ann. Ent. Soc.
      Amer.* *64:*1435-1448.

111. Koolman, J., J. A. Hoffman, and P. Karlson, 1973.
      Sulphate esters as inactivation products of ecdysone
      in *Locusta migratoria.* *H. S. Z. Physio. Chem.* *354:*
      1043-1048.

112. Krieger, R. I., P. P. Feeny, C. F. Wilkinson, 1971.
      Detoxication enzymes in the guts of caterpillars:
      An evolutionary answer to plant defense? *Science*
      *172:*579-581.

113. Leckstein, P. M. and M. Llewellyn, 1973. Effect of dietary amino acid on the size and alary polymorphism of *Aphis fabae*. *J. Insect Physiol*. *19*:973–980.

114. Loeb, J., 1915a. The salts required for the development of insects. *Jour. Biol. Chem*. *23*:431–434.

115. Loeb, J., 1915b. The simplest constituents required for growth and the completion of the life cycle in an insect (*Drosophila*). *Sci*. *41*:169–170.

116. Lukefahr, M. J. and D. F. Martin, 1966. Cotton-plant pigments as a source of resistance to the bollworm and tobacco budworm. *Jour. Econ. Ent*. *59*:176–179.

117. Maltais, J. B. and J. L. Auclair, 1962. Free amino acid and amide composition on pea leaf juice, pea aphid haemolymph, and honeydew, following the rearing of aphids on single pea leaves treated with amino compounds. *Jour. Insect Physiol*. *8*:391–400.

118. Mansingh, A., T. S. Sahota, and D. A. Shaw, 1970. Juvenile hormone activity in the wood and bark extracts of some forest trees. *Canad. Ent*. *102*:49–53.

119. Maxwell, F. G., J. N. Jenkins and W. L. Parrott, 1967. Influence of constituents of the cotton plant on feeding, oviposition, and development of the boll weevil. *J. Econ. Ent*. *60*:1294–1297.

120. Maxwell, F. G., J. N. Jenkins, and W. L. Parrott, 1972. Resistance of plants to insects. *Adv. Agron*. *24*:187–265.

121. Maxwell, F. G., H. N. Lafever, and J. N. Jenkins, 1966. Influence of the glandless genes in cotton on feeding, oviposition, and development of the boll weevil in the laboratory. *Jour. Econ. Ent*. *59*:585.

122. Meyer, H. J. and D. M. Norris, 1974. Lignin intermediates and simple phenolics as feeding stimulants for *Scolytus multistriatus*. *J. Insect Physiol*. *20*:2015–2021.

123. Mittler, T. E., 1973. Aphid polymorphism as affected by diet. *in* Perspectives in Aphid Biology, A. D. Lowe, Ed., pp. 65–75. Bull. 2. Ent. Soc. N.Z.

124. Mittler, T. E. and R. H. Dadd, 1966. Food and wing determination in *Myzus persicae*. *Ann. Ent. Soc. Amer*. *59*:1162–1166.

125. Mittler, T. E. and O. R. W. Sutherland, 1969. Dietary influences on aphid polymorphism. *Ent. Exp. Appl*. *12*:703–713.

126. Montgomery, M. E. and H. Arn, 1974. Feeding response
      of *Aphis pomi*, *Myzus persicae*, and *Amphorophora
      agathonica* to phlorizin. *J. Insect. Physiol.* 20:
      413-421.
127. Nakanishi, K., H. Moriyama, T. Okauchi, S. Fujioka,
      and M. Koreeda, 1972. Biosynthesis of α- and β-
      ecdysones from cholesterol outside the prothoracic
      gland in *Bombyx mori*. *Science 176*:51-52.
128. Nault, L. R. and W. E. Styer, 1972. Effects of
      sinigrin on host selection by aphids. *Ent. Exp.
      and Appl. 15*:423.
129. Neville, P. F. and T. D. Luckey, 1971. Bioflavonoids
      as a new growth factor for the cricket, *Acheta
      domesticus*. *J. Nutr. 101*:1217-1224.
130. Norris, D. M., J. E. Baker, T. K. Borg, S. M.
      Ferkovitch, and J. M. Rozental. An energy trans-
      duction mechanism in chemoreception by the bark
      beetle, *Scolytus multistriatus*. Contrib. Boyce
      Thomp. Instit. 24:263-274.
131. Oliver, B. F., F. G. Maxwell and J. N. Jenkins, 1970.
      Utilization of glanded and glandless cotton diets
      by the bollworm. *J. Econ. Ent. 63*:1965-1966.
132. Oliver, B. F., F. G. Maxwell and J. N. Jenkins, 1971.
      Growth of the bollworm on glanded and glandless
      cotton. *J. Econ. Ent. 64*:396-398.
133. Parr, J. C. and R. Thurston, 1972. Toxicity of
      nicotine in synthetic diets to larvae of the tobacco
      hornworm. *Ann. Ent. Soc. Amer. 65*:1185-1188.
134. Parrott, W. L., F. G. Maxwell, J. N. Jenkins, and J.
      K. Mauldin, 1969. Amino acids in hosts and nonhosts
      of the boll weevil, *Anthonomus grandis*. *Ann. Ent.
      Soc. Amer. 62*:255-261.
135. Pearl, R., 1926. A synthetic food medium for the
      cultivation of *Drosophila*. Preliminary note.
      Jour. Gener. Physiol. 9:513-519.
136. Randolph, P. A., J. C. Randolph, and C. A. Barlow,
      1975. Age-specific energetics of the pea aphid,
      Acyrthosiphon pisum. *Ecology 56*:359-369.
137. Reese, J. C., 1975. Effects of allelochemics on the
      black cutworm, *Agrotis ipsilon*. Ph.D. Dissert.,
      University of Wisconsin.
138. Reese, J. C., L. M. English, T. R. Yonke, and M. L.
      Fairchild, 1972. A method for rearing black cut-
      worm. *J. Econ. Ent. 65*:1047-1050.

139. Rehr, S. S., D. H. Janzen, and P. P. Feeny, 1973. L-dopa in legume seeds: A chemical barrier to insect attack. *Science* *181*:81–82.

140. Reichstein, T., J. von Euw, J. A. Parsons and M. Rothschild, 1968. Heart poisons in the Monarch butterfly. *Science* *161*:861–866.

141. Retig, N. and I. Chet, 1974. Catechol-induced resistance of tomato plants to *Fusarium* wilt. *Physiol. Plant Pathol.* *4*:469–475.

142. Riddiford, L. M., 1967. Trans-2-hexenal: Mating stimulant for polyphemus moths. *Science* *158*:139–141.

143. Robbins, W. E., J. N. Kaplanis, J. A. Svoboda, and M. J. Thompson, 1971. Steroid metabolism in insects. *Ann. Rev. Ent.* *16*:53–72.

144. Robbins, W. E., J. N. Kaplanis, M. J. Thompson, T. J. Shortino, C. F. Cohen, and S. C. Joyner, 1968. Ecdysones and analogs: Effects on development and reproduction of insects. *Science* *161*:1158–1160.

145. Robinson, T., 1974. Metabolism and function of alkaloids in plants. *Science* *184*:430–435.

146. Roller, H., K. H. Dahm, C. C. Sweely and B. M. Trost, 1967. The structure of juvenile hormones. *Angew. Che. (Int. Edit.)* *6*:179–180.

147. Ryan, C. A. and T. R. Green, 1974. Proteinase inhibitors in natural plant protection. *Rec. Adv. Phytochem.* *8*:123–140.

148. Salama, H. S. and A. M. El-Sharaby, 1972. Giberellic acid and B-sitosterol as sterilants of the cotton leafworm, *Spodoptera littoralis* Boisduval. *Experientia* *28*:413–414.

149. Sauer, H. H., R. D. Bennett, and E. Heftmann, 1968. Ecdysterone biosynthesis in *Podocarpus elata*. *Phytochem.* *7*:2027.

150. Schaeffers, G. A. and M. E. Montgomery, 1973. Influence of cytokinin (N^6Benzyladenine) on development and alary polymorphism in strawberry aphid, *Chaetosiphon fragaefolii*. *Ann. Ent. Soc. Amer.* *66*:1115–1119.

151. Schmialek, P., 1963. Metamorphosehemmung von *Tenebrio molitor* durch Farnesylmethylather. *Zeit f. Naturfor.* *18*B:513.

152. Schreiber, K., 1958. Uber einige Inhaltsstoffe der Solanaceen und ihre Bedeutung fur die Kartoffelkaferresistenz. *Ent. Exp. & Appl.* *1*:28–37.

153. Schroeder, L. A., 1971. Energy budget of larvae of
     Hyalophora cecropia (Lepidoptera) fed *Acer negundo*.
     *Oikos 22:*256-259.

154. Schroeder, L. A., 1972. Energy budget of cecropia
     moths, *Platysamia cecropia* (Lepidoptera;Staurniidae),
     fed lilac leaves. *Ann. Ent. Soc. Amer. 65:*367-372.

155. Seamans, H. L. and E. McMillan, 1935. The effect of
     food plants on the development of the pale western
     cutworm (*Agrotis orthogonia* Morr.). *Jour. Econ.
     Ent. 28:*421-425.

156. Self, L. S., F. E. Guthrie and E. Hodgson, 1964. Adap-
     tation of tobacco hornworms to the ingestion of
     nicotine. *Jour. Insect Physiol. 10:*907-914.

157. Self, L. S., F. E. Guthrie and E. Hodgson, 1964.
     Metabolism of nicotine by tobacco-feeding insects.
     *Nature 204:*300.

158. Shaaya, E., 1969. Studies on the distribution of
     ecdysone in different tissues of *Calliphora
     erythrocephula* and its biological half-life. *Z.
     Naturforsch. 24B:*718-721.

159. Shany, S., B. Gestetner, Y. Birk and A. Bond, 1970.
     Lucerne saponins. III. Effect of lucerne saponins
     on larval growth and their detoxification by various
     sterols. *J. Sci. Food Agric. 21:*508-510.

160. Shaver, T. N. and M. J. Lukefahr, 1969. Effect of
     flavonoid pigments and gossypol on growth and
     development of the bollworm, tobacco budworm, and
     pink bollworm. *J. Econ. Ent. 62:*643-646.

161. Shaver, T. N., M. J. Lukefahr and J. A. Garcia, 1970.
     Food utilization, ingestion, and growth of larvae of
     the bollworm and tobacco budworm on diets containing
     gossypol. *J. Econ. Ent. 63:*1544-1546.

162. Shaver, T. N. and W. L. Parrott, 1970. Relationship
     of larval age to toxicity of gossypol to bollworms,
     tobacco budworms, and pink bollworms. *J. Econ. Ent.
     63:*1802-1804.

163. Shigematsu, H., H. Moriyama and N. Arai, 1974. Growth
     and silk formation of silkworm larvae influenced
     by photoecdysones. *J. Insect Physiol. 20:*867-875.

164. Slama, K., 1969. Plants as a source of materials with
     insect hormone activity. *Ent. Exp. Appl. 12:*721-
     728.

165. Slama, K. and C. M. Williams, 1966a. "Paper factor"
     as an inhibitor of the embryonic development of the
     European bug, *Pyrrhocoris apterus*. *Nature 210:*329.

166. Slama, K. and C. M. Williams, 1966b. The juvenile
      hormone. V. The sensitivity of the bug, *Pyrrhocoris
      apterus*, to a hormonally active factor in American
      paper pulp. *Biol. Bull.* *130*:235.

167. Smissman, E. E., S. D. Beck, and M. R. Boots, 1961.
      Growth inhibition of insects and a fungus by indole-
      3-acetonitrile. *Sci.* *133*:462.

168. Smissman, E. E., J. B. Lapidus, and S. D. Beck, 1957.
      Isolation and synthesis of an insect resistance
      factor from corn plants. *Jour. Amer. Chem. Soc.*
      79:4697-4698.

169. Smith, D. S., 1959. Utilization of food plants by the
      migratory grasshopper, *Melanoplus bilituratus* (Walker)
      (Orthoptera:Acrididae) with some observations on
      the nutritional value of the plants. *Ann. Ent. Soc.
      Amer.* *52*:674-680.

170. Snyder, Karl D., 1954. The effect of temp. and food
      on the development of the variegated cutworm
      Peridroma margaritosa Haw. *Ann. Ent. Soc. of Amer.*
      47:603-613.

171. Soo Hoo, C. F. and G. Fraenkel, 1966a. The selection
      of food plants in a polyphagous insect, *Prodenia
      eridania*. *Jour. Insect Physiol.* *12*:693-709.

172. Soo Hoo, C. F. and G. Fraenkel, 1966b. The consump-
      tion, digestion, and utilization of food plants by
      a polyphagous insect, *Prodenia eridania* (Craner).
      Jour. Insect Physiol. *12*:711-730.

173. Sturchkow, B. and I. Low, 1961. Die Wirkung einiger
      Solanum-Alkaloidglykoside auf den Kartoffelkafer,
      Leptinotarsa decemlineata Say. *Ent. Exp. & Appl.4:*
      133-142.

174. Sutherland, O. R. W., N. D. Hood and J. R. Hillier,
      1975. Lucerne root saponins: a feeding deterrent
      for the grass grub, *Costelytra zealandica* (Coleoptera:
      Scarabaeidae). *N. Z. J. Zool.* *2*:93-100.

175. Svoboda, J. A., R. F. N. Hutchins, M. J. Thompson, and
      W. E. Robbins, 1969. 22-Trans-cholesta-5,22,24-trien-
      3B-ol-- an intermediate in the conversion of stig-
      masterol to cholesterol in the tobacco hornworm,
      Manduca sexta, (Johannson). *Steroids* *14*:469.

176. Svoboda, J. A., J. N. Kaplanis, W. E. Robbins, and
      M. J. Thompson, 1975. Recent developments in
      insect steroid metabolism. *Ann. Rev. Ent.* *20*:205.

177. Svoboda, J. A. and W. E. Robbins, 1968. Desmosterol
      as a common intermediate in the conversion of a
      number of C_{23} and C_{29} plant sterols to cholesterol
      by the tobacco hornworm. *Experientia 24:*1131.
178. Takemoto, T., S. Ogawa, and N. Nishimoto, 1967a.
      Studies on the constituents of *Achyranthis* radix.
      II. Isolation of the insect moulting hormones.
      Yak. Zass. (J. Pharm. Soc. Japan) *87:*1469-1473.
179. Takemoto, T., S. Ogawa, and N. Nishimoto, 1967b.
      Studies on the constituents of *Achyranthis* radix.
      III. Structure of Inokosterone. *Yak. Zass.*
      (J. Pharm. Soc. Japan) *87:*1474-1477.
180. Takemoto, T., S. Ogawa, N. Nishimoto and S. Taniguchi,
      1967c. Studies on the constituents of *Achyranthis*
      radix. IV. Isolation of the insect moulting
      hormones from Formosan *Achyranthes* spp. *Yak. Zass.*
      (J. Pharm. Soc. Japan) *87:*1478-1480.
181. Takemoto, T., S. Ogawa, N. Nishimoto, and K. Mue, 1967d.
      Studies on the constituents of *Achyranthis* radix.
      V. Insect hormone activity of ecdysterone and
      inokosterone on the flies. *Yak. Zass.* (J. Pharm.
      Soc. Japan) *87:*1481-1483.
182. Tarnopol, J. H. and H. J. Ball, 1972. A survey of
      some common midwestern plants for juvenile hormone
      activity. *J. Econ. Ent. 65a:*980-982.
183. Thompson, J. A., J. B. Siddall, M. N. Galbraith, D. H.
      S. Horn, and D. J. Middleton, 1969. The biosynthesis
      of ecdysones in the blowfly *Calliphora stygia.*
      *Chem. Commun. 1969:*669-670.
184. Thompson, M. J., J. N. Kaplanis, W. E. Robbins, and
      J. A. Svoboda, 1973. Metabolism of steroids in
      insects. *Adv. Lipids Res. 11:*219-265.
185. Thorsteinson, A. J., 1953. The chemotactic responses
      that determine host specificity in an oligophagous
      insect (*Plutella maculipennis* (Curt.) Lepidoptera).
      *Can. Jour. Zool. 31:*52-72.
186. Tipton, C. L., J. A. Klun, R. R. Husted, and M. D.
      Pierson, 1967. Cyclic hydroxamic acids and related
      compounds from maize. Isolation and characterization.
      *Biochem. 6:*2866-2870.
187. Todd, G. W., A. Getahun, and D. C. Cress, 1971. Resis-
      tance in barley to the greenbug, *Schizaphis graminum.*
      1. Toxicity of phenolic and flavonoid compounds and
      related substances. *Ann. Ent. Soc. Amer. 64:*718-
      722.

188. Uvarov, B. P., 1928. Insect nutrition and metabolism. *Trans. Ent. Soc. London 76*:255-343.

189. Vanderzant, E. S. and J. H. Chremos, 1971. Dietary requirements of the boll weevil for arginine and the effect of arginine analogues on growth and on the composition of body amino acids. *Ann. Ent. Soc. Amer. 64*:480-485.

190. Van Emden, H. F., 1972. Aphids as phytochemists *in* Phytochemical Ecology, J. B. Hartborne, Ed., Academic Press, N. Y. pp. 25-43.

191. Verschaffelt, E., 1910. The cause determining the selection of food in some herbiverous insects. Proc. K. Akad. Wetensch. *Amsterdam Sect. Sci. 13*:536-542.

192. Virtanen, A. I. and P. K. Hietala, 1955. 2(3)-Benzoxazolinone, an anti-fusarium factor in rye seedlings. *Acta Chem. Scand. 9*:1543-1544.

193. Virtanen, A. I. and P. K. Hietala, 1959. On the structure of the precursors of benzoxazolinones in rye plants. *Suom. Kemist. B 32*:252. (reprint filed as abstract).

194. Waldbauer, G. P., 1964. The consumption, digestion, and utilization of Solanaceous and non-solanaceous plants by larvae of the tobacco hornworm, *Protoparce sexta* (Johan) (Lepidoptera:Sphingidae). *Ent. Exp. & Appl. 7*:253-269.

195. Waldbauer, G. P., 1968. The consumption and utilization of food by insects. *Adv. Insect Physiol. 5*: 229-288.

196. Wardojo, S., 1969. Some Factors Relating to the Larval Growth of the Colorado Potato Beetle, *Leptinotarsa decemlineata* Say (Coleoptera:Chrysomelidae), on Artificial Diets. H. Veenman and Zonen N.V., Wageningen 75 pp.

197. Wensler, R. J. D., 1962. Mode of host selection by an aphid. *Nature 195*:830-831.

198. White, D., 1972. Effect of varying dietary amino acid and sucrose concentrations on production of apterous cabbage aphids. *J. Insect Physiol. 18*: 1241-1248.

199. Whittaker, R. H., 1970. The biochemical ecology of higher plants. *in* Chemical Ecology, E. Sondheimer and J. B. Simeone, Eds. Academic Press, pp. 43-70.

200. Whittaker, R. H. and P. P. Feeny, 1971. Allelochemics: Chemical interactions between species. *Science 171*: 757-770.

201. Wiklund, C., 1975. The evolutionary relationship
      between adult oviposition preferences and larval
      host plant range in *Papilio machaon* L. *Oecologia 18:*
      185-197.
202. Williams, C. M., 1970. Hormonal interactions between
      plants and insects. *in* Chemical Ecology, Eds.
      E. Sondheimer and j. B. Simeone. Academic Press,
      New York and London, pp. 103-132.
203. Yamada, H. and M. Kato, 1966. Chlorogenic acid as
      an indispensable component of the synthetic diet
      for the silkworm. *Proc. Jap. Acad. 42:*399.
204. Yang, R. S. H. and F. E. Guthrie, 1969. Physiological
      responses of insects to nicotine. *Ann. Ent. Soc.
      Amer. 62:*141-146.
205. Yang, R. S. H. and C. F. Wilkinson, 1972. Enzymic
      sulphation of p-nitrophenol and steroids by
      larval gut tissues of the southern army worm
      (*Prodenia eridania* Cramer.) *Biochem. J. 130:*487-
      493.

Chapter Three

MILKWEED CARDENOLIDES AND THEIR COMPARATIVE PROCESSING

BY MONARCH BUTTERFLIES (*Danaus plexippus* L.)

C.N. ROESKE AND J.N. SEIBER

Department of Environmental Toxicology

University of California, Davis, California

L.P. BROWER AND C.M. MOFFITT

Department of Biology

Amherst College, Amherst, Massachusetts

INTRODUCTION

    The milkweed family (Asclepiadaceae) comprises some 200 genera and 2500 species of perennial shrubs, herbs and vines distributed throughout the tropics and extending to temperate areas of the world. They include some highly

prized ornamentals and economically significant weeds, and
are generally characterized to the layman by the milky latex
they exude when a leaf or other organ is ruptured. Chemical
interest in the milkweeds has been stimulated by the use
of some plants in medicinal preparations to treat cancers,
tumors, and warts (Refs. in 54), as emetics, to treat
bronchitis (Refs. in 64), and as a source of digitalis-
like therapeutic agents (Refs. in 44). They are also known
for their poisonous nature, which has found advantageous
use in the preparation of arrow poisons, and also causes
occasional but extensive poisoning episodes among grazing
sheep and cattle in milkweed-infested rangelands[52,58].

Chemical and ecological interest in the milkweeds as
a source of poisons which serve a protective function in
insects has recently been forthcoming. It is now well
established, for example, that monarch butterflies of the
genus *Danaus* feed as caterpillars on species of milkweed
from which they accumulate protective quantities of
cardenolides (cardiac glycosides) in their body tissues[7,70,80]
If the caterpillar has accumulated sufficient cardenolide,
then the caterpillar, pupa, or adult will cause a bird
which eats it to vomit. This toxic response may be learned
by the predator, such that it avoids the prey in subsequent
encounters. In this chapter we will focus primarily on
the chemical ecology of the interactions between milkweeds,
monarch butterflies, and their vertebrate predators. Both
literature and research from our own laboratories will
serve as the basis for discussion. The milkweed-monarch
sequestering phenomenon should be of general interest,
since it has become increasingly apparent that food-derived
chemicals have several wide-ranging and important ecological
roles[6,23,28,34,84,91,108].

THE CHEMISTRY OF CARDENOLIDES

General. The cardenolides are a group of C_{23} steroid
derivatives which conform to the general structure:

I

They are characterized by the presence in the "genin" (where R=H in structure I) of (1) an α:β unsaturated γ-lactone (butenolide) ring attached at C-17, (2) a *cis* juncture of rings C and D, and (3) a 14-β hydroxy group. Differentiat- ing features include the configurations at C-3, C-5, and C-17, presence of additional oxygen substituents (usually hydroxy groups) at C-1,2,5,11,12,15,16, and **19** and in a few cases the presence of an additional olefinic double bond in the molecule.

The cardenolides usually occur in nature as glycosides ("cardiac glycosides"), attached through oxygen at C-3 of the genin to one or more sugar moieties. Over twenty sugars have been isolated from hydrolysis of cardenolides and only three of them occur in other plant glycosides. A common, but by no means inclusive, pattern is for the aglycone to be attached to one or more rare sugars and then to one or more glucose molecules. The seeds and leaves of many plants contain other glucosides and also the enzymes needed to hydrolyze the glucose units. Unless enzymatic hydrolysis is inhibited during extraction and isolation procedures, the glycosides isolated will retain only the rare sugars[33,78,94]. It is the degree of oxygenation in the steroid ring, and the nature of the glycosidic group which primarily determines cardenolide polarity.

Cardenolides have been isolated only from the Angio- sperms and are particularly abundant in the families Apocynaceae and Asclepiadaceae. The familiar digitoxin comes mainly from the Scrophulariaceae while strophanthidin

and ouabain occur primarily in the Apocynaceae. These
drugs are of the 5β series (*cis* A/B ring fusion) and, in
the case of digitoxin, are conjugated to three moles of
digitoxose, a 2,6-dideoxyhexose and in ouabain, to one
mole of rhamnose. This group includes several clinically
useful drugs for the treatment of atrial fibrillation and
related heart ailments. The cardiac activity of the
glycosides is due to the genin moiety, but the sugar provides important solubility and distribution characteristics
which increase the potency and duration of the effect on
the myocardium. Since the major toxic effect (cardiac
arrhythmias, emesis, visual disturbance) of these popular
medicinal agents are exerted at dosage levels near those
employed for the therapeutically useful actions on the
heart, numerous attempts have been made to uncover
cardenolides having a more advantageous therapeutic index.
These and other aspects of cardenolide chemistry, pharmacology and distribution have been summarized in several
excellent reviews[33,46,78,94,97-99].

   The Asclepiadaceae. Cardenolides have been isolated
from 12 genera of the Asclepiadaceae[46,94] and their presence noted in several others[1]. Particularly pertinent to
our review are the chemically related cardenolides of
Asclepias spp, *Calotropis procera*, and *Pergularia extensa*
(Table 1). The shrub *Calotropis procera*, which occurs
extensively in Asia and Africa, has attracted much attention
over the years probably from its popular use in folk medicine and in the preparation of arrow poisons. From the
plant latex, Hesse and his co-workers (see Table 1) sepa-
rated calotropin (II) ($C_{29}H_{40}O_9$), calactin (IV) ($C_{29}H_{40}O_9$),
calotoxin (V) ($C_{29}H_{40}O_{10}$), uscharidin (VI) ($C_{29}H_{38}O_9$),
uscharin (VII) ($C_{31}H_{41}O_8NS$), and voruscharin (VIII)
($C_{31}H_{43}O_8NS$), and suggested that these formed a new series
of cardenolides all with a common aglycone. The structure
of calotropin was formulated as II[21,22];

Calotropin

that is, calotropin possessed a *trans* A/B ring-fused genin, calotropagenin, common to all the latex-derived cardenolides and the unusual 4,6-dideoxyhexosone conjugated to the genin.

A major structural revision of calotropagenin and the associated glycosides was proposed recently[13,55,56] based on the physical and chemical similarity of the *Calotropis* cardenolides to gomphoside (III) isolated and identified

III
Gomphoside

from *Asclepias fruticosa*[15,19,20,105,107]. The revised
structures of the six latex cardenolides of *Calotropis
procera* and the more recently isolated proceroside (IX)
are given in structures (IV-IX)[14].

IV
Calactin and Calotropin
(configurational isomers)

VI
Uscharidin

VIII
Voruscharin

V
Calotoxin

VII
Uscharin

IX
Proceroside

Gomphoside (III) and the *Calotropis* cardenolides
(IV-IX) possess adjacent *trans* hydroxy groups at C-2 and
C-3 of the steroid nucleus which together participate in
bonding the hexosone, through hemiketal (at C-2) and acetal
(at C-3) bonds. This presence of two sugar-genin oxygen
bridges explains the stability of these cardenolides to
mild acid; this behavior is in marked contrast with that
of most of the known cardenolides and was a source of
considerable confusion in early work with these compounds.
It also clarified the placement of the "extra" hydroxy
group, originally assigned with no substantive evidence to
C-12[36]. Uscharidin (VI) has a carbonyl oxygen at C-3', and
may be converted to calactin-calotropin (IV) by partial
reduction[38,40]. Uscharin (VII) and voruscharin (VIII)
generate uscharidin (VI) by acid or mercuric salt-induced
hydrolytic cleavage of the thiazoline and thiazolidine
rings[39,40,42]. Proceroside (IX) is formulated as calactin-
calotropin with an additional hydroxy group, probably
located on rings C or D of the steroid nucleus[14].

   Asclepias curassavica analyzed by Santavy *et al.*,
(cited in 102) was found to contain an almost identical
array of cardenolides as those isolated from *Calotropis
procera* by Hesse and coworkers. Singh and Rastogi (1969)
isolated from *A. curassavica* an additional member (asclepin)
of the series represented by (IV-IX) which was identified
as 3-O-acetylcalotropin[95]. We have recently obtained
evidence, discussed in more detail in a subsequent section,
that calotropagenin-derived cardenolides are also distri-

buted among other *Asclepias* species of the western
United States.

The calotropagenin-derived cardenolides are by no
means the only ones encountered in this genus; from
Asclepias syriaca Bauer *et al.* (1961) and Masler *et al.*
(1962a and 1962b) isolated and identified uzarigenin,
syriogenin, desglucouzarin, syrioside and syriobioside.
These glycosides are formed in the more conventional manner;
the genins (syriogenin and uzarigenin) are linked to common
sugars (glucose and rhamnose) through a single oxygen bridge,
an acetal bond formed between the steroid alcohol (C-3) and
the hexose (C-1'). The structures of these genins and
their associated sugars as well as other genins and sugars
found in *Asclepias* species are presented in Figures 1 and
2. From *A. labriformis* and *A. eriocarpa* from the western
U. S. we have isolated three apparently new cardenolides
having an unusually high O/C ratio reflected in the tentative empirical formulas $C_{31}H_{39}O_{10}NS$, $C_{29}H_{36}O_{11}$, and
$C_{29}H_{38}O_{11}$. The structures of these compounds are presently
under study[96].

PLANT DISTRIBUTION AND CARDENOLIDE VARIATION

Geographic Distribution of Some Asclepias spp. The
milkweed family in North America, including Mexico, is
represented by several genera[110] and the genus *Asclepias*
contains 108 described species[111]. Widespread and abundant
North American species include *Asclepias syriaca* which
occurs from Maine to Virginia westward to North Dakota and
Kansas, and *A. speciosa* which has a western geographic
distribution from southern Manitoba to British Columbia in
the north, from Minnesota southward to northwestern Texas,
and westward to the Pacific Ocean. Extremely localized
and rare species include *A. masonii*, known only in the
vicinity of Magdalena Bay in Baja California, and *A.
labriformis*, apparently restricted to washes, sandstone
canyons, dry cliffs, and high flats of eastern Utah. Still
others are more or less widely distributed, yet extremely
specialized in their habitat. For example, in the east,
the subaquatic *A. incarnata* is found only in marshy areas;
A. exaltata is a deciduous forest border species; and *A.
amplexicaulis* only grows on sandy shores or associated with
old alluvial deposits. In Northern California, one of the
most specialized and infrequent milkweeds is *A. solanoana*

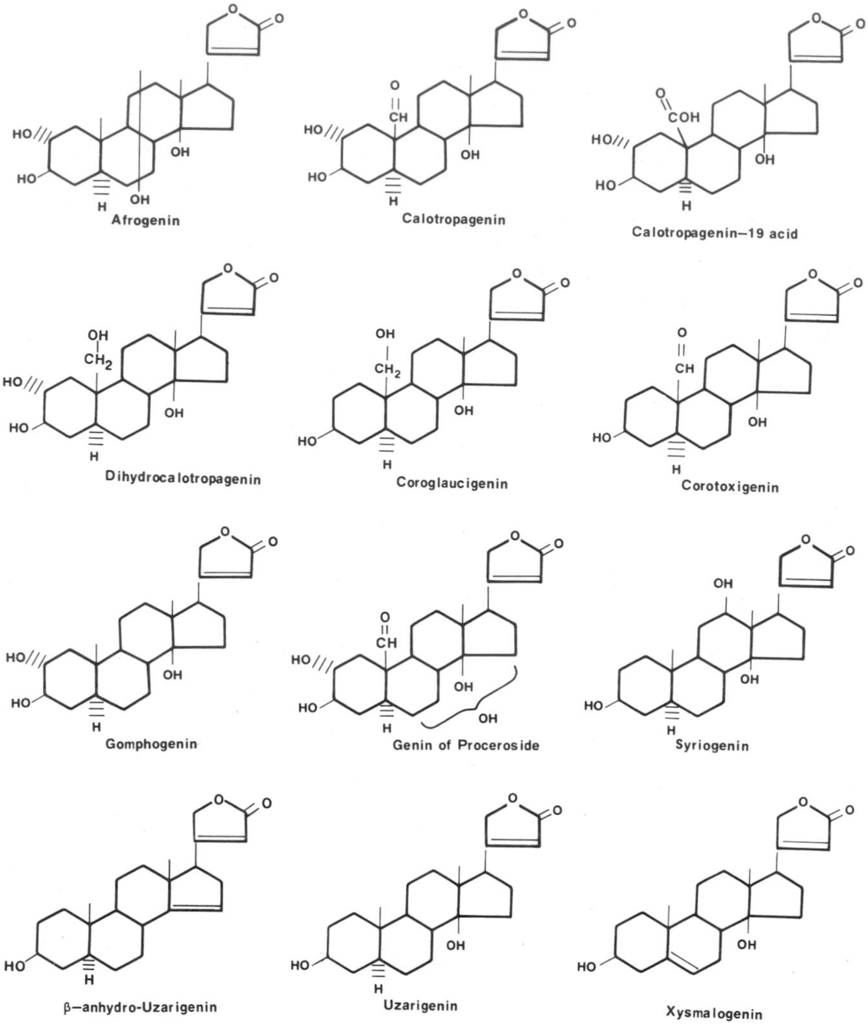

Figure 1. Structures of genins from *Asclepias* spp.,
Calotropis procera, and *Pergularia extensa*.

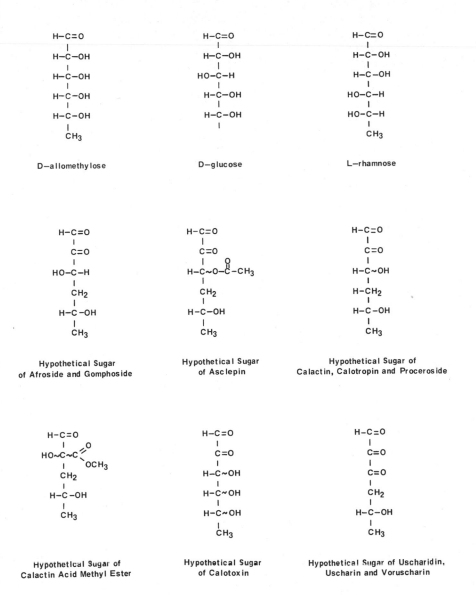

Figure 2.   Structures of sugars associated with genins from *Asclepias* spp, *Calotropis procera*, and *Pergularia extensa*.

which grows exclusively on hot, dry, south facing, ser-
pentine slopes in chaparrel communities.

  Cardenolide Variation Among Milkweeds. Extensive
quantitative and qualitative variation in cardenolide
content occurs both between and within single species of
milkweeds. This variation could be caused by genotypic
(*e.g.* geographic, ecotypic, and polymorphic) and/or by
phenotypicfactors (*e.g.* age of plant and part of plant
selected for analysis) and may be influenced by methods of
collection, handling, and analysis.

  For example, the search for cardenolides in *Calotropis
procera* resulted in an examination of the different parts
of the plant. The initial isolation of calactin (IV),
calotoxin (V), calotropin (IV), uscharidin (VI), uscharin
(VII), voruscharin (VIII), and calotropagenin
was achieved from the latex by Hesse and coworkers (see
Table 1). Bruschweiler *et al.* (1969b), working with both
the latex material and leaf and stem material found these
latex cardenolides along with three additional ones--
proceroside, syriogenin, and uzarigenin but
found no voruscharin or syriogenin in the leaf and stem
samples. Hassall and Reyle (1959) isolated calactin and
calotoxin from the root bark which they considered to be
the most convenient source of glycosides. Seeds from two
different locations, examined by Rajagopalan *et al.* (1955),
yielded calotropin and three additional cardenolides:
frugoside, coroglaucigenin, and corotoxigenin.

  It has been determined that cardenolide concentrations
in *Asclepias nivea* and *A. curassavica* are highest in
actively growing plant parts (Brower and Tahsler, unpublished). In both species, the apical leaves, buds and
flowers had the highest concentrations. For *A. nivea* the
buds and flowers contained 9.02 mg of cardenolide evaluated as digitoxin per gm dry plant material, the upper leaves
7.32 mg/gm, and the lower leaves 3.56 mg/gm. *A. curassavica* had uniformly lower concentrations; 6.27 mg/gm in the
buds and flowers, 4.87 mg/gm in the upper leaves, 2.38
mg/gm in the middle leaves, and 1.78 mg/gm in the lower
leaves. This is in qualitative agreement with the findings of Evans and Cowley (1972) with cardenolide variation
as a function of plant part in *Digitalis purpurea*. TLC
analysis of the different samples of *A. nivea* and *A.*

curassavica showed only slight variation in the relative proportions of the various cardenolides present among the samples from the different regions.

Qualitative variation in cardenolide content of *A. curassavica* is illustrated by the following analyses. Leaves and stems of Brazilian origin yielded five known cardenolides--uzarin, calotropagenin, coroglaucigenin, corotoxigenin, and uzarigenin[100,101]. Examination of specimens from India resulted in the isolation of four of these cardenolides and three additional ones-- asclepin, calactin (IV), and calotropin (IV)[93,95]. Santavy and coworkers isolated from *A. curassavica*, col- lected in Trinidad W.I., two of the cardenolides found in the Indian material, calactin and calotropin, and three additional ones, calotoxin V,uscharin (VII), and uscharidin (VI) (Ref. in 102). Rothschild *et al.*, (1975) reported that *A. curassavica* growing in England (of unknown origin) contained calotropin and uzarigenin and still another cardenolide, proceroside IX.It should be noted that all four examinations yielded either calotropagenin or its glycosides, and that uzarigenin was found in three of the four.

A. syriaca, a native North American species named after its introduction into Europe in colonial times, has also been studied by a number of investigators. Petricic (1967) in Yugoslavia found desglucouzarin, syriobioside and syrioside. These glycosides and their genins, uzarigenin and syriogenin were isolated. from the aerial parts of a European *A. syriaca*[2,59]. How- ever, leaves of *A. syriaca* growing in Switzerland were reported by Reichstein and co-workers to lack cardenolides (see ref. 88). From *A. syriaca* growing in Japan, Mitsu- hashi *et al.* (1970) isolated five cardenolides from the aerial parts, two of which were identified as uzarigenin and xysmalogenin. Five cardenolides of high to intermediate polarity were also observed in the leaves of this species growing in Ontario, Canada[26]; however, only one highly polar cardenolide was found in a sample from New York[88]. Our quantitative analysis of *A. syriaca* leaves from Massachusetts showed over a 40-fold variation in cardenolide concentration among different samples, from 0.006% to 0.264% (Table 2). Analysis of specimens from Canada indicate fifteen cardenolides in the seeds

Table 1

Summary of cardenolides of known structure identified from *Asclepias* spp, *Calotropis procera*, and *Pergularia extensa*

Plant	Glycoside	Genin	Sugar	References
Asclepias curassavica				
	asclepin	calotropagenin	3-0-acetyl-4,6-dideoxyhexosone	93,95
	calactin	calotropagenin	4,6-dideoxyhexosone	93,102
	calotoxin	calotropagenin	6-deoxyhexosone	102
	calotropin	calotropagenin	4,6-dideoxyhexosone	54,88,93,102
	proceroside	calotropagenin+OH	4,6-dideoxyhexosone	88
	uscharidin	calotropagenin	4,6-dideoxyhexosone	102
	uscharin	calotropagenin	3-spiro-Δ^3-thiazoline 4,6 dideoxyhexosone	102
	uzarin	uzarigenin	2, D-glucose units	93,101
		calotropagenin		93,100,101
		coroglaucigenin		93,100
		corotoxigenin		100
		uzarigenin		88,93,100
A. fruticosa				
	afroside	afrogenin	4,6-dideoxyhexosone	88,104,105,106
	desglucouzarin	uzarigenin	D-glucose	18
	frugoside	coroglaucigenin	D-allomethylose	17,47,48
	gofruside	corotoxigenin	D-allomethylose	17,47,48,49
	gomphoside	gomphogenin	4,6-dideoxyhexosone	20,88,104,105,107
		uzarigenin		18,63,88
A. glaucophylla				
	ascleposide	uzarigenin	D-allomethylose	68
		coroglaucigenin		67,68
		uzarigenin		67,68
		β-anhydro uzarigenin		68
A. lilacina				
	ascleposide	uzarigenin*	D-allomethylose	89
	frugoside	coroglaucigenin*	D-allomethylose	89
A. mellodora				
	desglucouzarin	uzarigenin	D-glucose	72,73
A. swynnertonii				
	ascleposide	uzarigenin*	D-allomethylose	51
	frugoside	coroglaucigenin*	D-allomethylose	51
	uzarin	uzarigenin*	2g D-glucose units	51
		corotoxigenin		51
A. syriaca				
	desglucouzarin	uzarigenin	D-glucose	2,59,72
	syriobioside	syriogenin	L-rhamnose + D-glucose	59,60,72
	syrioside	syriogenin	L-rhamnose + 2 D-glucose units	59,60,72
		syriogenin		59,60
		uzarigenin		2,59,62
		xysmalogenin		62

Table 1 (continued)

A. tuberosa	frugoside	coroglaucigenin*	D-allomethylose	72
	glucofrugoside	coroglaucigenin	D-allomethylose + D-glucose	72,74
Calotropis procera	calactin	calotropagenin	4,6-dideoxyhexosone	14,36,40,45
	calactin acid methyl ester#	calotropagenin	2-carbomethoxy-3,5-dideoxy-pentose	14
	calotoxin	calotropagenin	6-deoxyhexosone	14,36,40,45,57
	calotropin	calotropagenin	4,6-dideoxyhexosone	14,40,44,57,77
	calotropin-19-acid	calotropagenin-19-acid	4,6-dideoxyhexosone	14
	frugoside	coroglaucigenin*	D-allomethylose	77
	proceroside	calotropagenin+OH	4,6-dideoxyhexosone	14
	uscharidin	calotropagenin	4,6-dideoxyhexosone	14,39,40,42,57,41
	uscharin	calotropagenin	3-spiro-β^3-thiazoline 4,6-dideoxyhexosone	14,39,40,43,45,57,41
	voruscharin	calotropagenin	3-spiro-thiazolidine 4,6 dideoxyhexosone	14,41,42
		calotropagenin		14,44
		corotoxigenin		77
		syriogenin		14
		uzarigenin		14
Pergularia extensa	calactin	calotropagenin*	4,6-dideoxyhexosone	14,64
	calotoxin	calotropagenin*	6-deoxyhexosone	14,64
	calotropin	calotropagenin*	4,6-dideoxyhexosone	14,64
	proceroside	calotropagenin+OH	4,6-dideoxyhexosone	14
	uscharidin	calotropagenin*	4,6-dideoxyhexosone	14
	uscharin	calotropagenin*	3-spiro- 3-thiazoline 4,6-dideoxyhexosone	14
		coroglaucigenin		64
		corotoxigenin		64
		al-dihydrocalotropagenin		64
		uzarigenin		64

*Genins also isolate in free state from the plant.

@These three African species are considered botanically very similar (Sawlewicz et al., 1967 and Jaggi et al., 1967).

#Believed to be an artifact originating during extraction from uscharidin and methanol (Bruschweiler et al., 1969b).

Table 2

Gross cardenolide content of plant material from various
North American species of *Asclepias* listed as enumerated
by Woodson (1954)

Plants	Cardenolide content, calculated as mg of digitoxin per gm dried material@			
	Leaf or aerial		Seeds	
Subgenus I. Asclepias				
Series 1. Incarnata				
1. *incarnata*	nil – 0.28	(2)	0.70 – 5.42	(82)
7. *curassavica*	2.08 – 4.14	(10)		
8. *nivea*	5.44	(p)		
10. *fascicularis*	0.04* – 0.12	(1+2p)	0.51 – 0.73	(2)
13. *verticillata*	nil	(p)	nil	(p)
Series 2. Tuberosa				
17. *tuberosa*	nil – 0.06	(5)	nil –0.03	(38)
Series 3. Exaltata				
23. *exaltata*	nil – 0.70	(3)	1.59 – 7.24	(49)
26. *amplexicaulis*	nil – 0.22	(5)	2.64	(p)
Series 5. Syriaca				
36. *syriaca*	0.06 – 2.64	(16)	1.34 – 9.03	(160)
42. *linaria*	7.78*	(p)		
Series 6. Purpurascentes				
49. *speciosa*	0.149*	(p)	2.58 – 3.89	(2)
Series 8. Roseae				
63. *labriformis*	2.14*	(p)		
64. *erosa*	1.30*	(p)		
65. *eriocarpa*	1.13* – 5.20	(3+2p)	1.66 – 5.13	(2+p)
66. *masonii*	11.2* – 147.0	(p+1)		
67. *subaphylla*	7.65* – 18.8	(2p)		
68. *albicans*	1.79* – 55.1	(3p)		
69a. *vestita vestita*	3.05*	(p)		
69b. *vestita parishii*	1.25*	(p)		
Subgenus II. Podostemma				
74. *subulata*	3.21* – 29.6	(3p)		
Subgenus IV. Asclepiodella				
86. *cordifolia*	0.36* – 0.38	(1p)	3.06	(p)
Subgenus V. Acerates				
90. *viridiflora*	nil	(p)	0.51	(p)
Subgenus VI. Solanoa				
92a. *californica californica*	0.145*–0.271*	(2p)		
92b. *californica greenii*	0.166*	(p)		
93a. *cryptoceras cryptoceras*	0.252*	(p)		
94. *solanoana*	0.244*	(p)		

@Determinations were made using standard spectroassay method. Minimum and maximum values are given when possible, each representing a different plant or seed pod; the number of samples is given in ( ); if an individual sample is from more than one plant it is indicated by (p) as pooled. Data collected through 1974, seed data from Brower and Moffitt, in preparation.

*Denotes values determined on aerial samples after extraction and purification as described later.

but only five in the leaves[26]; our analyses of seeds from
Massachusetts shows a higher but narrower range of concentrations (0.134% to 0.903%, Table 2) than was observed for
the leaves (0.006% to 0.264%).

The separate analyses of *A. curassavica* and *A. syriaca*
illustrate the extensive variation of cardenolide content
within a single species and the similarities and differences
between species. Most of the glycosides identified from
A. curassavica had as their aglycone calotropagenin, and
those from *A. syriaca* had syriogenin. However, both of these
species contained glycosides derived from uzarigenin. The
presence of uzarigenin and/or its glycosides has been shown
in all but one of the ten plants listed in Table 1.
Uzarigenin, perhaps biogenetically the simplest cardenolide,
may prove to be a common genin in the genus *Asclepias*.

Several other North American asclepiads have been
examined for cardenolide content in our laboratories by
spectral and TLC procedures (Table 2). *A. masonii* and
A. tuberosa illustrate the extremes. A single sample of
A. masonii was 14.7% cardenolide by dry weight while five
samples of *A. tuberosa* ranged from nil to 0.006%. Quantitative variation between leaf and seed samples is evident,
and in most cases the seed concentrations are higher than
the corresponding leaf samples.

In our study, qualitative analysis of the cardenolides
in 22 different aerial plant samples, representing 14
different species of *Asclepias*, was carried out using TLC
(Figures 3 and 4). Ethanol extracts were cleaned by solvent
partitioning and gross cardenolide concentrations were
determined by spectroassay. Amounts equivalent to about 100
µg and 20 µg of digitoxin were then spotted, developed in
two solvent systems, sprayed with a cardenolide-selective
reagent[69]and photographed. The *A. eriocarpa* samples illustrate variation between two samples of the same species.
The sample labeled SL 353, collected in Kern County
(Southern California), contained about twice the cardenolide
concentration as the other sample (to the left of SL 353)
collected in Yolo County (Northern California). Qualitative
differences are also evident in the relative proportions of
different cardenolides present in the two samples. Slight
cardenolide variation is also apparent among the different
specimens of *A. californica* and *A. vestita* collected over

the geographic range which included two morphological subspecies for each species.

It appears that *A. vestita*, *A. subulata*, and *A.*
cordifolia have at least three major cardenolides in common.
The patterns of *A. californica*, *A. solanoana*, and *A.*
cryptoceras also resemble those of *A. vestita*
A. subulata, and *A. cordifolia*. A somewhat differerent
pattern is evident in samples of *A. albicans*, *A. masonii*,
and *A. linaria*. A third, completely different pattern
arises from samples of *A. eriocarpa*, *A. erosa*, and *A.*
labriformis. The sample of *A. curassavica* is used as a
reference. This plant, cultivated at Amherst College, was
grown from seeds obtained in Trinidad and is believed to be
similar in cardenolide content to that analyzed by Santavy
and coworkers (cited in 102). The general similarity of its
TLC pattern to those of *A. vestita*, *A. subulata*, *A.*
cordifolia, *A. californica*, *A. solanoana*, and *A. cryptoceras*
suggests that these species have some cardenolides in common.

At present no chemotaxomic classification based on
cardenolide content of *Asclepias* has been attempted. However, it is interesting to compare our quantitative (Table
2) and qualitative results (Figures 3 and 4) with Woodson's
(1954) botanical classification. The North American genus
Asclepias is divided into nine subgenera; subgenus 1,
Asclepias, is further separated into nine series. Series 1,
Incarnatae, contains 16 species, five of which are listed
in Table 2. Of these five plants, *A. incarnata*, *A. fascicu-*
laris, and *A. verticillata* have quantitatively less
cardenolides than *A. curassavica* and *A. nivea*. Qualitatively the cardenolides of *A. curassavica* and *A. nivea*
appear similar, but patterns from the first three species
were not clear due to low concentrations. Series 8, Roseae,
contains 13 species, seven of which we have examined. All
are rich in cardenolides but the range is great (0.125% to
14.7%) as is the qualitative cardenolide differences among
them. *A. labriformis*, *A. erosa* and *A. eriocarpa* and *A.*
masonii, *A. subaphylla* (not shown) and *A. albicans* form two
distinct TLC cardenolide patterns, both of which differ
significantly from that of *A. vestita*. Thus there may be
three different groups of cardenolides within this series.
Subgenus VI, Solanoa, includes *A. californica*, *A. crypto-*
ceras, and *A. solanoana*. The plant samples of these species
and subspecies which we have analyzed are qualitatively

similar and are somewhat lower in cardenolide concentration
(0.0145% to 0.0271%) than those in Series 8, Roseae. Cardenolide profiles in themselves may be of some taxonomic
value, but clearly a more complete examination of the genus
is in order before drawing definite conclusions.

Duffey (1970) and Duffey and Scudder (1972) have found
cardenolides in 18 *Asclepias* species. In addition to the 13
species we have examined, they also report positive results
for leaf samples of *A. arenaria, A. variegata,* and *A.
ovalifolia.*

Generalizing from the above, it appears that nearly
all of the *Asclepias* species contain cardenolides, but
with many examples of both qualitative and quantitative
differences among and within species. By correlating these
cardenolide analyses with distribution and density information from Woodson (1954), it appears that one group
(*A. amplexicaulis, A. exalta, A. incarnata, A. speciosa, A.
syriaca, A. tuberosa, A. verticillata,* and *A. viridiflora*)
comprises species having a generally low cardenolide concentration, wide geographic distribution, and fairly high
relative density within that distribution. A second group
(*A. albicans, A. eriocarpa, A. erosa, A. labriformis, A.
linaria, A. masonii, A. subaphylla, A. subulata,* and *A.
vestita)* is characterized by high cardenolide concentrations,
narrow distribution, and low density. The third group (*A.
californica, A. cordifolia,* and *A. solanoana*) possesses a
medium cardenolide concentration, narrow distribution, and
low density.

Because of these differences, it is perhaps appropriate
to ask what factors have influenced the cardenolide concentration, abundance and distribution of these milkweeds. The
answer will involve considerations of the importance of
cardenolides as herbivore deterrents in plant defense, the
physiological cost to the plants in cardenolide biosynthesis,
and the availability of other, noncardenolide defenses to
the plant, as well as nutritional and environmental
requirements and tolerances for each species. It is certainly premature to attempt to formulate the answer with
the limited data presently available, but further investigations, such as the ones described in the following
sections, may have some bearing on it.

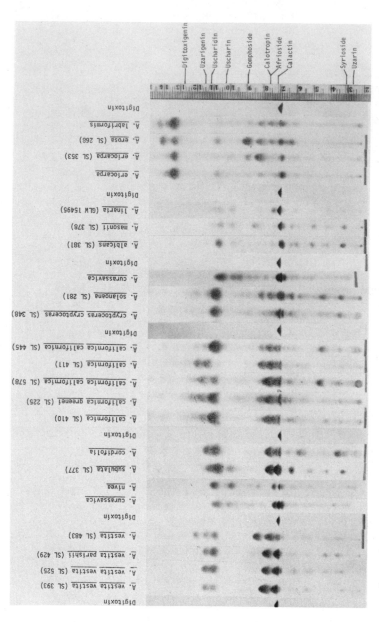

Figure 3. Thin layer chromatograms of cleaned extracts of some *Asclepias* spp and the standard digitoxin; mobilities of some other standards are also indicated. The silica gel plates were developed four times in chloroform–methanol–formamide (90:6:1, by volume). Chromatograms were aligned at the R_f of digitoxin, thus R_f values on either side may vary among the chromatograms.

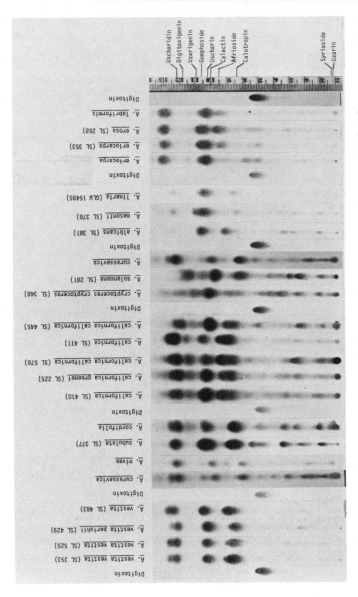

Figure 4. Thin layer chromatograms of cleaned extracts of some *Asclepias* spp and the standard digitoxin; mobilities of some other standards are also indicated. The silica gel plates were developed two times in ethylacetate-methanol (97:3, by volume). Chromatograms were aligned at the R_f of digitoxin, thus R_f values on either side may vary among the chromatograms.

SEQUESTERING AND STORAGE PHENOMENA

General Aspects. One major aspect of the rapidly developing field of chemical ecology is concerned with the roles which a wide array of plant chemicals play in ordering complex interactions of other species at various trophic levels in natural communities. They include allelochemics which function to regulate both intraspecific and interspecific competition in plants[66,82,108], and the utilization by certain insects of specific classes of plant chemicals as pheromones to attract mates or to regulate assembling behavior[37,75,109]. A particularly interesting and general phenomenon within this scope pertains to the sequestering of chemicals, from a lower level in the foodchain, for utility at higher trophic levels as predator deterrents. It is among the invertebrates, and especially the arthropods, that chemical defense seems to have reached its peak of diversification. It is particularly remarkable that many of the chemicals utilized in arthropod defense are compounds known to occur in plants[28]. While this fact in itself does not argue conclusively for the operation of general mechanisms of sequestering, it is highly suggestive. Examples of the diversity of plant-derived poisons and their immediate precursors used for defensive purposes by insects abound in reviews[28,90,108]. Rothschild alone (1972a) has tabulated examples from among 43 species of six orders shown to sequester and store plant toxins.

The argument for adaptive coevolution of insects and poisonous plants may be briefly stated as follows[4,5,31,53,86]: practically no plant, poisonous or otherwise, is immune to attack by plant-feeding insects. Some herbivorous species develop solely on poisonous plants and others may utilize poisonous and nonpoisonous species as food sources. In either case, benefits may accrue to the insects, particularly in their early stages, from the simple fact that the plants are avoided by large herbivores, and in some cases from more subtle behavioral clues associated with the poison which assist the insect in locating its chosen food plants. The poisons, particularly general biocides such as nicotine, may be handled in the plant-feeder by general detoxication processes of metabolism, conjugation, and elimination. It is implicit to this argument that some insects successfully adapted to specific poisons evolved the requisite detoxication ability, while others

have not. Alternately, some poisons, particularly those
selective towards vertebrates, may be sufficiently tolerated
by insects so that they are incorporated (sequestered and
stored) into the insect's body tissue. The biochemical
processes involved in sequestering are, for the most part,
unknown but they may be as simple as the partitioning of
low polarity chemicals, such as DDT, to lipid matter. Once
incorporated, these poisons act as deterrents to potential
predators in one or more ways. The poisons may themselves
have, or be associated with other chemicals having, some
distinctive property (*e.g.* bitterness) such that predators
learn to associate taste with unacceptability, and thus
avoid the potential prey. More efficiently, defended prey
may adopt distinctive colorations (*i.e.* become aposematic)
such that predators learn to avoid them on sight. In the
latter case, protection may be extended to nonpoisonous
prey species through their adoption of similar coloration--
the phenomenon of mimicry. It is this adoption of the
aposematic life style and evolution of mimics which
characterizes certain butterflies, aphids, grasshoppers,
and other insects which sequester and store poisons from
cardenolide-bearing food plants.

   One of the more extensive studies related to this
phenomenon to date was carried out with the grasshopper
Poekilocerus bufonius[79,102]. This North African species
probably feeds exclusively on Asclepiad plants in nature.
It is aposematic, and has a poison gland from which the
contents may be ejected as a spray at the immature hopper
stage, or as a foamy froth in the adult stage. This secretion contains histamine and one-cat lethal dose of a
digitalis-like toxin. The latter occurs in similar concentrations in the hemolymph. Analysis of eggs, egg pods,
cast skins and the dried secretion of hoppers and adults
fed on the Asclepiads *Calotropis procera* and *Asclepias
curassavica* revealed the presence of two major cardenolides,
calactin and calotropin, and small amounts of uscharidin
(Table 3). Approximately 10 times less cardenolide was
found in specimens reared on a nonpoisonous diet--strong
evidence that they are indeed plant-derived. It is particularly interesting that of the seven closely related
cardenolides (IV-IX), only calactin and calotropin were
present in appreciable quantities. Several possible
explanations were offered for the absence of other cardenolides apparently present in the plant. The insect could,

Table 3

Cardenolide content of *Poekilocerus bufonius* poison gland secretion, eggs, egg pod secretion, and cast skins: (Reichestein, 1967b)

Material, number of animals and stages, and diet	Total amounts found in mg* (and estimated amounts per animal or egg)		
	Calactin	Calotropin	Uscharidin
Dried secretion, 197 adults on Asclepiadaceae	28 (0.2)	present (0.1)	absent
Dried secretion, 1150 hoppers stages III-V on Asclepiadaceae	138.5 (0.14)	11.5 (0.06)	trace (0.01)
Dried secretion, 217 hoppers stages III-V on nonpoisonous diet	0.9 (0.01)	0.7 (0.01)	
Dried secretion, 200 hoppers stages I-II parents already on nonpoisonous diet	0.1 (0.002)	0.05 (0.001)	
19.8 gm eggs (*ca* 1900) from 16 animals on Asclepiadaceae	18 (0.01)	present (0.01)	0.1 (0.0005)
29.8 gm sand with secretion from 16 egg pods	trace	trace	
5.6 gm cast skins (exuviae)	23.8	9.5	trace

*Isolated crystals in all cases except where identified only by paper and thin-layer chromatography, in which cases they are denoted simply as "present".

for example, selectively absorb calactin and calotropin while metabolizing, excreting, or converting *in vivo* the other cardenolides to calactin and/or calotropin. The latter is a particularly distinct possibility for uscharidin, uscharin, and voruscharin, since these three may be readily transformed in the laboratory to calotropin by simple chemical reactions.

Some experimental predators, such as white mice and the European jay (*Garrulus glandarius* L.), learned to avoid *P. bufonius* only after experiencing disagreeable side effects following ingestion. These include, in the case of the jay, the characteristic vomiting intoxication. Some other predators which are much less sensitive to cardenolides, such as the European hedgehog (*Erinaceus europaeus* L.), consumed *P. bufonius* with impunity.

A parallel study with monarch butterflies (*Danaus plexippus*) yielded results with some striking similarities[70,79,80]. In adults, fed as larvae on *Asclepias curassavica*, calactin and calotropin were two of the major cardenolides isolated, and uscharidin, uscharin, and voruscharin were absent. Since relatively large amounts of calotropagenin, calotoxin, and two unidentified cardenolides were also present along with traces of four others, it was suggested that the monarch was less selective than *P. bufonius*. The monarch has no poison gland and the cardenolides were isolated from the whole insect. Protection occurs, it was implied, through ingestion of the prey from which the predator "learns", from its subsequent intoxication, to reject other members of the species.

It should be pointed out that the analysis on *P. bufonius* was primarily concerned with the cardenolide components of the defensive secretion which is only one of the four lines of defense listed by von Euw and coworkers for this insect. A second, which is more analogous to that in the monarch, but was not investigated, involves cardenolides in the body tissues. It may well be that calactin and calotropin were selectively stored in the poison gland of *P. bufonius*, but other cardenolides present in the food source could have been stored in other body tissues.

Support for this view comes from studies of live specimens of *Lygaeus kalmii kalmii* (Stoal), *L.kalmii angustomar-*

ginatus (Parshley), *Oncopeltus fasciatus* (Dallas) and *O.*
sandarachatus (Say) taken on *Asclepias* spp[92]. These
specimens revealed cardenolide presence in the hemolymph,
in secretions from the middorsal abdominal glands of the
nymphs and the ventral metathoracic glands of the adults,
and particularly high concentrations in the secretions of a
special series of dorsolateral glands. From the latter it
was inferred that these lygaeids use the chemicals for
defensive purposes in much the same way as does *P. bufonius*.
Duffey and Scudder (1974) showed that *O. fasciatus* selectively stores polar cardenolides in the dorsolateral space fluid
when reared on *Asclepias syriaca* seeds. However, analysis
of the whole insects and their haemolymph indicated the
presence of nonpolar cardenolides, which were absent from
the dorsolateral complex, but were probably among the
cardenolides found in the seeds. None of the cardenolides
were identified: evidence for this was based primarily on
chromatographic profiles. The supposition was made that
metabolic changes of nonpolar cardenolides to polar ones
was at least in part responsible for the accumulation of
the latter. This was based on the finding that *O. fasciatus*
fed on a diet of sunflower seeds, to which digitoxin was
added, contained two polar cardenolides, apparently
metabolites of digitoxin. However, administration of
ouabain, a more polar compound than digitoxin, resulted in
storage without metabolic alteration. A similar conclusion
might be reached from the data of Feir and Suen (1971), which
showed that *O. fasciatus* fed on *A. syriaca* seeds contained
one cardenolide of seven observed in the food source, plus
three others not found in the source. It should be pointed
out that in neither study were the *A. syriaca* cardenolides
nor the supposed metabolites identified thereby confirming
this supposition. Furthermore, neither digitoxin nor
ouabain have been reported in *A. syriaca* or, in fact, any
other asclepiad, so they might be questioned as adequate
models when considering selective sequestering and storage
in relation to the asclepiad host plants. An alternate
possibility, that polar cardenolides present in the *A.*
syriaca food source have been selectively sequestered and
concentrated in *O. fasciatus* more than the other plant
cardenolides, can not be excluded based on the evidence
presented. Nevertheless, the selective sequestering and
storage of more polar cardenolides in *O. fasciatus* is in
keeping with the results cited above for *P. bufonius* and
D. plexippus; uscharidin, uscharin, and voruscharin, three

cardenolides inferred to be present in the *Calotropis procera* and *A. curassavica* food source are generally absent in the adult insects which fed upon them. Chromatographic data **reveals** these compounds to be less polar than calactin and calotropin, the selectively sequestered and stored chemicals.

Selectivity in the sequestering process might also result from the insect's feeding habits. The aposematic aphid, *Aphis nerii*, contained cardenolides when reared on both *Nerium oleander* and *Asclepias curassavica*[87]. The three cardenolides found in the aphids reared on oleander occur in the plant; however, oleandrin, a major oleander component, was absent in the aphid and it was inferred that this was due to its absence in the phloem on which the aphids were believed to feed.

A similar explanation was invoked when finding that the aphids reared on *A. curassavica* stored proceroside and calotropin but not calactin. This hypothesis, though interesting, has little experimental support since in this study, as in the previously cited studies with *P. bufonius* and *D. plexippus*, the food plants, on which the insects had fed, were not analyzed. The presence of an entire group of cardenolides was only assumed from their presence in other plants of that species. It might be that the *A. curassavica*, upon which the aphids were reared, was similar to the plants growing in the Botanical Gardens at Oxford, that is, plants containing proceroside, calotropin and uzarigenin but lacking calactin[88]. Thus again the observed selective storage could be alternately explained by preferential storage of the more polar cardenolides. Such ambiguities illustrate the necessity of analyzing the food plant upon which the insects have been reared when investigating selective sequestering and storage.

Dynamics of Cardenolide Sequestering and Storage by the Monarch Butterfly. To the avian predators cardenolides sequestered and stored during larval feeding impart to the adults a degree of unpalatability which is dependent both upon cardenolide type and concentration[4,9,11,12]. Considering two somewhat extreme cases, Brower *et al.* (1967 and 1968) and Brower (1969) found that larvae reared on *Asclepias curassavica*, a species rather rich in cardenolides, produced adults unpalatable to blue jays while those reared

on a noncardenolide bearing milkweed (*Gonolobus rostratus*)
or cabbage produced palatable adults. A palatability
spectrum, with its origins in the food source, was
postulated for monarchs.

This theory received support from a cardenolide content
survey analysis of monarchs from natural populations in
eastern North America[10], and a comparison of both cardenolide concentration and emetic potency to blue jays of
monarchs from Massachusetts and California[11]. Migrating
east coast butterflies netted in Massachusetts were found
to have a higher mean concentration of cardenolides than
those sampled from the overwintering sites in California
(Fig. 5). In addition the Massachusetts samples showed

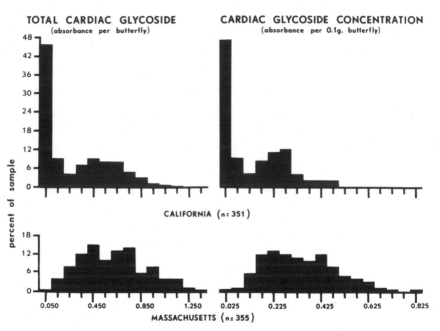

Figure 5. Variation in cardiac glycoside content of
monarch butterflies from California and Massachusetts. The
west coast monarchs are bimodal, with nearly half containing little to immeasurable amounts of cardiac glycosides.
The east coast monarchs are unimodal and have a substantially higher mean cardenolide content. An extract of 0.1 g of
butterfly in 5 ml of alcohol which gives an absorbance of
0.200 by our standard method is equivalent to a 3.16 x 10^{-5}
M solution of digitoxin, or 241µg per 0.2 g butterfly
(Brower and Moffitt, 1974).

a normal distribution of cardenolides in contrast to the California population of monarchs which was dimorphic; specifically, 47% of the individuals examined contained little or immeasurable amounts of cardenolides. However, the emeticity of the California monarchs were 4.6 to 6.5 times greater than the Massachusetts monarchs within the same absorbancy ranges (Fig. 6). These differences were believed to reflect the differences in *Asclepias* floras in the eastern and western states.

From examination of the limited and somewhat hetero-genous data summarized in Table 4, it appears that quanti-tative cardenolide variation in the Massachusetts and California populations probably cannot be entirely due to quantitative variation among food plants. Leaves of two northeastern species, A. *amplexicaulis* and A. *incarnata*, show similar concentrations of cardenolides; however, butterflies reared on A. *amplexicaulis* contain cardenolides generally lower in concentration than the leaves, while those fed on A. *incarnata* have a higher concentration. In California, monarchs reared on A. *eriocarpa* stored cardeno-lides equal to the already high concentration of cardeno-lides found in the leaves; those reared on A. *cordifolia* effectively increased the concentration. Among the Cali-fornia butterflies examined, it appears that A. *fasciular-is*-reared monarchs contain very little cardenolides and due to the widespread distribution of this plant in the west[111], it is probable that monarchs feeding on it contribute, in part, to the 47% of monarchs extremely low or lacking in cardenolides.

In order to explore the large differences in emetic potency of cardenolides present in the California and Massachusetts populations, some of the remaining extracts from the gross concentration determinations[1] were examined by TLC after purification by solvent partition. The Massachusetts sample appeared quite homogenous in cardeno-lide patterns in contrast to the California sample. The TLC patterns produced by the Massachusetts butterflies could be arranged into two groups only slightly different in pattern (Fig. 7). A similar treatment of the California sample resulted in at least five distinctive patterns (Fig. 8). Thus the monarchs netted in Massachusetts appear to have fed on asclepiads which were qualitatively similar in storable cardenolides but qualitatively different from

Table 4

Cardenolide content of the leaves of various *Asclepias* species
and of adult monarch butterflies reared on these plants

	Cardenolide content, calculated as mg of digitoxin per gm dried material@		
Northeastern plant species	Leaf		Butterfly
A. amplexicaulis	nil - 0.22	(5)	nil - 0.12 (26)
A. exaltata	nil - 0.70	(3)	1.34 - 1.97 (3)
A. incarnata	nil - 0.28	(2)	C.28 - 1.27 (84)
A. syriaca	0.06 - 2.64	(16)	1.08 - 4.53 (83)
California plant species			
A. cordifolia	0.38	(1)	1.46 - 2.64 (12)
A. eriocarpa	2.58 - 5.20	(3)	2.80 - 5.36 (14)
A. fascicularis	0.12	(1)	nil - 0.57 (10)
A. speciosa	0.15*	(p)	1.34 - 2.23 (7)

@Determinations were made using the standard spectroassay method. Minimum
and maximum values are given when possible, each representing a different
sample or individual; the number of samples is given in ( ); if a plant
sample is from more than one plant it is indicated by (p) as pooled.
Both male and female butterflies are included in the range except for
A. exaltata and *A. incarnata* which represent males only. Data collected
through 1974, data from Northeastern materials from Brower and Moffitt
in preparation.

*Value determined on aerial sample after extraction and purification as
described later.

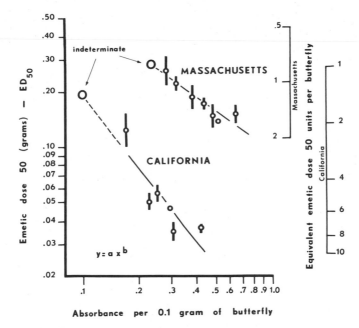

Figure 6. Relationship of cardiac glycoside concentrations and ED_{50} values. The lines extending from the small open circles are the 95% confidence limits of the means; the dotted lines represent indeterminate to nonemetic concentrations. Emetic dose is inversely related to cardiac glycoside concentrations, thereby proving that the palatability spectrum in these wild monarch butterflies is casually related to a spectrum in cardenolide concentration. Cardenolides in the California butterflies, however, are far more emetic than those in the Massachusetts butterflies, no doubt reflecting the completely distinct *Asclepias* species on the two coasts from which the wild larvae sequester the poisons. The ranges in equivalent emetic dose 50 units per blue jay predator per butterfly are shown on the right side of the graph (Brower and Moffitt, 1974).

most of the California species, while the California monarchs fed on *Asclepias* species differed greatly qualitatively among themselves in storable cardenolides. The uniformity of the Massachusetts sample undoubtedly accounts for the relatively clear cut relationship between concentration and number of ED_{50} units reported for these butterflies and the high degree of cardenolide variation in the California

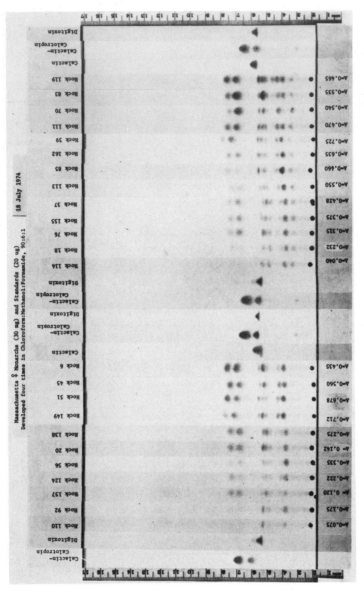

Figure 7. Thin layer chromatograms of cleaned extracts (equivalent to 30 mg, dry weight) of twenty-four individual female monarch butterflies caught during fall migration in the floodplain of the Connecticut River in Hockanum, Massachusetts and of calactin, calotropin and digitoxin standards (20 ug). Chromatograms were developed four times in chloroform–methanol–formamide (90:6:1, by volume). Chromatograms were aligned at the origin.

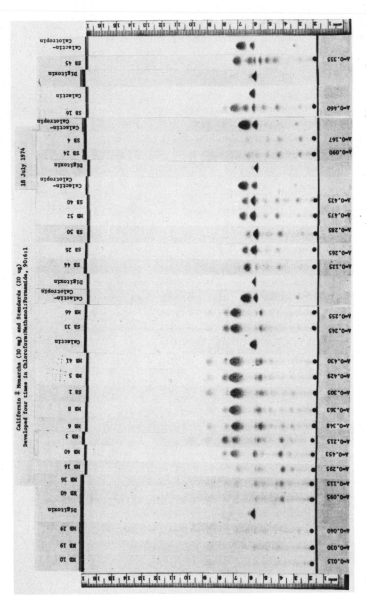

Figure 8. Thin layer chromatograms of cleaned extracts (equivalent to 30 mg, dry weight) of twenty-four individual female monarch butterflies taken from overwintering colonies in California at Muir Beach (MB), Natural Bridges State Park at Santa Cruz (NB), and at the Academy of Music at Santa Barbara (SB) and of calactin, calotropin and digitoxin standards (20 ug). Chromatograms were developed four times in chloroform–methanol–formamide (90:6:1, by volume). Chromatograms were aligned at the origin.

sample accounts for the scattered points in this relationship (Fig. 6).

The TLC patterns of the Massachusetts butterflies
(Fig. 7) show that calactin and calotropin are not present
in these butterflies but appear to be present in at least
two of the patterns found among the California butterflies
(Fig. 8). Since it is known that butterflies reared on *A.
curassavica* readily sequester and store these two cardenolides[80], it seems reasonable to assume that the asclepiads,
fed on by these Massachusetts larva and some of the
California larva, lacked calactin and calotropin.

By comparing the cardenolides present in several
western species of *Asclepias* (Fig. 3) with the cardenolides
stored by the overwintering California monarchs (Fig. 8),
it appears that the butterflies reflect simplified cardenolide patterns relative to those in the plants. The most
striking feature is that the monarchs have not stored
cardenolides above an R_f of 0.48 ($R_{digitoxin}=1.85$) even
though nearly every plant examined contained major cardenolides above this R_f. For the Massachusetts sample, no
cardenolides appear in the adult monarch above an R_f of
0.40 ($R_{digitoxin}=1.54$). This selective storage is consistent with our feeding studies in which *Asclepias
cordifolia, A. curassavica, A. eriocarpa, A. fascicularis,
A. nivea, A. speciosa, A. syriaca, Calotropis procera* and
Gomphocarpus sp. were used as food sources: no cardenolides were stored above the R_f of 0.48 in this solvent
system. Thus it appears to be generally true that the
adult monarch stores cardenolides only within a certain
polarity range. A more complete description of the quantitative dynamics of sequestering and storing will be presented in the subsequent section.

From cardenolide analysis of individual butterflies in
a migrating population, it may be possible to match individuals with their food sources for distinct geographic
regions. Studies are now underway which should help us
'fingerprint' most of the sampled overwintering California
monarchs from different sites. Combining the TLC patterns
of butterflies reared in captivity on known food plants, the
TLC patterns of overwintering butterflies, a knowledge of
the abundance and distribution of the various *Asclepias*
species and the dynamics of their roles as food plants for

both the migrating generations of monarchs and the preced-
ing generations, will help clarify the delicate interactions of this insect with its specific food plants and
the migration patterns of the monarch along the west coast.

DETAILED EXAMINATION OF CARDENOLIDE SEQUESTERING AND
STORAGE BY THE MONARCH BUTTERFLY (*Danaus plexippus*)
REARED ON *Asclepias curassavica* AND *Gomphocarpus* SP.

To better understand the relationship between the
monarch and its food plants, and the resulting palatability
spectrum, a detailed study was carried out with monarch
butterflies (*D. plexippus*) reared on two milkweeds--the
neotropical American *A. curassavica* and the African
Gomphocarpus sp. The former has been shown to produce
butterflies six times more emetic than the latter[12].
Samples of leaf material, adult butterflies and larval frass
from each plant were analyzed using three separate techniques, each yielding information related to sequestering
and storage phenomena. These techniques were: (1) a
spectroassay to determine the gross concentration of cardenolides present in each of the samples; (2) a bioassay
to determine the emetic dose fifty (ED_{50}) of the samples
to the blue jay; (3) TLC to separate and to qualitatively
analyze the cardenolides in each sample.

*Methods.   Spectrophotometric determination of
cardenolides.* One of the characteristic features of cardenolides is the presence of the α,β-unsaturated γ-lactone
ring attached to C-17 of the steroid nucleus (I). The
interaction between this structural feature of the cardenolide and polynitrated aromatic compounds, in the presence
of base, results in the formation of a colored complex.
The intensity of the absorbance of this colored complex
can be used as a measure of the concentration of cardenolides in an unknown, by comparison with that from solutions
of known concentrations of a standard such as digitoxin.
Absorbance values are converted to an equivalent molar
concentration of digitoxin, and then to equivalent gram
amounts of digitoxin. Reference to a readily available
standard such as digitoxin allows comparison of cardenolide
concentrations under different conditions, at different
times, and by different investigators, whereas direct
comparison of absorbancy values does not. The extinction

coefficient of cardenolides present in a sample may vary
from that of digitoxin, but this variation among extinction
coefficients for nine different cardenolides by our method
has been shown to be relatively minor (Moffitt and Brower,
as cited by Brower and Glazier, 1975, footnote 17). Thus
on a molar basis, calculations of concentration of cardenolide mixtures, based on the absorbancies of known standards,
give a fairly good estimate. When the equivalent molar
concentrations are converted to equivalent gram amounts,
a substantial variation would be observed if several
different standards were used. Thus this general method
is a convenient and accurate way to quantitate unknown
cardenolide mixtures. This spectroassay method has been
developed and utilized with previous butterfly samples[8,10,11].

Cardenolide-induced emesis and the blue jay bioassay.
Emesis was chosen as a bioassay parameter since it is the
intoxicating but nonlethal response to the ingestion of
cardenolide-laden monarchs from which the astute predator
learns to avoid the insect. Thus emetic determinations
allow speculation as to the effectiveness of the incorporated cardenolides as predator deterrents. Emeticity of
cardenolide-laden butterflies appears from previous discussions to be directly related to cardenolide concentration and is a recognized side effect of these drugs[65].
The physiology and pharmacology of vomiting induced by
cardenolides is not thoroughly understood but the following
summary is provided by Borison and Wang (1953): "...The
digitalis glycosides elicit emesis by acting at more than
one receptor site. The central site of emetic action is
the medullary chemoreceptor trigger zone. A local irritant
action on the gastrointestinal tract is probably insig-
nificant under most circumstances. While the cardiac
innervation is not essential to the emetic reflex, its
participation has not been excluded. The peripheral site(s)
of emetic action remain to be discovered." Most studies
upon which Borison and Wang based their summary utilized
purified compounds. In our assays using only mechanically
altered samples, perhaps the local irritation of the
gastrointestinal tract is more significant. Their summary
points to receptors in three different organs. To reach
the receptors in the brain and heart, cardenolides must be
absorbed and translocated prior to interaction with the
receptor itself. Thus the emeticity of cardenolides will

likely depend upon a number of factors: their ability to
be absorbed through the gut lining, the extent to which
they are translocated, their distribution throughout the
body, and finally their affinity for the receptors.
The relative emetic potency of cardenolide mixtures
administrated orally could result from differences in any
of the above factors.

Absorption and distribution of cardenolides is
presumably related to their structural features. For
example, the cardenolides present in *Digitalis* tinctures
or powders are adequately absorbed from the intestinal
tract following oral administration of this drug; however
ouabain, a highly polar cardenolide, is so irregularly and
poorly absorbed that its oral administration is dangerous[65].
Distribution studies in rats showed that the concentration
of the glycoside digitoxin in the heart was about four
times higher than that of the aglycone digitoxigenin.
However, in the brain the relative concentrations were
reversed, the concentration of the aglycone being about
ten times higher than that of the glycoside[81]. Such
differences in distribution could explain the observation
of Chen *et al.* (1938) with cats, that the aglycone
possessed a greater emetic potency than the parent glycoside. Differences in the emetic potency of cardenolides
such as those observed by Parsons and Summers (1971) could
also be due to differences in affinity for the receptor, which again could be influenced by structural
features and configuration as it is for cardiac potency[33,99].
Parsons and Summers have determined that calactin is about
three times as emetic as digitoxin to cats, and that for
these cardenolides and two others, vomiting occurs at
about 40% of the dose which would cause cardiac arrest.

Antagonistic or synergic effects among the cardenolides,
or between the cardenolides and other chemicals present
in mixtures could also occur. The emetic action of apo-
morphine is known to be synergistic with that of ouabain,
while nicotine under certain circumstances is antagonistic
to the emetic action of some cardenolides[3]. Catecholamine
depleting drugs like reserpine, tetrabenazine or syrosingo-
pine also have an antagonistic effect upon emesis in cats
produced by peruvoside and ouabain[35].

The bioassay technique of determining ED_{50}'s in the

blue jay (*Cyanocitta cristata bromia* Oberholser, Corvidae)
has also been well established and utilized in previous
studies[10,11]. Dried samples were ground in the manner
described for the spectroassay.[10] The "up and down" sequential analysis[24] was used to determine the ED_{50}'s. The
birds used for determinations were weighed, and deprived
of food but not water for about 2 hours. The dosages were
determined based on the weight of the birds, reducing the
birds' weight by one gram for each hour of deprivation.
The desired dose of dry material was packed into a #2
gelatin capsule, sealed, and forced down the bird's
esophagus into the gizzard. The bird was then placed in
a clean cage and observed for one hour through a one-way
mirror.

Ninety-two birds were used for a total of 104 force
feedings between 8 May - 30 June 1970. Of these 92
birds, 81 were captured in Hampshire and Franklin Counties,
Massachusetts, during the fall of 1969 through the spring
of 1970. Twelve of the 81 birds were force fed twice
during the course of the experiment with at least one
month intervening between the first and second sessions.
The remaining 11 birds had been captured and used several
months previously in force feeding experiments, loaned to
the University of Massachusetts for behavioral experiments,
and then returned to Amherst College and reused for single
force feedings during the course of the experiment. Of the
104 forced feedings, 91 fell within the dosage ranges
necessary for calculating the ED_{50} values (Table 6).

In a previously reported study of the emetic properties
of monarchs reared on these two plants[12], the butterfly
stock was obtained in Trinidad, W.I. In the neotropics
the monarch is smaller and has a slightly different color
pattern than the North America subspecies, and the observed
differences in the ED_{50} values of this and the previous
studies are not surprising.

*Thin layer chromatography, cardenolide purification
and recovery*. Utilizing this third technique, we were
able to separate cardenolide mixtures for both qualitative
and quantitative analyses. This procedure (see details
in fig. 3 and 4) separates these substances according to
varying polarity (the less polar the substance, the further
the migration). The presence of cardenolides on the plate

is visualized by spraying the plate with a solution of a polynitrated aromatic compound and alkali which react with the cardenolides to form a colored complex. Since R_f values vary slightly from one analysis to another, digitoxin is run as a standard and the R_d is determined.

Large quantities of coextracted fats, pigments, waxes, etc. commonly encountered in biological samples can alter the binding ability of the adsorbent and thus change cardenolide R_f values. To minimize this, and to eliminate pigments which could interfere in the visualization, the samples were partially purified before TLC analysis (Fig.9).

Preparative TLC was also done on silica gel. Seventy-five percent of the total sample was streaked on the plate with subsequent development in ethyl acetate-methanol (95:5 v/v). One edge was sprayed to visualize the major bands (A-F for *A. curassavica* and A'-G' for *Gomphocarpus* sp (Fig. 10)). The unsprayed bands were scraped off, eluted with 10% MEOH in MeCl$_2$, filtered and taken to dryness under N$_2$. Cardenolide concentration was determined spectrophotometrically.

Extraction efficiencies and recoveries through each step of the cleanup were determined on *A. curassavica* leaf material (1967 Trinidad plant stock TD-1-*As.c.*), *A. curassavica*-reared monarch butterflies (1966 Florida butterfly stock FLA-4) and frass from larvae reared on *A. curassavica* (1969 Massachusetts stock B-2). Four 5 gm plant samples, four 1.25 gm butterfly samples, and four 5 gm frass samples were extracted separately in ethanol using a Nillens Polytron. Three samples of each of the plant, butterfly, and frass were extracted separately with ethanol in a shaker water bath (0.1 gm/5ml@75-80°C for 1 hr). Gross cardenolide concentrations were determined by spectroassay and showed the heated extraction method to be more efficient. The extraction efficiencies by the Polytron method were 77% for the plant samples, 86% for the butterfly samples and 67% for the frass samples when compared with the heated ethanol extraction. TLC comparisons of the ethanol extracts obtained by both methods showed no qualitative differences in cardenolide profile.

Duplicate samples of plant, butterfly and frass materials were carried through the cleanup procedure and

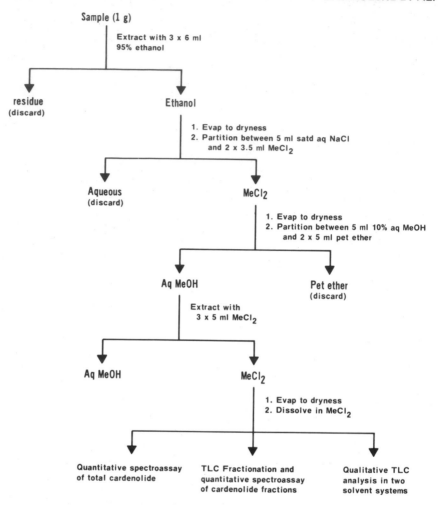

Figure 9. Extraction and cleanup scheme

gross cardenolide concentrations were determined by
spectroassay on each fraction (Fig. 9). The largest losses
for the plant and frass samples occurred in the petroleum
ether layer, where 51% of the plant cardenolides and 53%
of the frass cardenolides were lost. From the nature of
this particular partition process, particularly since salt
was added to the aqueous layer, losses to the petroleum
ether layer probably occurred primarily among the less

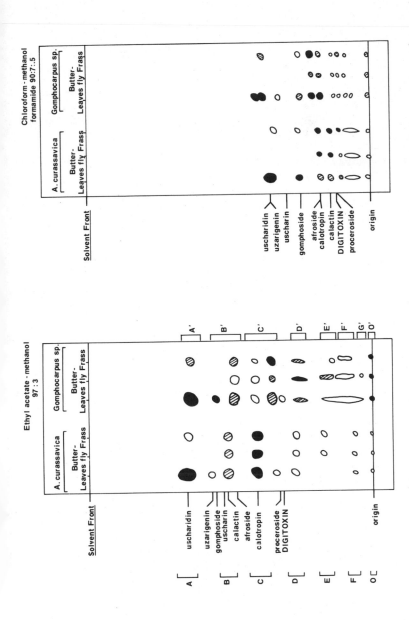

Figure 10. Redrawn thin-layer chromatograms of cleaned up extracts of *A. curassavica* and *Gomphocarpus* sp. leaves, and derived butterflies and frass. Mobilities of standards and areas scraped for spectroassay are marked. Both plates were developed twice in the solvents indicated. Relative spot intensities correspond to ● = dark, ◍ = medium, ○ = light.

polar cardenolides. The aqueous frass fraction contained
17% of the cardenolides originally present in the ethanol
extract; TLC indicated these were highly polar cardenolides
occurring mostly at the origin and in the regions of bands
E and F. The greatest losses of the butterfly sample (24%)
occurred in the aqueous layer and was composed of highly
polar cardenolides, those occurring at the origin and in
the regions of bands E and F. The overall recoveries
through this cleanup were thus poor--only 16% of the cardenolides present in the initial ethanol extract of the plant
samples remained in the methylene chloride layer; for the
butterfly samples 22% and for the frass samples 13% remained.
By totaling the cardenolide concentrations in all fractions
(*i.e.*, aqueous, petroleum ether, aqueous methanol and
methylene chloride), 71% of the cardenolides present in the
plant extract, 59% for the butterfly samples, and 87% for
the frass samples could be accounted for.

     Recoveries were also estimated by another means. The
concentration of cardenolides in the plant, butterfly and
frass samples determined after cleanup (listed in Table 7--
Columns A, B, and D) were compared with the values for the
ethanolic extracts in Table 5 (Column C). The following
recoveries through the extraction and cleanup were calculated
for *A. curassavica* materials: plant, 29%; butterflies, 27%;
frass, 17%; and for *Gomphocarpus* sp materials: plant, 25%;
butterfly, 30%; and frass, 13%. These figures are only
estimates since the spectroassay determinations and TLC
analysis used separate materials. Since the other recovery
figures were not determined simultaneously with the samples
used for TLC fractionation, they too must be regarded as
estimates. Combining these two determinations it appears
that recoveries through the cleanup procedure, for materials
from *A. curassavica* and *Gomphocarpus* sp, may have ranged
between 16% and 29% for plant, 22% and 30% for butterfly,
and 13% and 17% for frass samples.

     Recoveries from the preparative TLC were determined by
comparing the amounts of cardenolide streaked on the plate
with the amounts recovered through summation of the bands
(Table 7--Columns A, B, and D). The average recovery for
A. curassavica samples was 50% and *Gomphocarpus* sp 54%. No
attempt was made to determine where or how these losses
occurred.

We have taken no steps to prevent chemical breakdown during extraction, cleanup and TLC procedures other than the simple precautions of storing extracts in a freezer and evaporating volatile solvents under N_2. Reichstein and coworkers performed solvent extractions in a CO_2 atmosphere to minimize oxidative degradation[80]. We have not determined the extent to which breakdown might have occurred through our methods. The accountabilities of 59% to 87% of the cardenolides through the cleanup procedure and about 50% through the TLC isolation may be due in part to degradation. It is also possible that the high proportion of polar cardenolides in the butterfly sample is due in part to chemical breakdown.

Materials. The milkweed plants used to obtain the ED_{50} and spectroassay data were grown in the Amherst College greenhouse from seeds obtained in Trinidad, W.I. (*A. curassavica*) and in Uganda, Africa (*Gomphocarpus* sp) as described.[12] The seeds originally described as from Uganda were obtained from cultivated plants growing in Kampala, Uganda. The *A. curassavica* plants used for the TLC analysis were similarly grown, but from a new set of seeds obtained in Trinidad[9]. The *Gomphocarpus* sp plants were all propagated from cuttings of one surviving plant of the original stock described above.

For the ED_{50} tests and the spectroassays, the butterflies were the second generation of a stock originally obtained as larvae at Munroe Station on the Tamiami Trail, Collier Co., Florida on 23 January 1970 (Florida Stock F-2). The butterflies were reared in a controlled environment room under 12 hour light-12 hour dark conditions at 78°F and 45% RH. Through the fourth instar, the larvae were fed plucked leaves. After shedding their skins to the fifth instar, but before being allowed to eat further, the larvae were transferred to potted plants of either *Gomphocarpus* sp or *A. curassavica*. Intact stems of the plants were led through the side of narrow gauge nylon bags. Each bag was tied at the top and the open end was fitted around the top of a tractor funnel which served as a frass catcher. The funnel was lined with aluminum foil and a silk organza bag was attached. The silk bag was lowered into a large jar containing P_2O_5 and the jar was sealed against the bottom of the funnel mouth. As the larvae fed upon the leaves in the cylinder above, the frass dropped down through the funnel

into the organza bag where it immediately began desiccating.
The frass was removed, twice daily, and kept over P_2O_5 until
completely dry.

New plants (with flowers and buds removed) were brought
from the greenhouse daily, thus the larvae ate mainly
leaves, but also some stem material. As the potted plants
were placed in the bags, a duplicate set of plants was set
up next to them. Each time the larvae were given fresh
plants, the leaves of the previous duplicate set were
removed and dried for 16 hours in a forced draft oven at
60°C. In all, 34 monarchs were reared on *Asclepias
curassavica* and 37 on *Gomphocarpus* sp between 26 February -
2 March 1970. When the larvae completed feeding, they
were removed to one-half pint clear plastic containers,
where they attached to the lids and pupated. After harden-
ing for two or more days the lids with attached pupae were
set on screen cylinders to allow room for the emerging
adults to spread their wings and dry prior to being frozen[12].
All butterflies were dried for 16 hours at 60°C in a forced
draft oven on 21 April 1970 and then stored over Drierite
until used.

For the TLC study, the butterflies were the first
generation of a stock obtained in Massachusetts in August
1973 (Massachusetts Stock H-1) and were reared in September
1973. Unlike those described above, these insects were
reared individually through the fifth instar, on plucked
leaves rather than in the nylon cylinders on intact leaves.
Fresh leaves were added daily and the frass removed. This
was dried in the forced draft oven along with sets of
fresh plucked leaves. The rearing, hatching, drying, and
storage procedures were as described above. All the
material was subsequently packed in plastic bags, sealed
in a desiccant and shipped to the University of California
at Davis for analysis.

Results. A simplified view of the processes which
might be involved in cardenolide processing between the
ingestion of the leaf material, the egestion of frass and
the emergence of the adult butterfly is summarized in
Figure 11. Cardenolides ingested by a monarch larva feeding
on an asclepiad plant may be (1) transported across the gut
wall and thus sequestered into the larva, (2) passed
through the gut and egested in the frass in the same form in

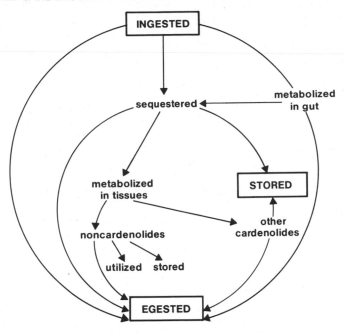

Figure 11. Possible processing of ingested cardenolides during larval development.

which they were ingested in the leaf, or (3) altered in the
gut and then sequestered or egested. The cardenolides which
are sequestered by the larva *via* routes 1 or 3 have three
possible fates: (1) they may be stored, as sequestered,
through pupation to the adult butterfly; (2) they may be
egested in the form sequestered, or (3) they may be metabolized in the hemolymph or tissues of the insect to yield
noncardenolides or other cardenolides. These noncardenolides, in turn, may be stored to the adult stage, utilized
by the larva, or egested. Alternately, the metabolically
altered cardenolides may be stored to the adult stage or
egested by the larva. In our nomenclature, cardenolides
that have been sequestered by the larva are those which have
passed through the gut wall; assimilated cardenolides are
those which have been sequestered and retained within the
larva; stored cardenolides are those which have been sequestered, assimilated and retained to the adult stage.

Cardenolide concentrations and emeticity. Table 5
Column A shows the spectrometrically determined absorbance

Table 5

Spectrophotometric absorbances and cardenolide concentrations (equivalent to digitoxin) of ethanolic extracts of the leaves of *Asclepias curassavica* and *Gomphocarpus* sp, of adult monarch butterflies (*Danaus plexippus*) reared on these two plants and of their larval frass.

Material analyzed	A Spectrophotometric absorbance per 0.1 gm dry weight	B Cardenolide Concentration (M x 10⁻⁵) per 0.1 gm dry weight	C mg of cardenolide per 0.1 gm dry weight
Asclepias curassavica			
Leaf	0.615	10.08	0.3857
Butterfly	0.510	8.33	0.3186
Frass	0.520	8.50	0.3250
Gomphocarpus sp			
Leaf	0.530	8.67	0.3314
Butterfly	0.170	2.66	0.1016
Frass	0.445	7.25	0.2771

measurements from which the equivalent molar concentrations
(Column B) and equivalent mg amounts of cardenolides per
0.1 gm of dry material (Column C) have been calculated.
The concentrations of cardenolides in *A. curassavica* leaf,
butterfly, and frass materials are, respectively, 1.16
times, 3.13 times, and 1.17 times greater than in the
corresponding *Gomphocarpus* sp material. For both plants,
the cardenolide concentration in the frass is 0.84 relative
to that of the leaf (a 16% decrease); however, the cardenolide concentration of the butterfly material relative to
the leaf material for the *Gomphocarpus*-reared butterflies
is much less: 0.31 as compared to 0.83 for the *A.
curassavica*-reared individiuals. This is the first line
of evidence that the monarch is more efficient at handling
the cardenolides in *A. curassavica*.

Table 6 summarizes the results of the emetic dose fifty
(ED_{50}) tests in which the same powdered dry leaf, butterfly,
and frass materials were force-fed to blue jays. The ED_{50}
values and 95% confidence limits are in Column A, calculated
as described[12,24]. Based on the ED_{50} values are the number
of ED_{50} units per 0.1 gm of dry powdered material, corrected
for the mean weight (85 gm) of the blue jays (Column B).
The number of ED_{50} units per 0.1 gm in *A. curassavica* leaf,
butterfly and frass materials are higher than the corresponding *Gomphocarpus* sp values as they were for cardenolide
concentrations. However, the differences, respectively
3.4 times, 8.2 times, and 4.9 times, are even greater.
The differences in emeticity is expressed absolutely in
Column C which gives the number of ED_{50} units per mg of
cardenolide (*i.e.*, emetic potency). For the leaf, butterfly and frass, the cardenolides in *A. curassavica* materials
are 2.9 times, 2.6 times, and 4.2 times more emetic than
those in the corresponding *Gomphocarpus* materials. In
other words, per mg of cardenolide, the cardenolides in
and derived from *A. curassavica* appear to be intrinsically
more emetic than those originating from *Gomphocarpus* sp.
Thus the larger number of emetic units per 0.1 gm in the
A. curassavica materials is due not only to greater amounts
but also to greater emetic potency of the cardenolides
from this plant. This latter fact is evidence that the
cardenolides in the two plants differ qualitatively as well
as quantitatively, a conclusion supported by the qualitative
TLC analysis (discussed below). In addition, Column C
indicates that the butterflies reared on *A. curassavica*

Table 6

Emetic dose fifty and emetic potency of leaves of *A. curassavica* and *Gomphocarpus* sp, of adult monarch butterflies (*Danaus plexippus*) reared on these two plants and of their larval frass.

Material analyzed	A			B	C
	Emetic dose fifty (ED50) as gm of dry powder per 100 gm jay			No. of ED50 units per 0.1 gm of dry material per 85 gm jay@	Emetic Potency
	No. of birds tested	ED50	95% confidence limits		No. of ED50 units per mg cardenolide@@
Asclepias curassavica					
Leaf	14	0.021	0.018 to 0.24	5.60	14.52
Butterfly	14	0.026	(0.023 to 0.028)*	4.52	14.19
Frass	12	0.016	0.014 to 0.019	7.35	22.62
Gomphocarpus sp					
Leaf	14	0.071	0.057 to 0.088	1.66	5.01
Butterfly	23	0.212	0.175 to 0.258	0.55	5.41
Frass	14	0.078	0.066 to 0.093	1.51	5.45

@Equals 0.1 gm divided by ED50 in gm (Column A), corrected for the mean weight of the blue jays 85 gm).
@@Equals ED50 units (Column B) divided by concentration (Table 5, Column C).
*Estimate subject to error because $NB-A^2/N^2 = 0.25$ (Dixon and Massey, 1957).

have an emetic potency approximately equal to that of the leaves; however, the emetic potency of their frass is greater, by a factor of 1.56 (22.62/14.52). For *Gomphocarpus* sp, the emetic potency of all materials is relatively similar. The two most obvious explanations for the increase in emetic potency observed in the frass, from larva reared on *A. curassavica*, are selective sequestering of cardenolides with lower emetic potency leaving those of higher emetic potency concentrated in the frass, and/or metabolically altering of the egested cardenolides to more emetic ones.

Summarizing, Tables 5 and 6 show that (1) the two species of milkweed differ in the amounts and emetic potencies of their contained cardenolides, (2) the butterflies, when reared as larvae on the two plants, reflect these differences, and (3) the larvae on both plants store cardenolides for the adult butterfly that have about the same emetic potency as those originally in the leaves; whereas, the larva on *A. curassavica* sort or alter the cardenolides so that those egested have a higher emetic potency than those originally in the leaves.

Qualitative and quantitative TLC analysis. TLC mobilities have been used qualitatively to characterize cardenolides occurring in the asclepiad, *Calotropis procera* L.[102], to ascertain the number of cardenolides stored during feeding experiments with the large milkweed bug, *Oncopeltus fasciatus* (Dallas)[32], and to determine the number of cardenolides present in various *Asclepias* species from North America[26].

The chromatograms in Fig. 10 show the qualitative TLC results from our study. The major cardenolides from both plants are of low and intermediate polarity (A, B, and C in *A. curassavica* and A', B', C', and D' in *Gomphocarpus* sp). The chromatograms clearly demonstrate that none of the major high R_f cardenolides (A in *A. curassavica* and A' in *Gomphocarpus* sp) are stored by the butterflies. From both plant sources, cardenolides intermediate in polarity (R_f regions corresponding to B and C in *A. curassavica* , and C', D', and E' in *Gomphocarpus* sp) are the ones stored in greatest quantities in the butterflies. The frass material, from larva reared on both plants, contains small amounts of the relatively nonpolar cardenolide (A in *A.*

Table 7

Cardenolide amounts in equivalent µg of digitoxin for TLC fractions* of the leaves of Asclepias curassavica and Gomphocarpus sp, adult monarch butterflies (Danaus plexippus) reared on these two plants and of their larval frass

Bands	Leaf	Butterfly		Frass	
	A µg per 0.1 gm dry weight	B µg per 0.1 gm dry weight	C relative to leaf@	D µg per 0.1 gm dry weight	E relative to leaf@
Asclepias curassavica					
A	19.8	nil	0.0	2.6	0.13
B	6.1	9.6	1.57	5.2	0.85
C	18.2	16.8	0.92	11.0	0.60
D	2.7	3.4	1.26	2.5	0.93
E	2.4	3.8	1.58	1.9	0.79
F	1.2	3.4	2.83	1.0	0.83
O	0.1	nil	0.00	0.2	2.00
R	0.1	nil	0.00	nil	0.00
I	6.6	4.1	0.62	3.0	0.45
TOTAL	57.2	41.1	0.72	27.4	0.48
Amount prior to TLC fractionation	113.0	85.1		54.8	
Recovery through TLC fractionation@@	51%	48%		50%	
Total recovery through extraction, cleanup and TLC@@@	15%	13%		8%	

Gomphocarpus sp

A'	14.1	nil	0.00	3.6	0.26
B'	9.6	1.0	0.10	4.2	0.44
C'	8.9	2.4	0.27	6.7	0.75
D'	8.3	4.6	0.55	2.4	0.29
E'	0.4	2.2	5.50	0.4	1.00
F'	0.3	0.4	1.33	0.2	0.67
G'	0.2	1.4	7.00	nil	0.00
O'	0.1	0.1	1.00	nil	0.00
R'	nil	nil	0.00	nil	0.00
I'	6.4	3.6	0.56	2.6	0.41
TOTAL	48.3	15.7	0.33	20.1	0.42

Amount prior to TLC fractionation	82.9	30.8		37.2
Recovery through TLC fractionation	58%	51%		54%
Total recovery through extraction, cleanup and TLC@@@	15%	15%		7%

*TLC fractions designated in Figure 10. Bands 0 and 0' correspond to 0.2 cm on each side of the origin; I and I', to the remaining silica gel between the origin and 0.1 cm above band A or A'; and R and R' to that between 0.1 cm above A or A' and the solvent front.

@Determined by dividing amount in butterfly (Column B) or frass (Column D) material by amount in leaf (Column A) material.

@@Determined by dividing total amount recovered from TLC bands by amount prior to fractionation then multiplying by 100.

@@@Determined by dividing total amount recovered from TLC bands by amount in ethanolic extracts given in Table 5 Column C then multiplying by 100.

curassavica and A' in *Gomphocarpus* sp); but, again the major cardenolides appear to be those of intermediate polarity.

Thus storage selectivity of cardenolides observed previously with the grasshopper, *Poekilocerus bufonius*, fed on *Calotropis procera* and *Asclepias curassavica*[102], the monarch butterfly reared on *Asclepias curassavica*[80], and the aphid, *Aphis nerii*, fed on *A. curassavica* and *Nerium oleander*[87] was also found in the present study.

The major cardenolides of *A. curassavica*, those in bands A, B and C, were identified as uscharidin, calactin and calotropin, respectively, by comparison of Rf's with authentic standards. Furthermore, the relative quantities in the plant are in substantial agreement with those found from an independent examination of *A. curassavica* from Trinidad W.I. (Reichstein, personal communication). Calactin and calotropin are two of the principal cardenolides stored in the adult monarch butterfly, again in agreement with earlier studies[80]. The identities of the other cardenolides occurring in lesser concentrations in *A. curassavica* were not pursued.

Gomphocarpus sp contains a different array of cardenolides (Fig. 10). The major one(s) in band A' was not identified. Matching Rf values with those of authentic standards indicated the possible presence of gomphoside and afroside among the minor components; these were isolated previously from *Gomphocarpus fruticosus (Asclepias fruticosa)* (see references in Table 1). Calotropin, if present at all, occurs in only small amounts.

Table 7--Column A, gives the quantitative determinations of cardenolide amounts as equivalent μg of digitoxin per 0.1 gm dry material in each TLC band and the sum of these amounts for the leaf material of both *A. curassavica* and *Gomphocarpus* sp. Below the total amounts is listed the amount prior to TLC. The last figure in this column is an approximation of the total recovery through extraction, cleanup and TLC fractionation and was obtained by comparing the total cardenolide amounts from the band to that of the ethanolic extracts in Table 5--Column C. The total cardenolides in *A. curassavica* leaf after processing is only 15% of that in the ethanolic extract

(Table 5--Column C). Similarly Column B gives the cardenolide amounts in the TLC bands, the total isolated and the recoveries of the butterfly material, and Column D shows these values for the frass samples. These amounts are an estimate in quantitative terms of the qualitative observations (Fig. 10). The cleanup procedure yielded samples devoid of TLC-interfering pigments and other coextracted material, so that little uncertainty was associated with the final quantitative determinations.

Columns C and E of Table 7 list the relative amounts of cardenolides in each TLC band of the butterfly and frass materials, relative to that of the leaf material for each plant. From the data, we conclude that the decrease in total cardenolide amounts, relative to the leaf material, is not due to a uniform decrease in each TLC band. For example, the decrease in relative cardenolide amounts of the butterfly material (Column C) appears to be due to the complete lack of cardenolides in the region of A while cardenolides in band B, D, E and F show an increase in amounts, relative to the leaf material. The same general pattern was observed in frass material from *A. curassavica* (Column E).

The large decrease in relative cardenolide amounts in the *Gomphocarpus*-reared butterfly material (0.33) is apparently due to the absence of cardenolides in band A' and decreases in the relative amounts in bands B' and C'. There was, however, an increase in the relative amounts in bands E', F' and G'. The largest decreases in relative cardenolide amounts in the frass, for these larvae, occurred in bands A' and D'.

The TLC data lead to the following conclusions: (1) the cardenolides of *A. curassavica* and *Gomphocarpus* sp have approximately the same polarity, evidenced by their similar TLC R_f's, but differ qualitatively; (2) major leaf cardenolides A and A' are not stored by monarchs reared on these plants; (3) storage selectivity is further evidenced by the preponderant occurrence of cardenolides of intermediate polarity (B and C from *A. curassavica*, C', D', and E' from *Gomphocarpus* sp) in the adult, though this tendency is considerably greater for *A. curassavica*-reared individuals than those from *Gomphocarpus* sp; and (4) from the amounts of individual cardenolides in the leaf, butterflies and frass, it appears likely that metabolic transformation or degradation occurs, particularly from the complete absence of A and A' in the adults and only small smounts in the larval frass.

Comparative utilization of food plants. To explore further the storage phenomenon and two of its major components inferred in the previous results--selective sequestering and metabolic transformation--the utilization efficiency of the two plants, by monarchs, should be considered. The utilization of different asclepiad food plants by larvae of danaid butterflies has been studied[29,61]. Particularly useful in evaluating the comparative utilization of food plants are indices which measure (1) approximate digestibility, (2) conversion of digested food, and (3) conversion of ingested food. These indices and other useful parameters are defined and their limitations outlined by Waldbauer (1968).

Experiments to determine the utilization of *A. curassavica* and *Gomphocarpus* sp by the monarch larvae have been completed by Brower and Moffitt (in preparation) and the results are summarized in Table 8--Column A. This table presents the total weight of leaf material ingested through all instars for larvae on both *A. curassavica* and *Gomphocarpus* sp, the weight of the adult butterflies and the total weight of larval frass produced. The ingested material, not accounted for through egestion as larval frass or as adult butterfly, is categorized as that metabolized or lost in other ways.

Table 8, Column A shows that an average larva, reared on *A. curassavica*, will consume 1.35 gm of leaf, emerge as an adult butterfly weighing 0.21 gm and produce 0.68 gm of frass, leaving by subtraction, 0.46 gm unaccounted for. Similarly the average larva on *Gomphocarpus* sp will ingest 2.13 gm of leaf, emerge as an adult butterfly weighing 0.21 gm and produce 1.25 gm of frass leaving 0.67 gm unaccounted for. This indicates that the larvae on *Gomphocarpus* sp consume 1.58 times as much leaf material as the larva on *A. curassavica* while achieving the same dry weight as an adult, but produce substantially more (1.84 times) frass.

These differences in plant utilization can perhaps best be expressed by using the values in Table 8, Column A to calculate the values shown in Table 9, Part A. These

Table 8

Consumption and utilization of leaf material and contained cardenolides by monarch butterfly larvae (*Danaus plexippus*) reared on *Asclepias curassavica* and *Gomphocarpus* sp. The leaf processing data (Column A) are from Brower and Moffitt, in preparation

	A Mean dry wt (gm) of material processed per monarch	B Total cardenolide (mg) in material processed per monarch*	C No ED_{50} units per 85 gm blue jay in amount of material processed per monarch@
Monarchs reared on *Asclepias curassavica*			
Leaf	1.35	5.21	75.6
Butterfly	0.21	0.67	9.5
Frass	0.68	2.21	50.0
Metabolized or other	0.46@@	2.33@@@	
Monarchs reared on *Gomphocarpus* sp			
Leaf	2.13	7.06	35.4
Butterfly	0.21	0.21	1.2
Frass	1.25	3.46	18.9
Metabolized or other	0.67@@@	3.39@@@	

*Obtained by multiplying mean dry weight of material processed per monarch (Column A) by cardenolide concentration (Table 5 Column C).

@Obtained by multiplying mean dry weight of material processed per monarch (Column A) by number of ED_{50} units (Table 6 Column B).

@@Of this 0.46 gm, molted exoskeletons weighed 0.02 gm, adult eclosion fluid (meconium) weighed 0.02 gm, leaving by subtraction approximately 0.42 gm lost to metabolism.

@@@Obtained by subtraction.

Table 9

Utilization efficiencies of leaf material and contained
cardenolides by monarch butterfly larvae (*Danaus plexippus*)
reared on *Asclepias curassavica* and *Gomphocarpus* sp.*

	A. curassavica	*Gomphocarpus* s
A. Utilization of food plant		
a. Approximate digestibility of ingested food@	50%	41%
b. Efficiency of conversion of digested food@@	31%	24%
c. Efficiency of conversion of ingested food@@@	16%	10%
B. Utilization of cardenolides		
a. Assimilation efficiency of ingested cardenolides@	58%	51%
b. Storage efficiency of assimilated cardenolides@@	22%	6%
c. Storage efficiency of ingested cardenolides@@@	13%	3%

*Figures in Part A are based on the values in Table 8 Column
A, figures in Part B are based on the values in Table 8
Column B; calculations were obtained as follows:

@Approximate digestibility and Assimilation efficiency=
[(amount ingested - amount egested) ÷ amount ingested]x 100.

@@Efficiency of conversion of digested food and Storage efficiency of assimilated cardenolides=[amount in adult butterfly
÷ (amount ingested - amount egested)] x 100.

@@@Efficiency of conversion of ingested food and Storage effi
ciency of ingested cardenolides =(amount in adult butterfly
÷ amount ingested) x 100.

are (a) approximate digestibility of ingested food, (b) efficiency of conversion of digested food to body substance, and (c) efficiency of conversion of ingested food to body substance[103]. These indices show that larvae on A. *curassavica* are more efficient than those on *Gomphocarpus* sp in digesting the ingested leaf material (50% vs. 41%); more efficient in converting the digested leaf material to butterfly (31% vs. 24%); and overall more efficient at converting ingested leaf material to butterfly (16% vs. 10%). *Asclepias curassavica* is clearly the more efficient food plant for monarchs.

Comparative processing of the total cardenolide content. By combining the data in Table 8, Column A with the gross cardenolide concentration determinations in Table 5, Column C, the actual amounts of cardenolides ingested, stored in the butterfly, egested in the frass and metabolized or otherwise excreted are listed in Column B (Table 8). Similarly Table 8, Column C lists the number of ED_{50} units ingested, stored in the butterfly and egested in the frass.

An average larva, on A. *curassavica,* ingests 1.35 gm of leaf during its larval stages resulting in a total consumption of 5.21 mg of cardenolide worth 75.6 ED_{50} units. The adult butterfly weighs 0.21 gm and contains 0.67 mg of cardenolides possessing 9.5 ED_{50} units. This larva produces 0.68 gm of frass which contains 2.21 mg of cardenolides worth 50.0 ED_{50} units, leaving 0.46 gm of consumed leaf and 2.33 mg of cardenolide unaccounted for, presumably metabolized or excreted in another manner. Table 8 gives the same information for the average larva reared on *Gomphocarpus* sp.

In comparing the utilization of cardenolides in the two food plants, by the monarch butterfly larva, it is again useful to calculate efficiencies similar to those applied to food utilization. The data in Table 8, Column B have been used to calculate these efficiencies (Table 9, Part B): (a) assimilation efficiency of ingested cardenolides, (b) storage efficiency of assimilated cardenolides, and (c) storage efficiency of ingested cardenolides. The first two, of these three parameters, are only approximations since cardenolides converted, in the gut, to noncardenolides are considered assimilated as well as cardenolides present in the molted exoskeletons and meconium. The assimilation efficiency will depend upon the larva's ability to sequester cardenolides and also its ability to retain the sequestered cardenolides. The storage efficiency of ingested cardeno-lides (c) will depend both upon the assimilation efficiency of ingested cardenolides (a) and the storage efficiency of assimilated cardenolides (b).

With regard to comparative processing of cardenolides, the total cardenolide amounts processed by the larvae, on both food plants, was calculated (Table 8, Column B). Even though *Gomphocarpus* sp leaves have a lower concentration of cardenolides (Table 5), the fact that the monarch consumes more leaves results in their ingesting 1.36 times more cardenolides than they do on *A. curassavica* (7.06 mg vs. 5.21 mg). However, only 3% of the total amount of cardeno-lides in the leaves is successfully stored by adults; whereas, on *A. curassavica* the adult stores 13% [Table 9, Part B (c)]. In terms of gross amounts stored, the *A. curassavica*-reared monarchs contain 0.67 mg of cardenolides compared to 0.21 mg for the *Gomphocarpus*-reared adults, or 3.2 times as much (Table 8, Column B). In other words, because the monarchs must eat so much more of the *Gompho-carpus* leaves, they ingest about 13% more cardenolides than they do on *A. curassavica*, but end up with only 31% as much cardenolides in the adult butterflies. Therefore it appears that *Gomphocarpus* is less efficient, not only as a foodplant, but also as a source of cardenolides for the butterflies. Butterflies reared as larvae, on both plants, store only small amounts of ingested cardenolides. The cardenolides, which were not stored, were about equally egested (42% for *A. curassavica*-reared monarchs and 49% for *Gomphocarpus*-reared monarchs) and metabolized (45% and 48%, respectively) for both plants (Table 8, Column B).

It appears that the differences in storage efficiencies of ingested cardenolides [Table 9, Part B(c)] is due to slight differences in assimilation efficiencies (a) and larg differences in storage efficiencies of assimilated cardeno-lides (b). Qualitative similarities and differences between the cardenolides in *A. curassavica* and *Gomphocarpus* sp could probably effect both. The polarity of the cardenolides in these two plants are qualitatively similar (Fig. 10). If polarity is important in determining assimilation, then it could be predicted that, on a gross level cardenolides from these two plants would be assimilated with similar efficien-

cies. Differences in the larval storage efficiencies of
assimilated cardenolides is probably due to qualitative
differences in cardenolides of the two species. Qualitative
differences, affecting the storage efficiencies of assimilated cardenolides, could result from differences in the
ability to store the assimilated cardenolides and/or differences in the ability to metabolize the assimilated cardenolides to noncardenolides. Qualitative differences are also
evident from the R_f values. Cardenolides with similar
polarities may vary in their structural features and/or
configuration; these variations could account for the differences in storage efficiencies. Clearly more isolation and
structural identification must be done before accepting
this explanation.

Table 8, Column C lists the number of ED_{50} units present in the total amount of leaf ingested by the larva, the
butterfly, and the frass egested for both plants. The *A.
curassavica*-reared butterflies contain 13% of the emeticity
present in the leaves they ingested and the corresponding
value for the *Gomphocarpus* butterflies is 3%. These values
are identical to the storage efficiency of cardenolides
[Table 9, Part B (c)]. The percentages of cardenolides
stored, exactly parallels the number of ED_{50} units present
in the butterflies reared on these two milkweeds; especially,
the emetic potency of the leaf and butterfly material for
each plant is similar (Table 6, Column C).

Thus the food consumption data has allowed us to
conclude that (1) the larvae are about equally efficient in
assimilating cardenolides from *Asclepias curassavica* and
Gomphocarpus sp, but (2) the larvae on *A. curassavica* are
more efficient at storing cardenolides and (3) only a
relatively small percent of the ingested cardenolides are
stored in the adults reared as larvae on both plants.

Comparative processing of individual cardenolides. By
combining the food plant utilization data of Table 8, Column
A with amounts of cardenolides isolated from each TLC band
(Table 7, Columns A, B, and D), the actual amounts of
cardenolides in each TLC band ingested by the larva, stored
in the butterfly, and egested in the frass were calculated
for each plant (Table 10). Using the amounts of cardenolides
in the bands of the butterfly (Column B) and the actual
amount ingested (Column A), the storage efficiencies were
calculated (Column C). Similarly the egestion efficiencies

Table 10

Consumption and utilization of cardenolides in TLC regions from *Asclepias curassavica* and *Gomphocarpus* sp by monarch butterfly larvae (*Danaus plexippus*)

	Leaf	Butterfly		Frass	
Monarchs reared on *Asclepias curassavica* Bands	A μg of cardenolides ingested*	B μg of cardenolides stored@	C Storage efficiency@@@	D μg of cardenolides egested@@	E Egestion efficiency#
A	267	0	0%	18	7%
B	82	20	24%	35	43%
C	246	35	14%	75	30%
D	36	7	19%	17	47%
E	32	8	25%	13	41%
F	16	7	44%	7	44%
O	1	0	0%	1	100%
R	1	0	0%	0	0%
I	89	9	10%	20	22%
TOTAL	770	86	11%	186	24%
Monarchs reared on *Gomphocarpus* sp					
A'	300	0	0%	45	15%
B'	204	2	1%	53	26%
C'	190	5	3%	84	44%
D'	177	10	6%	30	17%
E'	9	5	56%	5	56%
F'	6	1	17%	3	50%
G'	4	0	0%	0	0%
O'	2	0	0%	0	0%
R'	0	0	0%	0	0%
I'	136	8	6%	33	24%
TOTAL	1028	31	3%	253	25%

*Determined by multiplying the mean dry weight of the leaf ingested (Table 7 Column A) by cardenolides isolated from leaf material (Table 7 Column A).

@Determined by multiplying the mean dry weight of the butterfly (Table 8 Column A) by the cardenolides isolated from the butterfly material (Table 7 Column B).

@@Determined by multiplying the mean dry weight of frass egested (Table 8 Column A) by cardenolides isolated from frass material (Table 7 Column D).

@@@Storage efficiency determined by dividing μg of cardenolides stored (Column B) by μg of cardenolides ingested (Column A) and multiplying by 100.

#Egestion efficiency determined by dividing g of cardenolides egested (Column D) by g of cardenolides

were calculated (Column E).

The average larva on *A. curassavica* will ingest 267 g of cardenolides in the region of A, will store 0 µg in the adult butterfly (Column B), and the larva will egest 18 g (Column D). This reflects a storage efficiency of 0% (Column C) and an egestion efficiency of 7% (Column E). Similar calculations were made for each of the other regions in *A. curassavica* and *Gomphocarpus* sp. Thus from the total of the bands, *A. curassavica*-reared butterflies store 11% of the cardenolides they ingested and those reared on *Gomphocarpus* sp store 3%.[1]

---

[1]The total amount of cardenolides ingested in all bands by the average larva on *A. curassavica* is 770 µg (Column A). This figure represents only 15% of the true amount ingested (Table 8, Column B) because the total recovery for *A. curassavica* leaf material through extraction, cleanup, and TLC fractionation when based on these figures in only 15% (see Table 7, Column A). The total cardenolides in the TLC bands of the *A. curassavica*-reared butterfly (Column B) is only 13% of that listed in Table 8, Column B because its total recovery was 13%. These considerations also apply to the *A. curassavica* frass material and all the *Gomphocarpus* sp materials. A total of 86 µg of cardenolides stored in the *A. curassavica*-reared butterfly is calculated by summing the TLC bands (Table 10, Column B); this reflects a storage efficiency of 11% (Column C). This value differs from the value calculated on gross concentrations [Table 9, Part B (c)] due to the absolute differences in total recovery between the leaf material and butterfly material, *i.e.*, 15% vs. 13% (Table 7, Columns A and B). Similarly the egestion efficiency for the total of TLC bands in the frass sample is 24% whereas calculations based on gross cardenolide concentrations (Table 8, Column B) give a value of 42% [(2.21÷5.21) x 100] about 1.8 times as great because the total recovery of leaf material was 1.8 (15%÷8%) times that of frass material (Table 7, Columns B and D).

Comparing the storage efficiencies of individual cardenolides (Table 10, Column C) it is apparent that the butterflies do not store the cardenolides with uniform efficiency. The *A. curassavica*-reared butterfly stores no cardenolides in the region of A, stores 24% of those ingested in the region of B, 14% of C, etc. The absolute amounts of cardenolides stored in the regions of B and C are greater than in any other TLC region for the *A. curassavica*-reared butterfly even though the storage efficiencies are only equal to or less than those of the other TLC regions. The frass material from the *A. curassavica*-reared larva reflects a similar pattern.

The *Gomphocarpus*-reared butterflies (Table 10, Column B) show a different pattern than those reared on *A. curassavica*. Again no cardenolides are stored in the region of A' and the major cardenolides, stored by the *Gomphocarpus*-reared butterfly, are more polar (regions C', D', and E'). For example, in the *A. curassavica*-reared butterfly the cardenolides in region B comprise 23% [(20÷86) x 100] of the total cardenolides stored but for the *Gomphocarpus* sp-reared butterfly, cardenolides in the same region (B') comprise only 6% of the total stored. Furthermore the *Gomphocarpus* sp-reared monarchs store cardenolides in regions B', C', and D' with much less efficiency than those reared on *A. curassavica* (Column C). The contrast in storage efficiency suggests basic differences in processing of individual cardenolides of these two plants having similar polarities. These differences, in turn, probably account for the overall difference in storage efficiency of the butterflies whose larva were reared on these two plants. The relative amounts of cardenolides found in the frass, from the *Gomphocarpus* sp-reared larva and their egestion efficiencies (Table 10, Columns D and E) are similar to those for the frass from *A. curassavica*-reared larvae.

Butterflies reared as larvae on both plants reflect different storage patterns: both are selective and show no storage of cardenolides in the regions of A and A' (Fig. 10 and Table 10, Column B). Furthermore, since only 7% of the ingested cardenolides in the region of A from *A. curassavica* and 15% of the ingested cardenolides in band A' from *Gomphocarpus* sp can be accounted for in the frass (Table 10, Column E), metabolic degradation or transformation to other cardenolides must be inferred.

Summarizing, Table 10 shows that larvae reared on
Asclepias curassavica and *Gomphocarpus* sp are processing
the cardenolides of the individual TLC bands such that they
are egested in the larval frass and stored in the adult
butterfly with different efficiencies. It is also apparent
that the adult monarch of *A. curassavica*-reared larva is
more efficient in storing the cardenolides of intermediate
polarity than the adult monarch of *Gomphocarpus*-reared larva.

We have inferred metabolic transformation of cardenolides, and selective sequestering from the foregoing data.
Absolute proof must, however, await the outcome of feeding
studies in which individual cardenolides of these plants
are fed to larvae in a medium known to lack cardenolides.
In lieu of this much-needed proof, we can further support
these inferences from the known processing of other cardenolides, and other steroids, in insects and other organisms.

Duffey and Scudder (1974) indicated that the large
milkweed bug, *Oncopeltus fasciatus*, could metabolize
digitoxin to at least two (more polar) cardenolide metabolites. Brower and Weeks (unpublished) found monarch butterfly larvae to be capable of metabolically altering digitoxin
and ouabain. Thus metabolic alteration of some cardenolides
is clearly possible for these two insects. It should be
noted, however, that neither digitoxin nor ouabain are
natural constituents of *Asclepias* species.

The metabolic processing of other steroids in insects
has been reviewed by Robbins *et al.*, (1971). While all
insects require a dietary or exogenous source of sterols,
they do have the capacity to sequester, then utilize, these
sterols to fulfill their functional needs. In addition, some
insects have the capacity metabolically to alter the sterols
prior to their utilization or excretion. It would then
seem reasonable that an insect such as *Danaus plexippus*
which sequesters and utilizes cardenolides ostensibly for
protective purposes, also has developed metabolic capa-
cities to alter the ingested cardenolides.

Selective sequestering may be inferred from the selective storage observed in this study. It is known that
cardenolides are unequally absorbed from the human gastrointestinal tract and that the more polar cardenolides are
generally not as well absorbed as the less polar ones. For

the series lanatoside C, digoxin and digitoxin, extent of
absorption was estimated as 10, 50 and 100%, respectively[65].
Selective sequestering of cardenolides in the monarch
butterfly larva could be governed by polarity in a similar
way.

The absence of nonpolar cardenolides (*i.e.*, for *A.
curassavica* inclusive of the polarity range of A and the
solvent front) in the adult monarch butterfly, observed in
this and other feeding studies as well as among wild monarchs, suggests it is a general occurence and not limited to
a particular cardenolide. In a study of different stages
of the development of *Danaus plexippus*, Brower and Tomashow
(unpublished) observed that the selective storage pattern
apparent in the adult was present throughout the life
cycle. These observations seem to favor selective sequestering over metabolic alteration as an explanation for the
absence of A and A' in the *A. curassavica*- and *Gomphocarpus*-
reared butterflies. Thus the suggestions that the monarch
is capable of selectively sequestering and metabolically
altering cardenolides is not without precedence.

Selective storage within the adult monarch butterfly.
Our investigations of monarch butterflies reared as larvae
on *A. curassavica* and *Gomphocarpus* sp showed that they are
definitely selective in the cardenolides they store. Brower
and Glazier (1975) analyzed the selective storage of the
cardenolides present in the body parts of the adult monarch
reared on *A. curassavica*.

The monarch, unlike the grasshopper, *P. bufonius* and the
milkweed bug *O. fasciatus*[27,102] apparently has no specific
glands or organs which serve to hold or excrete cardenolides
as a protective function. They have, however, been found to
concentrate cardenolides of differing emetic potencies in
their different body parts in a manner hypothesized to
maximize the effectiveness of these compounds as deterrents.
The concentration, number of ED_{50} units, and relative emetic
potency for the abdomen, wings, and thoraces of both male
and female monarchs reared on *A. curassavica* is given in
Figure 12. The butterflies have the highest concentration of
cardenolides in the wings, the abdomen being intermediate
in concentration and the thoraces being lowest. The relative
emetic potencies for each part show that the cardenolides
contained in the wings are least emetic while those of the
abdomen have the greatest emetic potency and that the females

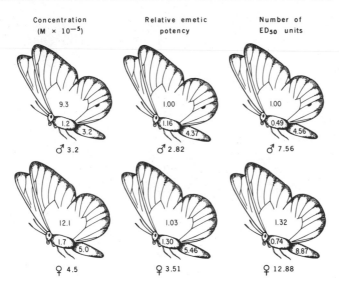

Figure 12. Cardiac glycosides concentration (left) relative emetic potency (center), and ED_{50} units (right) for monarch butterflies reared on *A. curassavica* (for explanation see p. 156-57).

contain cardiac glycosides which are more emetic than the males. Because the females have a higher concentration of the more potent cardiac glycosides, they contain more ED_{50} units in each body part than do the males[9].

A preliminary examination of purified extracts of abdomen, wings, and thorax has shown that these parts differ in the proportions of cardenolides of different polarities present. Thus the differences in emetic potency observed for the different body parts might be explained by differences in the individual cardenolide content, and would be amenable to quantitative study in the same manner as the leaf, frass and adult samples previously discussed.

Selective storage, selective sequestering and metabolic alteration of cardenolides may represent processes by which the monarch manipulates the ingested cardenolides to minimize any deleterious effects, while maintaining them for their benefit. Deleterious effects of cardenolides upon the monarch are as yet unknown, but they have many **adverse** effects upon the human system[65] including cyctotoxicity[50,54].

Summary of comparative cardenolide processing by monarchs on two food plants. In summarizing the study of comparative processing of cardenolides by the monarch (*Danaus plexippus*) we have shown (1) that *Asclepias curassavica* and *Gomphocarpus* sp differ qualitatively and quantitatively in their cardenolide content and in their emetic potency; (2) that butterflies when reared as larvae on these two plants reflect these differences; (3) that the larvae, when reared on each plant, have selectively sequestered and/or altered the cardenolides so that the butterflies show a selective storage pattern, have an emetic potency about equal to that of the leaves they ingested, and in the case of *A. curassavica* the frass egested has a greater emetic potency than the ingested leaf material; and (4) from a previous study it is known that butterflies reared on *A. curassavica* have selectively stored cardenolides in different concentrations and of different emetic potencies within different body tissues. From the food utilization data we were further able to conclude that (5) larvae on both plants assimilate cardenolides with about equal efficiency but (6) they differ in their efficiency of storing cardenolides, and (7) on each plant the larvae differ in their efficiency of storing and egesting individual cardenolides of different polarities.

CONCLUSION

In this chapter we have considered a single group of chemicals, cardenolides, within a single plant family, Asclepiadaceae; and the utilization of these chemicals for defensive purposes principle by a single insect species, *Danc plexippus*. This utilization of cardenolides is perhaps the most apparent and thus the most studied interaction between the monarch and its host plant. There are assuredly many more subtle interactions which allow the monarch to locate and select a host plant and then to oviposit. Cues which act as feeding stimulants and components which effect food utilization, growth rate and development are all aspects of this unique relationship. While we have touched on just one group of chemicals for one specific utilization by the monarch, these chemicals may have other purposes, and other chemicals from the plant may also serve some defensive purposes. The interactions between the milkweed, the

monarch and its predators is an intriguing study involving chemistry, biology and ecology.

Acknowledgements. We would like to thank Professors
T. Reichstein, T. R. Watson, and G. Russo for supplying
cardenolide standards; S. Lynch and R. F. Keeler for
providing *Asclepias* samples. We are grateful to the
following Amherst College people for assistance at various
stages in this work: Howard Uman, Susan Swartz, Lorna
Coppinger, Homer Bessonette, George McIntyre, Barbara
Tiffany, and Helen Sullivan. Research support has been
provided by the U. S. National Science Foundation, most
recently by Grant GB 35317x, BMS 75-14265, and BMS
75-14266 to the Amherst College and UC Davis groups.

REFERENCES

1. Abisch, E. and T. Reichstein. 1962. Chemical orienta-
        tion investigation of some Asclepiadaceae and
        Periplocaceae. *Helv. Chim. Acta, 45:*2090.
2. Bauer, S., L. Masler, O. Bauerova and D. Sikl. 1961.
        Uzarigenin and desglucouzarin from *Asclepias
        syriaca* L. *Experientia, 17:*15.
3. Borison, H.L. and S.C. Wang. 1953. Physiology and
        pharmacology of vomiting. *Pharmac. Rev., 5:*193.
4. Brower, L.P. 1969. Ecological chemistry. *Scient.
        Amer., 220:*22.
5. Brower, L.P. 1970. Plant poisons in a terrestrial food
        chain and implications for mimicry theory. *In:Bio-
        chemical Coevolution,* Proceedings of the Twenty-
        Ninth Annual Biology Colloquim, 1968. K.L. Chambers
        (ed), Oregon State University Press, Corvallis,
        Oregon, p. 69.
6. Brower, L.P., and J.V.Z. Brower. 1964. Birds, butter-
        flies and plant poisons:a study in ecological
        chemistry. *Zoologica, 49:*137.
7. Brower, L.P., J.V.Z. Brower and J.M. Corvino. 1967.
        Plant poisons in a terrestrial food chain. *Proc.
        Nat. Acad. Sci., 57:*893.
8. Brower, L.P., M. Edmunds and C.M. Moffitt. 1975.
        Cardenolide content and palatability of a population
        of *Danaus chrysippus* butterflies from West Africa.
        *J. Ent. (A), 49:*183.

9. Brower, L.P. and S.C. Glazier. 1975. Localization of
    heart poisons in the monarch butterfly. *Science*,
    188:19.

10. Brower, L.P., P.B. McEvoy, K.L. Williamson and M.A.
    Flannery. 1972. Variation in cardiac glycoside
    content of monarch butterflies from natural
    populations in eastern North America. *Science*,
    177:426.

11. Brower, L.P. and C.M. Moffitt. 1974. Palatability
    dynamics of cardenolides in the monarch butterfly.
    Nature, *249*:280.

12. Brower, L.P., W.N. Ryerson, L.L. Coppinger and S.C.
    Glazier. 1968. Ecological chemistry and the pala-
    tability spectrum. *Science*, *161*:1349.

13. Brüschweiler, F., K. Stöckel and T. Reichstein. 1969a.
    Calotropis-glycosides, presumed partial structure.
    Helv. Chim. Acta, *52*:2276.

14. Brüschweiler, F., W. Stöcklin, K. Stöckel and T.
    Reichstein. 1969b. Glycosides of *Calotropis
    procera* R. Br. *Helv. Chim. Acta*, *52*:2086.

15. Carman, R.M., R.G. Coombe and T.R. Watson. 1964. The
    cardiac glycosides of *Gomphocarpus fruticosus* (R.
    Br.) IV. The nuclear magnetic resonance spectrum
    of gomphoside. *Aust. J. Chem.*, *17*:573.

16. Chen, K.K., E.B. Robbins and H. Worth. 1938. The
    significance of sugar component in the molecule of
    cardiac glycosides. *J. Am. Pharm. Ass.*, *27*:189.

17. Chernobai, V.T. 1957. Cardiac glycosides of
    Gomphocarpus fruticosus. I. Glycosides of the
    seed. *Med. Prom. S.S.S.R.*, *11*:38. [*Chem. Abstr.*,
    52:8465a]

18. Chernobai, V.T. and N.F. Komissarenko. 1971. Cardeno-
    lides from *Gomphocarpus fruticosus* and partial
    synthesis of uzarigenin glycosides. *Khim. Prir.
    Soedin.*, *7*:445. [*Chem. Abstr.*, *75*:141103w]

19. Coombe, R.G. and T.R. Watson. 1962. Structure of
    β-anhydrogomphogenin. *Chem. and Ind.*, *1962*:1724.

20. Coombe, R.G. and T.R. Watson. 1964. The cardiac glyco-
    sides of *Gomphocarpus fruticosus* R. Br. III.
    Gomphoside. *Aust. J. Chem.*, *17*:92.

21. Crout, D.H.G., R.F. Curtis, C.H. Hassall and T.L. Jones.
    1963. The cardiac glycosides of *Calotropis procera*
    Tetrahedron Lett., *1963*:63.

22. Crout, D.H.G., C.H. Hassall and T.L. Jones. 1964. Car-
    denolides. Part VI. Uscharidin, calotropin, and
    calotoxin. *J. Chem. Soc.*, *1964*:2187.

23. Dethier, V.G. 1970. Chemical interactions between
     plants and insects. *In:Chemical Ecology*, E.
     Sondheimer and J.B. Simeone (eds), Academic Press,
     New York, Chapter 5.

24. Dixon, W.J. and F.J. Massey, Jr., 1957. *Introduction
     to Statistical Analysis*, McGraw-Hill, New York.

25. Duffey, S.S. 1970. Cardiac glycosides and distaste-
     fulness: some observations on the palatability
     spectrum of butterflies. *Science, 169:*78.

26. Duffey, S.S. and G.G.E. Scudder. 1972. Cardiac glyco-
     sides in North American Asclepiadaceae, a basis for
     unpalatability in brightly coloured Hemiptera and
     Coleoptera. *J. Insect Physiol., 18:*63.

27. Duffey, S.S. and G.G.E. Scudder. 1974. Cardiac glyco-
     sides in *Oncopeltus fasciatus* (Dallas) (Hemiptera:
     Lygaeidae). I. The uptake and distribution of
     natural cardenolides in the body. *Can. J. Zool.,
     52:*283.

28. Eisner, T. 1970. Chemical defense against predation in
     arthropods. *In: Chemical Ecology*, E. Sondheimer and
     J. B. Simeone (eds), Academic Press, New York,
     Chapter 8.

29. Erickson, J.M. 1973. The utilization of various
     Asclepias species by larvae of the monarch butter-
     fly, *Danaus plexippus. Psyche, 80:*230.

30. Evans, F.J. and P.S. Cowley. 1972. Cardenolides and
     spirostanols in *Digitalis purpurea* at various stages
     of development. *Phytochem., 11:*2971.

31. Feeny, P. 1975. Biochemical coevolution between plants
     and their insect herbivores. *In: Coevolution of
     Animals and Plants*, L.E. Gilbert and P.H. Raven
     (eds), University of Texas Press, Austin, p. 3.

32. Feir, D. and J. Suen. 1971. Cardenolides in the milk-
     weed plant and feeding by the milkweed bug. *Ann.
     Entomol. Soc. Am., 64:*1173.

33. Fieser, L.F. and M. Fieser. 1959. *Steroids*, Reinhold,
     New York, Chapter 20.

34. Fraenkel, G.S. 1959. The raison d'etre of secondary
     plant substances. *Science, 129:*1466.

35. Gaitonde, B.B. and S.N. Joglekar. 1975. Role of
     catecholamines in the central mechanism of emetic
     response induced by peruvoside and ouabain in cats.
     *Br. J. Pharmacol., 54:*157.

36. Hassall, C.H. and K. Reyle. 1959. Cardenolides. Part
     III. Constitution of calotropagenin. *J. Chem. Soc.,
     1959:*85.

37. Hendry, L.B., J.K. Wichmann, D.M. Hindenlang, R.O.
    Mumma and M.E. Anderson. 1975. Evidence for
    origin of insect sex pheromones: presence in food
    plants. *Science, 188*:59.

38. Hesse, G., H. Fasold and W. Geiger. 1959. African
    arrow poisons. X. Calotropin from uscharidin.
    Liebigs Ann. Chem., 625:157.

39. Hesse, G. and H.W. Gampp. 1952. African arrow poisons.
    VI. The heterocyclic part of uscharin. *Chem.
    Ber., 85:* 933.

40. Hesse, G., L.J. Heuser, E. Hutz and F. Reicheneder. 1950
    African arrow poisons. V. Relationships between
    the most important poisons of *Calotropis procera*.
    Liebigs Ann. Chem., 566:130.

41. Hesse, G. and G. Lettenbauer. 1957. A second sulfur-
    containing compound from the milky-juice of
    Calotropis procera. *Angew. Chem., 69*:392.

42. Hesse, G. and G. Ludwig. 1960. African arrow poisons.
    XIV. Voruscharin, a second sulfur-containing
    heart poison from *Calotropis procera* L. *Liebigs
    Ann. Chem., 632*:158.

43. Hesse, G.and K. Mix. 1959. African arrow poisons.
    IX. Structure and partial synthesis of uscharin.
    Liebigs Ann. Chem., 625:146.

44. Hesse, G. and F. Reicheneder. 1936. African arrow
    poison calotropin. I. *Liebigs Ann. Chem., 526*:
    252.

45. Hesse, G., F. Reicheneder and H. Eysenbach. 1939.
    African arrow poisons. II. Heart poisons in
    Calotropis latex. *Liebigs Ann. Chem., 537*:67.

46. Hoch, J.H. 1961. *A Survey of Cardiac Glycosides and
    Genins,* Univ. of South Carolina Press, Charleston,
    South Carolina.

47. Hunger, A. and T. Reichstein. 1952a. Frugoside, a
    second crystallized glycoside from the seeds of
    Gomphocarpus fructicosus. *Helv. Chim. Acta, 35*:429.

48. Hunger, A. and T. Reichstein. 1952b. The constitution
    of gofruside and frugoside. *Helv. Chim. Acta, 35*:
    1073.

49. Keller, M. and T. Reichstein. 1949. Gofruside, a
    crystalline glycoside from the seeds of *Gomphocarpus
    fructicosus* (L) R. Br. *Helv. Chim. Acta, 32*:1607.

50. Kelly, R.B., E.G. Daniels and L.B. Spaulding. 1965.
    Cytotoxicity of cardiac principles. *J. Med. Chem.,
    8:*547.

51. Jaggi, K., H. Kaufmann, W. Stocklin and T. Reichstein. 1967. Glycosides of *Asclepias swynnertonii* S. Moore. roots. *Helv. Chem. Acta, 50:*2457.

52. Kingsbury, J.M. 1964. *Poisonous Plants of the United States and Canada,* Prentice-Hall, Englewood Cliffs, N.J., p. 269.

53. Krieger, R.I., P.P. Feeny and C.F. Wilkinson. 1971. Detoxication enzymes in the guts of caterpillars: an evolutionary answer to plant defenses? *Science, 172:*579.

54. Kupchan, S.M., J.R. Knox, J.E. Kelsey and J.A.S. Renauld. 1964. Calotropin, a cytotoxic principle isolated from *Asclepias curassavica* L. *Science, 146:*1685.

55. Lardon, A., K. Stockel and T. Reichstein. 1969. Gomphogenin—partial synthesis and structure of calotropagenin. *Helv. Chem. Acta, 52:*1940.

56. Lardon, A., K. Stockel and T. Reichstein. 1970. Partial synthesis of 2α, 3β, 19-triacetoxy-14β-hydroxy-5α-card-20(22)-enolide. Additional proof of calotropagenin structure. *Helv. Chim. Acta, 53:*167.

57. Mahran, G.H., M.M. Rizkallah and A.H. Saber. 1971. A phytochemical study of *Calotropis procera* (Ait) R. Br. growing in Egypt. *Bull. Fac. Pharm. Cario Univ., 10:*1.

58. Marsh, C.D. and A.B. Clawson. 1924. The woolly-pod milkweed(*Asclepias eriocarpa*) as a poisonous plant. United States Department of Agriculture Bulletin. No. 1212.

59. Masler, L., S. Bauer, O. Bauerova, and D. Sikl. 1962a. Cardiac glycosides from *Asclepias syriaca.* I. Isolation of cardiac active steroids. *Collect. Czech. Chem. Commun., 27:*872.

60. Masler, L., S. Bauer, O. Bauerova and D. Sikl. 1962b. Cardiac glycosides from *Asclepias syriaca.* II. Structure of syriogenin and its glycosides. *Collect. Czech. Chem. Commun., 27:*895.

61. Mathavan, S. and R. Bhaskaran. 1975. Food selection and utilization in a Danaid butterfly. *Oecologia, 18:*55.

62. Mitsuhashi, H., K. Hayashi and K. Tomimoto. 1970. Studies on the constituents of the Asclepiadaceae plants. XXVIII. Components of *Asclepias syriaca* L. *Chem. Pharm. Bull., 18:*828.

63. Mitsuhashi, H. and M. Kurumi. 1968. Constituents of Asclepiadaceae plants. XXIII. Components of *Gomphocarpus fruticosus. Shoyakagaku Zasshi, 22:* 86. [*Chem. Abstr., 71:*88459f]

64. Mittal, O.P., Ch. Tamm and T. Reichstein. 1962.
     Glycosides of *Pergularia extensa* (Jacq) N.E. Br.
     *Helv. Chem. Acta, 45:*907.

65. Moe, G.K. and A.E. Farah. 1970. Digitalis and allied
     cardiac glycosides: *In: The Pharmacological Basis
     of Therapeutics,* L. S. Goodman and A. Gilman
     (eds), Macmillan, New York. Chapter 31.

66. Muller, C. H. 1970. The role of allelopathy in the
     evolution of vegetation. *In: Biochemical Co-
     evolution,* Proceedings of the Twenty-Ninth Annual
     Biology Colloquium, 1968. K.L. Chambers (ed),
     Oregon State University Press, Corvallis, Oregon,
     p. 13.

67. Nascimento, J.M. 1964. Cardenolides in leaves of
     *Asclepias glaucophylla. Rev. Port. Quim., 6:*97.
     [*Chem. Abstr., 64:*5549f]

68. Nascimento, Jr., J.M., Ch. Tamm, H. Jager and T.
     Reichstein. 1964. Glycosides of *Asclepias
     glaucophylla* Schlechter roots. *Helv. Chem. Acta,
     47:*1775.

69. Neher, R. 1969. TLC of steroids and related compounds.
     In: Thin-Layer Chromatography, A Laboratory Handbook
     E. Stahl (ed), Springer-Verlag, New York, p. 311.

70. Parsons, J.A. 1965. A digitalis-like toxin in the
     monarch butterfly, *Danaus plexippus* L. *J. Physiol.
     178:*290.

71. Parsons, J.A. and R.J. Summers. 1971. Cat assay for th
     emetic action of digitalis and related glycosides
     (digitoxin, digoxin, lanatoside C, ouabain, and
     calactin). *Br. J. Pharmacol., 42:*143.

72. Petricic, J. 1967. Cardenolides of some species of
     the genus *Asclepias. Farm. Glas., 23:*3. [*Chem.
     Abstr., 67:*8686g]

73. Petricic, J. 1966a. Desglucouzarin, the principal
     cardenolide glycoside of *Asclepias mellodora* St.Hi
     *Naturwissenschaften, 53:*332.

74. Petricic, J. 1966b. Cardenolides from the roots of
     *Ascelpias tuberosa. Arch. Pharm., 299:*1007.
     [*Chem. Abstr., 66:*76274a]

75. Pliske, T.E. and T. Eisner. 1969. Sex pheromone of
     the queen butterfly: biology. *Science, 164:*1170.

76. Rabitzsch, G. and U.Tambor. 1969. Method for the
     quantitative determination of cardiac glycosides
     and genins of the cardenolide type with 2,4,2',4'-
     tetranitrodiphenyl. *Pharmazie, 24:*262.

77. Rajagopalan, S., Ch. Tamm and T. Reichstein. 1955.
       The glycosides of seeds of *Calotropis procera* R.
       Br. *Helv. Chem. Acta, 38*:1809.
78. Reichstein, T. 1967a. Cardenolide-and pregnanglycosides.
       Naturwissenschaften, 54:53.
79. Reichstein, T. 1967b. Cardiac glycosides as defensive
       substances in insects. *Naturwiss. Rundsch., 20*:
       499.
80. Reichstein, T., J. von Euw, J.A. Parsons, and M.
       Rothschild. 1968. Heart poisons in the monarch
       butterfly. *Science, 161*:861.
81. Repke, K. 1963, Metabolism of cardiac glycosides. *In:*
       New aspects of cardiac glycosides, Proceedings of
       the first international pharmacological meeting,
       W. Wilbrandt and P. Lindgren (eds), Macmillan,
       New York, Volume 3, p. 47.
82. Rice, E.L. 1974. *Allelopathy.* Academic Press,
       New York.
83. Robbins, W.E., J.N. Kaplanis, J.A. Svoboda and M.J.
       Thompson. 1971. Steroid metabolism in insects.
       In: Annual Review of Entomology., R.F. Smith and
       T.E. Mittler (eds), Palo Alto, Calif., Volume 16,
       p. 53.
84. Robinson, T. 1974. Metabolism and function of
       alkaloids in plants. *Science, 184*:430.
85. Rothschild, M. 1972a. Secondary plant substances and
       warning colouration in insects. *In: Insect/Plant*
       Relationships, Symposia of the Royal Entomological
       Society of London, H.F. van Emden (ed), Blackwell
       Scientific Publications, Oxford, p. 59.
86. Rothschild, M. 1972b. Some observations on the
       relationship between plants, toxic insects, and
       birds. *In: Phytochemical Ecology,* J.B. Harborne
       (ed), Academic Press, New York. Chapter 1.
87. Rothschild, M., J. von Euw and T. Reichstein. 1970.
       Cardiac glycosides in the oleander aphid, *Aphis*
       nerii. J. Insect Physiol., 16:1141.
88. Rothschild, M., J. von Euw, T. Reichstein, D.A.S. Smith,
       and J. Pierre. 1975. Cardenolide storage in
       Danaus chrysippus (L.) with additional notes on
       D. plexippus. Proc. R. Soc. Lond. B., 190:1.
89. Sawlewicz, L., E. Weiss and T. Reichstein. 1967. Car-
       denolide and pregnanglycosides of *Asclepias lilacina*
       Weimarck roots. I.Isolation. *Helv. Chim. Acta,*
       50:504.

90. Schildknecht, H. 1971. Evolutionary peaks in the
     defensive chemistry of insects. *Endeavor, 30*:136.
91. Schoonhoven, L.M. 1972. Secondary plant substances and
     insects. *In: Structural and Functional Aspects of
     Phytochemistry, Recent Advances in Phytochemistry*,
     V.C. Runeckles and T.C. Tso (eds), Academic Press,
     New York; Volume 5, p. 197.
92. Scudder, G.G.E. and S.S. Duffey. 1972. Cardiac glyco-
     sides in the Lygaeinae (Hemiptera:Lygaeidae).
     Can. J. Zool, 50:35.
93. Singh, B. and R.P. Rastogi. 1969. Chemical investiga-
     tion of *Asclepias curassavica* Linn. *Indian J.
     Chem., 7*:1105.
94. Singh, B. and R.P. Rastogi. 1970. Cardenolides--
     glycosides and genins. *Phytochem.. 9*:315.
95. Singh, B. and R.P. Rastogi. 1972. Structure of
     asclepin and some observations on the nmr spectra
     of *Calotropis* glycosides. *Phytochem., 11*:757.
96. Seiber, J.N., J.M. Benson, C.N. Roeske and L.P. Brower.
     1975. Qualitative and quantitative aspects of milk
     weed cardenolide sequestering by monarch butter-
     flies. Paper presented at 170th National Meeting
     of the American Chemical Society, Division of
     Pesticide Chemistry (Pest 103), Chicago, August.
97. Tamm, Ch. 1956. New developments in the field of
     glycosidal cardiac poisons--fundamentals and the
     aglycons. *Fortschr. Chem. Org. Naturstoffe, 13*:
     137.
98. Tamm, Ch. 1957. New results from the field of glyco-
     sidic heart poisons; sugar and glycoside.
     Fortschr. Chem. Org. Naturstoffe, 14:71.
99. Tamm, Ch. 1963. The stereochemistry of the glycosides
     in relation to biological activity. *In: New
     Aspects of Cardiac Glycosides*, Proceedings of the
     first international pharmacological meeting, W.
     Wilbrandt and P. Lindgren (eds), Macmillan, New
     York, Volume 3, p. 11.
100. Tschesche, R., D. Forstmann and V.K.M. Rao. 1958.
     Cardenolide components of *Asclepias curassavica*
     L. *Chem. Ber., 91*:1204.
101. Tschesche, R., G. Snatzke and G. Grimmer. 1959.
     Calotropagenin from *Asclepias curassavica* L.
     Naturwissenschaften, 46:263.

102. von Euw, J., L. Fishelson, J.A. Parsons, T. Reichstein, and M. Rothschild. 1967. Cardenolides (heart poisons) in a grasshopper feeding on milkweeds. *Nature, 214*:35.

103. Waldbauer, G.P. 1968. The consumption and utilization of food by insects. *In: Advances in Insect Physiology,* Volume 5, J.W.L. Beament, J.E. Treherne and V.B. Wigglesworth (eds), Academic Press, New York, p. 229.

104. Watson, T.R. 1966. The cardiac glycosides of *Asclepias fruticosa. Colloq. Int. Centre Nat.Rech. Sci.* No. *144*:173.

105. Watson, T.R. and S.E. Wright. 1954. The Cardiac glycosides of *Gomphocarpus fruticosus* (R. Br.). *Chem. and Ind., 1954*:1178.

106. Watson, T.R. and S.E. Wright. 1956. The cardiac glycosides of *Gomphocarpus fruticosus* R. Br. I. Afroside. *Aust. J. Chem., 9,* 497.

107. Watson, T.R. and S.E. Wright. 1957. The cardiac glycosides of *Gomphocarpus fruticosus* R. Br. II. Gomphoside. *Aust. J. Chem., 10*:79.

108. Whittaker, R.H. and P.P. Feeny. 1971. Allelochemics: chemical interactions between species. *Science, 171*:757.

109. Wood, D.L. 1970. Pheromones of bark beetles. *In: Control of Insect Behaviour by Natural Products,* D.L. Wood, R.M. Silverstein, and M. Nakajima (eds), Academic Press, New York, p. 301.

110. Woodson, Jr., R.E. 1941. The North American Asclepiadaceae. I. Perspective of the genera. *Ann. Mo. Bot. Gard., 28*:193.

111. Woodson, Jr., R.E. 1954. The North American species of *Asclepias* L. *Ann. Mo. Bot. Gard., 41*:1.

Chapter Four

TOWARD A GENERAL THEORY OF PLANT ANTIHERBIVORE CHEMISTRY

DAVID F. RHOADES

Department of Zoology
University of Washington, Seattle, Washington

REX G. CATES

Department of Biology
University of New Mexico, Albuquerque
New Mexico

INTRODUCTION

Much research has been conducted in an attempt to fit the so-called plant secondary substances into the general framework of plant metabolism. Though some success has been achieved in this area, e.g., chlorogenic acid as a regulator of plant metabolic systems under stress[21], most such studies have met with a conspicuous lack of success. Possible metabolic roles for the plant alkaloids as intermediate

metabolites have probably received the greatest such attention and it was concluded at an early stage[44] that, since
a definitive metabolic role could not be assigned to alkaloids,
they best be considered as plant waste products. More recent
workers in the field of alkaloid biochemistry have concurred
with this evaluation[44,83] and the "waste product hypothesis"
has been expanded, principally by Muller (1969, 1970), to
include plant secondary substances in general.

On the other hand, a substantial literature has been
accumulating to implicate many plant secondary substances
as antibiotic agents in ecological interactions between plants
and their associated biota, including plant-plant interactions
(allelopathy)[76], plant-pathogen interactions[56] and plant-
herbivore interactions. Some secondary substances, particularly tannins and other phenolic compounds, have been
implicated as defensive in all three of these classes of
interactions. We will concentrate our attention on plant-
herbivore interactions, without inference as to the relative
importance of plant chemistry, to the three classes of
interactions.

Plant secondary substances that have been shown either
to have a negative effect on herbivore fitness (increased
mortality, lowered growth rates or fecundity), or to have a
deterrent effect on herbivore grazing activities include:
alkaloids, pyrethrins, rotenoids, long chain unsaturated
isobutylamides[43], cyanogenic glycosides[46], phytoecdysones
and juvenile hormone analogues[99], cardenolides and saponins[20],
sesquiterpene lactones[16,43], nonprotein amino acids[73], mustard
oil glycosides and isothiocyanates[29], oxalates, protoanemonin,
hypericin, fluoro-fatty acids, seleno-amino acids[48], 6-
methoxybenzoxazoline[10], gossypol[68], condensed tannin[33], phenolic resin and associated phenoloxidase[75] and proteinase
inhibitors of the soy-bean trypsin inhibitor type[98].

In other investigations some of these substances have
been found to exhibit zero effect on herbivore fitness[41,49]
or herbivore grazing patterns[65], and in some cases attractive
responses by some herbivores have been noted[89,95], leading to
the possibility of an ambivalent viewpoint as to the adaptive
significance of plant secondary substances in plant-herbivore
interactions. This viewpoint can be summarized as follows:
though in individual cases a positive effect on plant fitness
by repulsion of, or antibiotic action on, a particular

herbivore by a plant secondary substance has been noted,
cases of zero antibiosis, zero repulsion or indeed actual
attraction for other herbivores to these same substances,
negate the description of plant secondary substances as
defensive compounds. However, we subscribe to the viewpoint[25]
that cases of apparent zero or negative effect of secondary
substances on plant fitness, are best described in terms of
coevolution by specialized herbivores with the defensive
system resulting, in some cases, with the use by the herbivore of the substances as feeding cues and/or their
sequestration and incorporation into the herbivore's own
defensive system. In the total milieu of grazing and other
pressures on the plant, it is postulated that secondary
substances must have a net positive effect on plant fitness,
since if they did not, metabolic costs associated with their
production and sequestration should cause the elimination of
genotypes producing these substances from the plant population.

    This presentation is an attempt to bring together, into
one coherent framework, many facts and ideas from the literature and from our own work concerning attraction and
repulsion of herbivores, within and between plant distribution
of various classes of secondary substances, relative cost of
secondary substance classes to the time and energy budget of
the plant and notions of plant defense unrelated to chemistry
such as "escape in space and time" from herbivory. To set
the stage for the theoretical portion, we first present a
brief summary of our work with creosote bush.

ANTIHERBIVORE CHEMISTRY OF CREOSOTE BUSH

    Two species of creosote were studied[75], *Larrea tridentata*
(tetraploid), a dominant desert shrub of southwestern North
America, and *L. cuneifolia*, a South American desert species.
Creosote is characterized by the presence of large quantities
of diethyl ether soluble phenolic resin on the surface of
the leaves and terminal stem portions, ranging in concentration
for *L. tridentata* from 26% (d. wt.) for the spray tips, defined
as the first pair of immature folded leaves, to 11% (d. wt.)
for the rest of the leaves, here defined as mature leaves. For
L. cuneifolia the corresponding figures are 44% resin for tips
versus 15% for mature leaves. Over 90% of the resin is composed of phenolic constituents,the main component of which is
nordihydroguaiaretic acid (Fig. 1, V), a lignan catechol. The

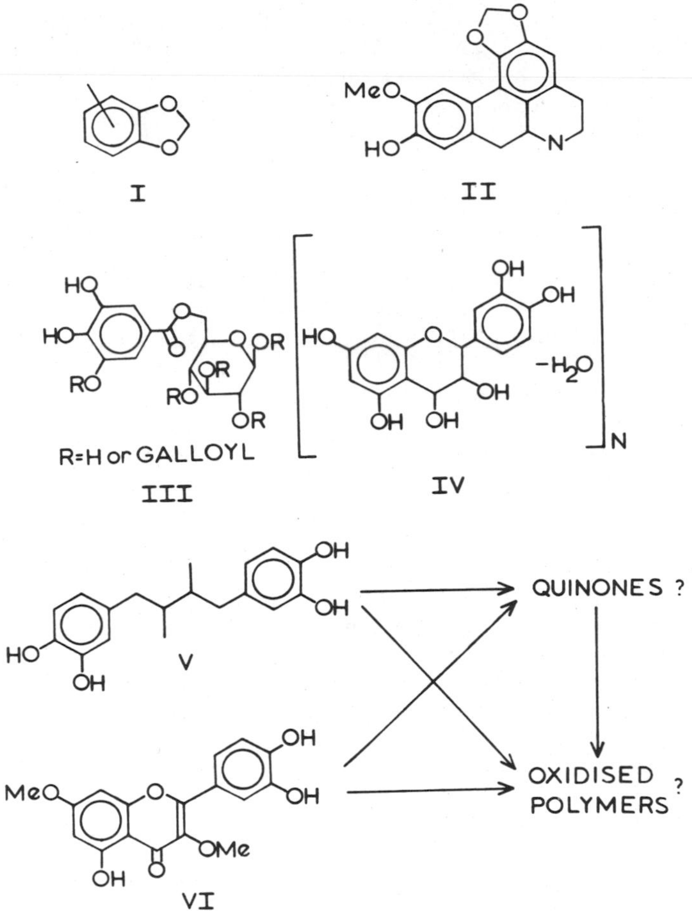

Figure 1. Structures of the methylenedioxyphenyl group I, actinodaphnine II, representative hydrolysable III and condensed IV tannins, nordihydroguaiaretic acid V and a representative *Larrea* leaf resin flavonoid 3,7-dimethylquercetin VI.

remainder is a complex mixture of partially O-methylated flavonoid phenolics (Fig. 1, VI)[23,82]. Grazing observations for three species of grasshoppers, a katydid and a geometrid moth larva on *L. tridentata* and a proscopiid grasshopper on *L. cuneifolia*, in the laboratory and in the field, showed

that most of the herbivores strongly preferred the less
resinous mature leaves. Palatabilities of mature leaves
and tips to the herbivores were compared (Table 1) as relative
palatability indices, defined as the percentage of mature
leaves present that were consumed during the experiment,
divided by the percentage of tips present that were consumed.
In this method, for equal palatability, tissues are eaten in
the same proportion as that in which they occur. A limited
number of field observations of grazing by the katydid *Insara
covilleae* (Tettigoniidae), a *L. tridentata* specialist, indicated (J. C. Schultz, personal communication) that this
herbivore may not discriminate between tips and mature leaves.
Removal of resin by diethyl ether wash followed by thorough
drying of the leaves at room temperature to remove traces
of diethyl ether and rehumidification, led to a randomization
of grazing by *Cibolacris parviceps* (Acrididae) and *Semiothisa
colorata* (Geometridae). In addition, extracted leaves were
more palatable than fresh unextracted leaves to *Cibolacris*
(P extracted/P fresh = 5.1), a generalized herbivore, but less
palatable than fresh leaves to *Semiothisa* (P extracted/P fresh
= 0.29), a *L. tridentata* specialist. In both cases P extracted and P fresh were significantly different from each other
at the 0.01 level (Student's t test). It was therefore concluded that the resin was repellent to the generalized herbivore *Cibolacris* at all concentrations, whereas to the monophagous *Semiothisa* it was repellent at high concentrations
but attractive at low concentrations.

A study of *L. tridentata* bushes either occupied or
unoccupied by *Ligurotettix coquilletti* (Acrididae), a terri-
torial grasshopper (D. Otte, unpublished), showed that the
preferred leaf tissue for *Ligurotettix* (the second leaf
below the tips) of occupied bushes had a significantly
(0.05, t test) lower average resin content (7.8 $\pm$ 2.4% d. wt.)
than unoccupied bushes (9.2 $\pm$ 2.1% d. wt.).

These results are consistent with a function for creosote
bush resin as an antiherbivore substance. Attraction of
Semiothisa to resin at low concentration suggests the use
of resin as a feeding cue by this specialized herbivore.

An investigation into the mechanism of resin action as
an antiherbivore substance stemmed from an initial observation that although *L. tridentata* resin exhibited a low
solubility in water it was quite soluble in dilute protein

TABLE I

Relative Palatabilities of mature leaves and tips of Creosote Bush to Insects

HERBIVORE	OVERALL DIETARY SPECIALIZATION	$\frac{\text{P MATURE}}{\text{P TIPS}}$*	RESIN $\frac{\text{MATURE}}{\text{TIPS}}$**	PLANT SPECIES
Cibolacris parviceps (Acrididae)	Generalist	81	0.40	*L. tridentata*
Semiothisa colorata (Geometridae)	Specialist	35	"	"
Ligurotettix coquilletti (Acrididae)	Fairly specialized	21	"	"
Bootettix punctatus (Acrididae)	Specialist	5	"	"
Insara covilleae (Tettigoniidae)	Specialist	~1	"	"
Astroma quadrilobatum (Proscopiidae)	Fairly specialized	3	0.35	*L. cuneifolia*

*$\frac{\text{P MATURE}}{\text{P TIPS}} = \frac{\text{\# Mature leaves eaten}}{\text{\# Mature leaves present}} \times \frac{\text{\# Tips present}}{\text{\# Tips eaten}} = $ Relative Palatability Index

**Dry weight comparison

All values of $\frac{\text{P MATURE}}{\text{P TIPS}}$ (except for *Insara*) are significantly (<0.01, Student's t test) greater than unity.

solutions, indicating that the resin interacted (complexed)
with proteins. Using an unbuffered diethyl ether/aqueous
extraction procedure it was shown[75] that although resin was
rapidly recovered in the ether layer during a resin/ether/
water extraction, resin was recovered much more slowly from
an ether/1.5% gelatin extraction. In addition, exhaustive
extraction of the proteinaceous layer with ether left
approximately 40% of the resin irreversibly complexed.
Partition of resin between diethyl ether and 1.5% solutions
of various solutes, buffered to approximately pH 7.0, showed
(Table 2) that the resin complexes with gelatin (degraded
collagen), casein (milk phosphoprotein) and edestin (hemp
seed protein) to a very similar degree. Complexing also
occurs with α–chymotrypsin (bovine endopeptidase) and with
starch, though more weakly than with the aforementioned
proteins. Disappearance of the interactions with gelatin at
8M urea concentration is consistent with stabilization of
the complex by hydrogen bonds. Lack of interaction with
hydrolyzed casein or glucose shows that stable complexes
are formed only with the polymeric forms of these substances.
The resin is functionally a tannin, though it does not fall
within the classic definition of a tannin. Whereas tannins
are hydrophilic polymeric phenols, that precipitate proteins
and starch, creosote resin is composed of lipophilic monomeric
phenols forming complexes that are quite water soluble. Like
a tannin, however, the resin inhibits proteolysis. Inhibition

Table 2. Partition of *Larrea tridentata* resin between 15 ml
of diethyl ether and 5 ml of a 1.5% solution of various
solutes in pH 7.0, 0.2 M sodium phosphate buffer.

Solute	pH	% Resin in Aqueous Phase
---	7.03	3.0 ± 0.9
Gelatin	7.02	72.9 ± 0.7
Casein	6.98	66.4 ± 0.8
Edestin	6.92	68.5 ± 0.7
α - Chymotrypsin	7.00	39.0 ± 1.0
Starch	7.01	16.5 ± 0.8
Gelatin/8 M Urea	7.60	2.7 ± 0.5
Hydrolyzed Casein	6.89	2.8 ± 0.5
Dextrose	7.00	2.4 ± 0.8

by *L. cuneifolia* resin of casein digestion was demonstrated[75] for both erepsin, a mixture of porcine proteolytic enzymes, and for the enzymes of *Astroma quadrilobatum* (Proscopiidae), an important *L. cuneifolia* herbivore[85]. Fig. 2 shows inhibition of casein/α-chymotrypsin and casein/erepsin digestion by *L. tridentata* resin.

We therefore propose that the mechanism of herbivore deterrence by creosote resin is basically the same as that claimed by Feeny (1970) for deterrence of winter moth larvae by the condensed tannin of the oak *Quercus robur*. On maceration of leaves by the herbivore the resin complexes with plant proteins, starch and perhaps also with herbivore digestive enzymes, to form complexes which are refractory to digestion. Creosote resin is a digestibility-reducing substance. The complexes are probably stabilized by hydrogen bonding between phenolic OH and, in the case of resin/protein complexes, peptide carbonyl, as has been shown to be the case for tannin/protein complexes[55].

A powerful phenoloxidase (PO) system present in the leaves of *L. tridentata* and *L. cuneifolia* may also be involved in herbivore deterrence[75]. As with resin concentrations, PO activity is higher in the young leaves than in the mature leaves. Introduction of PO plus creosote resin into casein/proteolytic enzyme systems shows an increase in inhibition of digestion over that obtained by introduction of resin alone. This enhancing effect of PO on digestibility-reduction by the resin may be due to the formation of reactive quinones (Fig. 1) which cross-link the proteins by covalent bonds, or alternatively, oxidative condensation of resin components may occur to give polymers of the condensed tannin type that are more effective digestibility-reducing substances than are the monomeric resin constituents.

WITHIN-PLANT DISTRIBUTION OF DIGESTIBILITY-REDUCING SUBSTANCES

An important question now arises. Why does creosote resin, which is of higher concentration in the young leaves than in the mature leaves, have an inverted within-plant distribution compared to the distribution of condensed tannin in oak[32], where tannin is of lower concentration in the young leaves than in mature leaves? This trend for tannin being more concentrated in the mature leaves than in the young

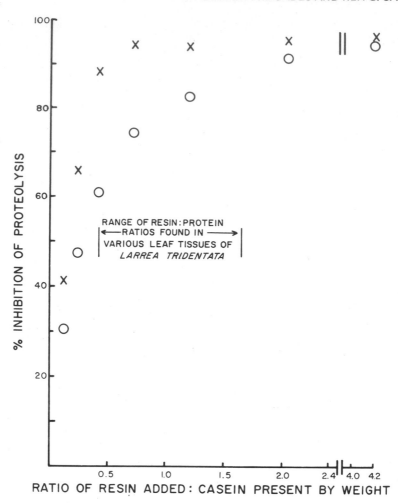

Figure 2.   Inhibition of casein proteolysis by the
diethyl ether soluble leaf resin of *Larrae tridentata*.   The
digestive system parameters were: 6 ml of 0.833% casein + 2
ml of 0.1% erepsin (O) (Nutritional Biochemical Corp.) or
0.0025% α–chymotrypsin (X) (type II, Sigma Chemical Co.),
enzymes and substrate dissolved in pH 7.0 0.2 M sodium
phosphate buffer, 2.0 hrs. digestion, 37°C, chloroform
antiseptic, termination by addition of 0.5 ml of trifluora-
cetic acid, free α-amino nitrogen estimated by the method
of Rosen (1957).[75]

leaves is also shown (Table 3) for bracken fern (*Pteridium*) and toyon (*Heteromeles*). Leaf protein digestibilities for young, intermediate and mature leaves within any given plant are best compared as tannin/protein and resin/protein ratios, since tissue protein (crude) contents, for oak, bracken and toyon are strongly and inversely dependent on leaf age.

The apparently divergent defensive strategy of creosote bush can be unified with that exhibited in the other plants, by a consideration of the effect of leafing phenology on tissue predictability and availability as a food resource to herbivores. The leafing pattern of oak, bracken, and toyon is epitomized by that of oaks. Oaks leaf out rapidly and, within a single plant, synchronously. Young oak leaves are available to herbivores for only a few weeks and are diffi- cult for herbivores to locate in both space and time[33]. Mature oak leaves, on the other hand, constitute a more predictable resource to herbivores. Creosote bush displays an unusual leafing pattern in which the youngest leaves are the most predictable resource. During moist periods leaves mature in sequence from the growing tips of the sprays. As the young leaves mature they are immediately replaced by still younger leaves at the tips. During dry periods, the leaves senesce and are shed, the oldest first, until under conditions of drought, only the youngest leaves remain. Thus, the youngest leaf age class is always present on the bush and sometimes is the only leaf age class present.

The generalization that emerges from the data in Table 3 is that the more predictable and available a leaf tissue is as a food resource to herbivores, the more heavily this tissue will be defended by digestibility-reducing systems.

WITHIN-PLANT DISTRIBUTION OF TOXINS

We will now examine the within-plant distribution of some other plant secondary substances that have been implicated as defensive against herbivores (Table 4). A natural division occurs between digestibility-reducing substances (Table 3), which act within the gut of the herbi- vore to reduce the availability of plant nutrients, and plant toxins (Table 4) which we define as substances which act on metabolic processes that are topologically internal

TABLE 3

WITHIN-PLANT DISTRIBUTION OF DIGESTIBILITY-REDUCING SUBSTANCES

| | LEAF AGE | | | | | | Plant Species | Growth Form | Ref. |
	Young		Intermediate		Mature				
	T	T/P	T	T/P	T	T/P			
TANNINS	0.5	0.017	1.2	0.085	5.0	0.35	*Quercus robur*	WP	32
	2.0	0.083	2.5	0.21	4.5	0.91	*Pteridium aquilinum*	HP	54
	7.0	0.5	10.5	1.7	11.0	1.7	*Heteromeles arbutifolia*	WP	22
	R	R/P	R	R/P	R	R/P			
PHENOLIC RESIN	26.2	1.64	12.3	0.70	8.4	0.52	*Larrea tridentata*	WP	75
	44.0	3.35	19.7	1.44	12.2	0.88	*L. cuneifolia*	WP	75

T = Tannin content   R = Resin Content   P = Protein (crude) content
WP = Woody Perennial   HP = Herbaceous perennial   (Dry weight percentages)

TABLE 4

WITHIN-PLANT DISTRIBUTION OF TOXINS (% DRY WEIGHT)

TOXIN	Leaf Age Classes			PLANT SPECIES	Growth Form	Ref.
	Young	Intermediate	Mature			
Alkaloids	1.5	1.3	1.0	Prosopis velutina	WP	17
"	1.8	1.7	0.9	P. chilensis	WP	17
"	1.6		0.7	P. flexuosa	WP	17
"	0.4		0.01	P. torquata	WP	17
"	0.39	0.34	0.12	Atropa belladonna	HP	44
"	H		L	Colchicum autumnale	HP	18
"	H		L	Solanum tuberosum	HP	18,84
"	H		L	Lupinus spp.	HP	67
"	H	I	L	Conium maculatum	B	66
"	H		L	Cicuta virosa	HP	66
"	H		L	Delphinium barbeyi	HP	48
"	H		L	Sarothamnus scoparius	WP	88
"	H		L	Datura stramonium	A	12
Cyanogenic Glycosides	4.8		2.5	Heteromeles arbutifolia	WP	22
HCN	0.03	0	0	Pteridium aquilinum	HP	54
"	0.45		L	Ximenia americana	WP	12

TABLE 4 - Continued

TOXIN	Leaf Age Classes			PLANT SPECIES	Growth Form	Ref.
	Young	Intermediate	Mature			
HCN	H		L	Acacia chiapiensis	WP	74
"	H		L	A. farnesiana	WP	74
"	H		L	Prunus spp.	WP	48
"	H		L	Pangium edule	WP	94
"	H		L	Sorghum halepense	HP	12
"	H		L	S. vulgare	A	12
"	H		L	Plantanus spp.	WP	69
"	H		L	Lotus corniculatus	HP	45
Mustard Oil Glycosides	0.12		0.07	Brassica spp.	A	97
Total Isothiocyanates	0.009		0.005	Brassica oleracea*	A	47
Non-protein Amino Acids						
Pipecolic acid**	H		L	P. velutina	WP	17
N-methyl-L-serine**	0.8	0.45	0.25	Dichapetalum cymosum	WP	27
Hypericin	0.060	0.035	0.026	Hypericum hirsutum	HP	72

Exceptions to the trend of decreasing toxin concentration with increasing leaf age

Nicotine***	1.00	2.92	5.36	Nicotiana tabacum	A	44
Saponins	0.47		0.70	Dioscorea tokoro	HP	1
"	L		H	Agrostemma githago	A	18
Protoanemonin	L		H	Ranunculus spp.	A,HP	48

TABLE 4 – Continued

*   Average of 11 commercial varieties, % fresh weight
**   Not definitely known to be toxic
***   Relative dry weight values

Growth Form

A = Annual; B = Biennial; HP = Herbaceous perennial; WP = Woody perennial
H = Highest concentration; I = Intermediate concentration; L = Lowest concentration

to the herbivore. Whereas digestibility-reducing substances
are of higher concentration in predictable leaf tissues,
toxins are in most cases of higher concentration in the
ephemeral leaf tissues (the youngest leaves in cases tabulated
This general trend for toxins to be concentrated in young
actively growing tissue has been previously noted[59]. Exceptions to the rule are provided by species of *Nicotiana*,
Dioscorea, *Agrostemma*, and *Ranunculus*. Ishaay and Birk (1965
have shown that saponins can function as inhibitors of
proteolytic enzymes and thus a dual function for these compounds as both toxins and digestibility-reducing substances
is likely. Similarly, such a dual function can be postulated
for the highly reactive and electrophilic N-alkylating agent,
protoanemonin, since Henderson (1971) has shown that the
catalytic activity of α-chymotrypsin is destroyed by N-
alkylation. Thus, of the four exceptions to the trend of
decreasing toxin concentration with increasing leaf age,
three can be explained because the compounds may have a dual
function. A dual function is also probable for mustard oil
glycosides since the active products of these compounds,
isothiocyanates, have long been known to combine with amino
groups in proteins, a reaction which has given rise to the
Edman amino end-group degradation of proteins[100]. Reduction
of food assimilation efficiency by a dietary mustard oil
glycoside, sinigrin, has been observed in the case of the
black swallowtail butterfly[29]. Mustard oil glycosides,
however, appear to be of highest concentration in the young
leaves. The distributional data for nicotine, in tobacco,
illustrates an additional possible correlation between
tissue age and toxin concentration. Nicotine distributional
data were obtained by taking single samples of leaves from
seedlings, immature plants, and mature plants[44], whereas
most of the data in Table 4 were obtained by sampling young,
intermediate, and mature leaves from plants of the same age.
It is thus possible that both leaf age and plant age are
important to leaf toxin distribution, at least for annual
plants and perhaps also for perennials, during their first
few years of growth. Alternatively, the exceptional distribution of nicotine in cultivated tobacco may be the result
of agricultural selection.

Seasonal distribution of both a toxin and a digestibilit
reducing system, in the same plant, is illustrated for bracke
fern in Fig. 35[4]. In the spring, the emerging fiddle-heads
contain high levels of cyanogenic glycosides which rapidly

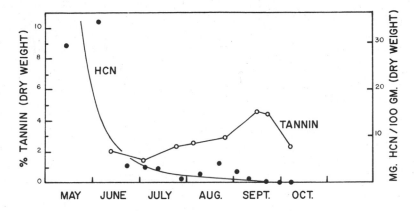

Figure 3.   Seasonal variation in the cyanide and tannin contents of bracken fern (*Pteridium aquilinum*) fronds[54].

decrease as the fronds mature.   Conversely, tannin content is low in the young tissue, increasing to maturity and decreasing as the fronds senesce. *Heteromeles* (Tables 3 and 4) shows a similar seasonal distribution for cyanogenic glycosides and tannin.   From our own work[17], of four *Prosopis* species examined, (Table 4) all contain the highest concentration of alkaloids in the young leaves.   In addition, the 70% acetone leaf extracts of all four *Prosopis* species contain digestibility-reducing substances, though the within-plant distributions of these substances are not known: *P. chilensis*, *P. alba* and *P. flexuosa* (precipitate obtained with 1.5% gelatin); *P. velutina* (no precipitate obtained with gelatin, but strongly inhibits digestion of casein by erepsin).

We believe that many plants, perhaps most plants, contain a "double or multibarreled" defensive system, with tissue predictability and availability as a food resource to herbivores being the main determinant of the defense type employed. From data so far available (Tables 3 and 4) it appears that the toxic defense mode is characteristic of ephemeral tissues, whereas, the digestibility-reducing mode is characteristic of predictable tissues.

PROPERTIES OF PLANT TOXINS

      The characteristics of a "good" toxin, from the plant's
viewpoint, should be that it is cheap to produce, highly
active in small amounts against physiological systems found
in animals but not in plants (to minimize autotoxicity
problems), have such physiochemical properties that the
compound can gain easy entry into the herbivore's body and
target cells, and, once in the herbivore, the toxin system
should be resistant to degradation or deactivation by
detoxification mechanisms. Known plant toxins satisfy these
criteria well. They are generally present and active in
small quantities, often less than 2% of plant material by dry
weight (Table 4). Many alkaloids, pyrethrins, and rotenoids
are nerve poisons[43] while cardiac glycosides (cardenolides)
and saponins[40] interfere with muscle action. Alkaloids and
oxalates interfere with liver and kidney function[15,48].
Insect hormone analogues[99]have a highly specific action
against insects. All these substances should present
minimal autotoxicity problems to the plants. Toxins active
against metabolic systems found in both animals and plants,
i.e., mustard oils[31], cyanide[46], protoanemonium[39] are
sequestered in the plant as inactive derivatives from which
the active toxins are released upon tissue damage.

      To gain access to the site of action in a herbivore,
toxins must possess properties that allow facile passage
across cell membranes. The rate of diffusion of molecules
across the unit membrane is inversely dependent on both
molecular size and molecular polarity[93]. Small molecules
such as oxygen, carbon dioxide and water, diffuse rapidly
across cell membranes, no matter what their polarity,
whereas large molecules such as biopolymers have low or
zero diffusion rates. For substances of intermediate
molecular weight (~100-1000), diffusion rates are controlled
mainly by molecular polarity or, to a first approximation,
by lipophilic nature, a phenomenon which is due mainly to
the lipoidal core of the membrane. Lipophilic substances
diffuse rapidly whereas hydrophilic substances diffuse
slowly. Hydrophilic substances of metabolic importance,
e.g., sugars and amino acids, cross the membrane either by
active transport or facilitated diffusion[93]. Plant toxins
should, therefore, be generally of low molecular weight
(most are below 500) and possess lipophilic properties,
which most of them do.

Exceptions do occur. Biopolymers satisfy neither the criterion for molecular size nor that for lipophilic nature, and although many potent biopolymeric toxins such as snake venom components are known, these substances are not expected as plant toxins unless some special adaptation to allow entry into the animal cells is also present. For instance the occurrence of oxalate and silica raphides[81] in the plant tissues may be such an adaptation, to aid toxin penetration, by abrasion of intestinal surfaces, or plant trichomes may inject the toxins into intestinal or external tissues of the herbivore[57]. Alternatively polypeptide toxins may be adapted to digest their way across cellular membranes, which may explain why the phytotoxins (a group of intermediate molecular weight range polypeptides) show a rapid diffusion rate across cellular membranes, in spite of their molecular size and polarity[48]. Toxins with detergent properties, e.g., saponins and cardenolides, may also diffuse by membrane disruption. The toxic, non-protein amino acids, found in many members of the Leguminosae[11], have high polarity. These compounds probably enter the herbivore's body by processes in which carrier molecules "mistake" these substances for normal dietary amino acids.

Resistance to degradation of plant toxins by herbivore detoxification systems, may be accomplished by the use of synergistic substances. Resistance of insect pests to a variety of chemically unrelated commercial insecticides can be reduced by simultaneous application of synergistic compounds containing the methylenedioxyphenyl (MDP) group (Fig. 1, I), such as piperonyl butoxide. These substances act by inhibition of insects' microsomal oxidase detoxification mechanism, in some unknown fashion[61].

Plant secondary substances containing the MDP group are common[60]. A role for any of these substances as synergistic agents for toxins present in the plant has not, to our knowledge, been established, but such a role has been postulated[50] for sesamin (an MDP compound) which occurs together with pyrethrins in pyrethrum flowers. In some cases, e.g., actinodaphine (Fig. 1, II), the MDP group occurs in fusion with an alkaloid nucleus. This leads to the speculation that in compounds of this type both active toxic principle and synergistic agent are contained within the same molecule.

PROPERTIES OF PLANT DIGESTIBILITY-REDUCING SYSTEMS

Digestibility-reducing substances on the other hand have properties which are quite different from those shown above for plant toxins. Since they are not required to cross cellular membranes, digestibility-reducing substances can be large or small molecules, of hydrophilic or lipophilic nature. The only requirement is that they disrupt digestive processes.

There are two types of plant protein digestibility-reducing substances known to date. The first type, which includes the hydrolyzable (Fig. 1, III) and condensed (Fig. 1, IV) tannins and creosote bush resin (Fig. 1, V and VI) are generalized protein complexing agents, which disrupt digestion by reducing availability of substrate peptide groups to the digestive enzyme and possibly by complexing with the enzyme itself. Generalism of action is observed, since the sole requirement for complex formation is the presence of peptide groups in the substrate. In contrast to plant toxins, generalized protein complexing agents are often present in large quantities (to over 60% dry weight) in the tissues in which they occur (Table 3)[14].

The second class of protein digestibility-reducing substances known to date have been found in a variety of edible crop plants: wheat, corn, barley, oats, rye, peas, beans, peanuts, potatoes, and beets[53,98]. These substances are specific inhibitors of proteolytic enzymes. Research has centered on the edible portions of plants, but in certain cases the inhibitors have been shown to be present in the leaves as well. The inhibitors which typically constitute less than 2% plant tissue, dry weight, are themselves polypeptides or small proteins which act by stoichiometric combination with the digestive enzyme, either at the active site or an allosteric site, and are often highly specific for a particular enzyme type. For instance, the soybean inhibitor is active against the endopeptidases, trypsin and chymotrypsin, but inactive against exopeptidases. The rapid denaturation of these inhibitors at 100°C is highly significant to their presence in human foodstuffs.

Inhibition of all enzyme-dependent digestive processes, by complex formation with the enzyme, is to be expected for tannins and creosote bush resin. However, it is conceivable

that adaptation of digestive enzymes, particularly in highly
specialized herbivores, could render their enzymes less
susceptible to complex formation. The lower degree of
complex formation between creosote resin and bovine α-
chymotrypsin than with other proteins (Table 2) may be an
example. A possible mechanism for this effect may be the
geometrical protection of peptide groups in the enzyme
molecule. Of course, adaptation of bovine enzymes specifically against the action of creosote resin is not to be
expected, since creosote is not a natural dietary component
for cattle. However, adaptation against generalized protein
complexing agents in grasses[9] could have resulted in "pre-
adaptation" for bovine enzymes against creosote resin. For
these reasons, we think that digestibility-reduction by
complex formation merely with the digestive enzyme should be
a less effective defense than complex formation with the
substrate or with both enzyme and substrate. For instance
with soybean trypsin inhibitor or other specific endopeptidase
inhibitors in crop plants described above, it seems likely
that adaptive modification of the trypsin-like enzymes of
specialized herbivores to a form that is not recognized by
the inhibitor, could render immunity to these herbivores.

Tannins[60] and creosote resin (Table 2) complex with
starch, thus starch digestibility-reduction by both enzyme
and substrate complex formation is to be expected, for these
substances. Inhibition of starch digestion by condensed
tannin has been demonstrated by Feeny (1970). Tannins
interact strongly with cellulose as shown by their low
or zero Rf values on paper chromatography[77], so inhibition
of cellulose digestion by both enzyme and substrate complex
formation is to be expected for tannin. Interaction between
creosote resin and cellulose has not been studied.

Another plant polysaccharide digestibility-reducing
system is known. Bacon and Cheshire (1971) have shown that
2-0-methylxylose and 2-0-methylfucose residues incorporated
into the chains of hemicellulose and other polymeric
carbohydrate leaf fractions of ash (*Fraxinus excelsior*),
horse chestnut (*Aesculus hippocastanum*), oak (*Quercus* spp.),
poplar (*Populus* spp.), sweet-chestnut (*Castanea sativa*),
sycamore (*Acer pseudoplatanus*), beech (*Fagus sylvatica*) and
in two marine flowering plants *Posidonia australis* and
Zostera marina, render the carbohydrates refractory to
enzymatic digestion by snail crop juice. Duff[7] has suggested

that the presence of methylated sugar residues might also
hinder biological degradation and cause the observed buildup of such methylated sugar rich polysaccharides in soil.
In addition 2-0-methylxylose and 2-0-methylfucose have been
shown to be leaf carbohydrate constituents in plum (*Prunus*
spp.)[3,4], lucerne (*Medicago sativa*)[6] and sisal (*Agave
sisalana*)[5].

The acidic leaf and wood oleoresins of conifers have
been implicated as defensive against insects. Most notably,
protection against bark beetles by mechanical "pitching
out" of the insects by wood resin has been established[91].
In addition, the volatile terpenoid fraction of conifer
resins has been shown to be toxic to a variety of insects[90].
The main constituents of conifer resins, however, are non-
volatile mono-, di-, and tri-terpene carboxylic acids, which
occur together with smaller quantities of fatty acids and
neutral non-volatiles[2,28]. Similarly, the acidic resins of
many tropical angiosperms, in particular tropical members of
the Leguminosae and Dipterocarpacae, are terpenoid based
and contain terpene carboxylic acids as major constituents[52].
We consider it likely that acidic resins are involved in
digestibility-reducing systems, in addition to their
mechanical and toxic protective properties. Partition of
the diethyl ether soluble leaf resin of *Cedrus deodora*
(Cupressaceae), between diethyl ether and 1.5% gelatin,
initially indicated that the resin was a protein complexing
agent. However, it was later found that if the aqueous
phase was carefully buffered, so that the proteinaceous
and control aqueous phases had similar pH (cf. Table 2), the
interactions largely disappeared. In addition, zero inhibition of a glucose/glucose-oxidase system was observed for
exuded *Cedrus deodora* wood resin in contrast to inhibition
of 60.1% and 58.5% obtained for *Larrea divaricata* (creosote)
resin and gallotannic acid respectively. Resins and gallotannic acid were added to the system in a quantity equal
to the weight of glucose oxidase present. It may thus be
tentatively concluded that neither the diethyl ether soluble
leaf resin nor the exuded wood resin of *Cedrus* are generalize
protein complexing agents of the tannin and creosote resin
type. This does not exclude these substances from participation in digestibility-reducing systems, however. Possibly,
acidic resins act in conjunction with phenolic constituents
not present in large quantities in the exudable resin or
diethyl ether extracts but nonetheless present in the

tissues[28], to form complexes with proteins or carbohydrates.
Free carboxyl groups in some hydrolysable tannins have
been shown to be important to complex stability[55]. Alternatively, acidic resins may act as "anti-buffering" agents.
Stability of protein/tannin complexes decreases rapidly above
~pH 8[55]. The same is true for protein/creosote resin
complexes. Alkaline gut pH has been cited as an adaptation
of the winter moth to circumvent the action of dietary hydrolyzable tannins[33] and high gut pH is quite common, especially
for lepidopteran herbivores. In the presence of large
quantities of acidic resins it should be very difficult or
costly for herbivores to buffer their gut pH into the
alkaline range.

BETWEEN-PLANT DISTRIBUTION OF TOXINS AND DIGESTIBILITY-REDUCING SUBSTANCES

Table 5 summarizes data thus far assembled, concerning
the occurrence of toxins and digestibility-reducing systems
in annual, herbaceous perennial, deciduous woody perennial,
and evergreen woody perennial plants. Data for tannin
distribution were calculated from Bate-Smith and Metcalf
(1957). Plants were recorded as tanniniferous, if they
gave either a positive histochemical reaction with iron salts,
or if a positive leucoanthocyanin reaction was observed.
Condensed tannins characteristically give a positive leucoanthocyanin reaction, but it should be pointed out that a
positive reaction is also to be obtained from all flavan-3,4,-
diol monomers, substances which have no true tanning properties. A division of plant species, into two groups, based
on tannin occurrence is apparent (Table 5). At one extreme
are the non-woody species: annuals (17% of species tanniniferous) and herbaceous perennials (14%), and at the other
are the woody species: deciduous woody perennials (79%) and
evergreen woody perennials (87%). This association between
the woody perennial habit and tannin occurrence has been noted
by Bate-Smith and Metcalf (1957), who in a discussion of this
phenomenon concluded that a possible explanation for this
correlation, namely a function for tannins as percursors in
the lignification process, was unlikely on biochemical grounds.
An alternative explanation, which we favor, is that tannin
occurrence is correlated to the predictability and availability
of the plants as a food resource to herbivores, in a manner
similar to that shown for the within-plant distribution of

Table 5. Distribution of plant toxins and digestibility-reducing substances among plants of various growth forms.

	A	HP	WPD	WPE	REF.

Digestibility-
  reducing Systems
   Non-specific:					

Tannins	17%(30)	14%(182)	79%(211)	87%(78)	8
Phenolic Resins	?	?	?	*	75
Specific	*	*	?	?	98
Refractory					
Carbohydrates	?	*	*	*	3-7
Toxins	*	*	*	*	48
Alkaloids					
(North America)	30%	------------20%----------			58

* = Present; A = Annuals; WPD = Woody Perennials (Deciduous)
? = Data not available; HP = Herbaceous Perennials; WPE = Woody Perennials (Evergreen)

tannins (Table 3). Occurrence of tannin-like phenolic resin in creosote bush, an evergreen woody perennial, is consistent with this correlation.

Specific proteinase inhibitors have, so far, been found only in annuals and herbaceous perennials, and thus seem to constitute the defense of ephemeral plants, corresponding to the generalized digestibility-reducing systems of predictable plants. That many ephemeral plants do indeed contain digestibility-reducing systems is tentatively indicated by preliminar‑ results (Rhoades and Cates, unpublished) which show that the 70% aqueous acetone leaf extracts of desert annuals (14 spp) contain proteolysis-reducing principles, the activity of which is no less on average, than that found for desert perennials (12 spp) when measured in the model casein/erepsin proteolytic system. It should be emphasized here that porcine erepsin is naive with respect to all the inhibitors involved in this study, since arguments have been made above that specific inhibitors should be less effective against co-evolved enzyme systems than should generalized inhibitors.   Refractory carbohydrates have so far been found only in perennial plants.

Toxins are known to be present in all plant groups. Levin (1975) has estimated that, for North American species, approximately 30% of the annuals and approximately 20% of the perennials contain alkaloids, indicating that perhaps annuals may rely more heavily on this toxic defense mode than do perennials.

To summarize, both within-plant (Tables 3 and 4) and between-plant (Table 5) distributions of toxins and digestibility-reducing substances, indicate that the ephemeral leaf tissues (youngest leaves in most cases) of all plants are defended primarily by toxic constituents. The predictable leaves (mature leaves in most cases) of predictable plants (woody perennials) are defended primarily by generalized digestibility-reducing systems. Within-plant distribution of specific digestibility-reducing substances found in ephemeral plants (annuals and herbaceous perennials) is not presently known, but we think it likely that they will be found to be of higher concentration in predictable leaf tissues than in ephemeral tissues, for these plants, by analogy with the distribution of generalized digestibility-reducing substances in predictable plants. Similarly, within-plant distribution of refractory polysaccharides is unknown, but since these substances appear to be characteristic of the more predictable plant groupings, we think it likely that they will also be characteristic of predictable leaves.

REASONS FOR THE OBSERVED DISTRIBUTIONS OF PLANT DEFENSIVE SYSTEMS

To clarify the remainder of the discussion, we will refer to herbivores mainly in terms of two groupings: generalists and specialists, with full realization that there is a gradation of herbivore type from extreme polyphagy to strict monophagy. We are thus referring to the opposite ends of this continuum.

We propose that ephemeral plants and ephemeral plant tissues escape from their herbivores in space and time, to a larger extent than do predictable plants and plant tissues,

since it is difficult for herbivores to locate ephemeral
food resources.

The average annual mortality of winter moth between the
time that the overwintering eggs are laid in the late fall
and a new population of larvae is established in the spring
is approximately 90%, largely caused by asynchrony between
larval hatching and bud-burst, demonstrating the importance
of escape in time, for young oak leaves[33]. If the larvae
hatch too early they cannot enter the buds; if they hatch
too late the young leaves have already matured sufficiently
that they are no longer a suitable food due to decreasing
nitrogen and increasing tannin content. Similarly, escape
in time for young leaves has been demonstrated for black
spruce/spruce budworm[36], balsam fir/spruce budworm[26], and
Quercus robur/Tortrix viridana[86]. Escape in space for an
ephemeral plant has been demonstrated by Orians (in press)
who showed that clumped individuals (average conspecific
interplant distance <1.5m) of the desert annual *Ibicella
parodii* (Martyniaceae) experienced much higher grazing
pressure and grazing induced mortality (68% killed) from a
specialized grazer, than isolated individuals (>10m, 33%
killed).

Furthermore, we think that escape in space and time is
more effective against specialist herbivores than against
generalist herbivores because specialist herbivores have no
alternative food source. Specialist herbivores must allocate
time and energy to a search for their host plant. The more
ephemeral the resource, the greater will be herbivore
mortality during the search. For a generalist herbivore, on
the other hand, the predictability and availability of any
individual resource is of less consequence, since a generalist
can opportunistically utilize whatever resource happens to
be available. In other words, predictability of resource
should select for specialism in the herbivore and ephemerality
of resource should select for generalism, *all other things
being equal* (Fig. 4; for explanation see remainder of text).

Since ephemeral plants and tissues escape from herbivores
to a greater extent than do predictable plants and tissues,
ephemeral plants and tissues should utilize defenses, including
chemical defenses (Fig. 5) which are less costly to produce
and store, but more easily circumvented, than those utilized
by predictable plants and tissues. Escape should be more

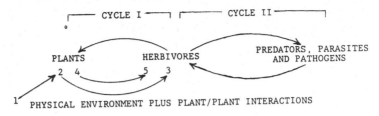

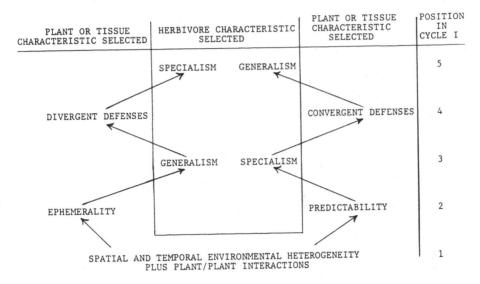

Figure 4. Coevolution between plants and herbivores.

effective against specialists than against generalists.
Therefore, the defenses evolved in ephemeral plants and
tissues should be directed particularly against generalist
herbivores and this should give, and has given, rise to a
divergent system of chemical defenses in such plants and
tissues (Fig. 4). Selective forces acting on generalist
herbivores should render these herbivores most capable of
accommodating the "average defensive chemistry" of their
host plant range. Thus plant species or individuals, whose
defensive chemistry deviates most widely from this average,
will be at a selective advantage compared to plants of more
common chemical type. This mode of disruptive selection
has been termed apostatic selection[19,70]. The extremely
diverse array of toxic defensive chemicals found in ephemeral

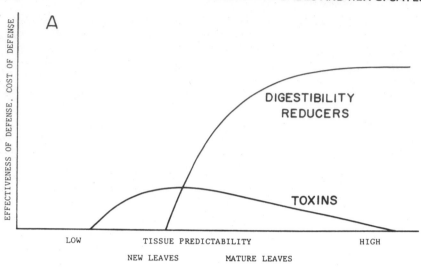

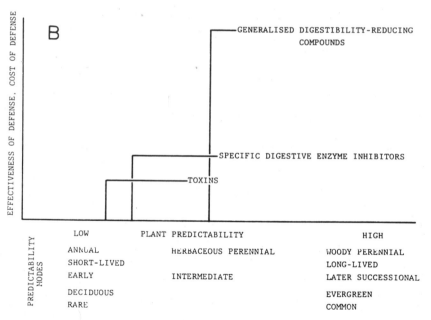

Figure 5. Postulated cost and effectiveness of chemical antiherbivore defenses of leaves, as a function of leaf (A) or plant (B) predictability to herbivores.

plants and ephemeral tissue is, we think, the result of this divergent selective process.

Conversely, the defenses of predictable plants and tissues should be directed particularly against specialist herbivores. In the extreme case where each predictable plant species (or predictable tissue) is experiencing grazing by its own specialist herbivore, the defense evolved in any one plant species (or tissue) will not be affected by the defenses evolved in other plant species (or tissue) in the community, in contrast to the condition for evolution of ephemeral plant defenses. We can therefore expect parallelism or convergence of chemical defense for predictable plants and tissues (Fig. 4). We propose that the defensive system, that has been converged upon, is the digestibility-reduction mode of defense. In contrast to the plethora of known plant toxins, the diversity of known digestibility-reducing systems is much lower. Of the major class of digestibility-reducing substances, namely tannins, there are only two main types: condensed and hydrolysable, hydrolysable tannins being further subdivided into gallo- and ellagitannins. A similar mechanism of digestibility-reduction involving phenolic hydroxyl hydrogen bonding is displayed by all tannins.

Specific proteinase inhibitors, present in ephemeral plants, and the postulated defense for the predictable tissues of these plants, may show a chemical diversity, which is intermediate between that shown for toxins and generalized digestibility-reducing substances.

We propose that toxic substances represent a "lower level" of chemical defense than do digestibility-reducing systems both in terms of cost to the time and energy budget of the plant, and in terms of effectiveness as a deterrent to herbivores (Figure 5). Since escape in space and time should be greater for ephemeral plants and tissues, the total level of all other defenses optimal for those plants and tissues should be less than those optimal for predictable plants and tissues, *all other factors being equal*. In a total description of the expected defensive levels for all plant tissues (e.g., flowers, seed, leaves, roots and structural tissues) factors other than tissue predictability must be taken into account. The most important of these is probably the loss in fitness accompanying loss of the tissue under consideration, or, in other words, the value of the particular tissue to the plant. To minimize the effect of these other variables the discussion has been restricted and will continue to be restricted to a single tissue type--leaves.

Toxic substances are cheaper for the plant to synthesize and sequester than are digestibility-reducing substances, for the following reasons:

(1) Toxins are present and active in small amounts, commonly less than 2% tissue dry weight (Table 4), whereas tannins are present and are effective only in high concentrations that often approach or exceed tissue protein concentration (Table 3). This argument does not apply to specific digestibility-reducing systems which are effective at concentrations similar to those of toxins, but see (3) below.

(2) Toxin activity is often restricted to metabolic systems, such as the neuromuscular system found in animals but not in plants. Where general metabolic toxins are employed (e.g., cyanide, isothiocyanates) the toxins are sequestered as inactive derivatives from which the active moieties are released, on tissue damage. Sequestration costs for toxins should therefore be minimal compared to those for tannins and resins, which, by virtue of their generalized protein-complexing properties must be completely segregated from cellular contents. Furthermore, sequestration of non-specific generalized protein-complexing agents may be limited in rapidly growing and differentiating tissues because dividing cells do not have large well organized vacuoles for compartmentalizing these compounds[30]. Again, the argument does not apply to specific digestibility-reducing substances, for which sequestration costs would be expected to be minimal.

(3) By virtue of their low molecular weight (commonly less than 500) many toxins can be transported, in the vascular system, to the deposition site from elsewhere in the plant. This is true of alkaloids which are often synthesized in the roots[44]. Specific digestibility-reducing substances and tannins cannot cross cellular membranes, due to their high molecular weight, and must be synthesized at the deposition site. The metabolic diversion resulting from on site synthesis of these substances, in ephemeral tissues, should slow growth rates and thus decrease the escape component of defense.

To exert its activity a toxin must traverse the gut membrane system of the herbivore, travel to the target cells, enter the cells, and finally, disrupt the metabolism of the cells[51]. "Detoxification" of the system, by the herbivore, can occur at any one of these levels. Physiological adaptations of the digestive system of herbivores may effectively reduce passage of the toxins across gut membranes. The low gut pH of tobacco herbivores[87] may be such an adaptation since protonated alkaloids, due to their polar nature, should exhibit much lower diffusion rates across membranes than the unprotonated form[93]. Herbivores, including both vertebrates[13], and insects[89], have a wide ability to detoxify foreign substances, by a combination of metabolism and excretion. Specialist herbivores have evolved detoxification mechanisms, of particularly high potency, against toxic substances present in their host plants. For example, the tobacco hornworm, *Manduca sexta*, can rapidly excrete large injected doses of nicotine[87]. Polyphagous (generalist) herbivores, on the other hand, deal with a broader spectrum of toxic substances and should have evolved a correspondingly broader based, lower intensity detoxification ability than have specialist herbivores. Kreiger et al. (1971) measured the microsomal activity of a variety of lepidopteran larvae against a synthetic, chlorinated insecticide. A gradation of activities was found in which Lepidoptera utilizing 11 plant families or more exhibited an average activity over 10 times that of Lepidoptera utilizing one plant family. This indicates, to us, that the more generalized the feeding habit of the insect, the higher is the likelihood that its microsomal system will show activity against a random foreign substance. The complementary experiment, measurement of activity against substances normally found in the food plants or of a similar chemistry, should show the opposite result.

Toxins may therefore confer little protection against specialist herbivores but should provide some protection against generalists. In fact, specialized herbivores in many cases utilize toxic substances, in their host plants, as location and feeding cues and exhibit an attractive response to these substances[72,88,95]. Sequestering of plant toxins by specialist insects for use as defensive or pheromonic substances is also common[24,80,101] (Roeske, this volume).

Digestibility-reducing systems, on the other hand, should provide protection against both generalist and specialist herbivores for the following reasons:

(1) The mode of action of generalized digestibility-reducing substances is to immediately bind proteins and other plant nutrients when the cells are disrupted. Chemical reactions begin before the leaves enter the gut, and consequently, there are fewer pathways by which the herbivore can "interfere" with the system, when compared to those available against the toxic system. For nutrients possessing low intrinsic digestibility, e.g., refractory polysaccharides, the system is in operation before the arrival of the herbivore

(2) Tannins, and creosote bush resin, are very general and nonspecific protein complexing agents. Tannins complex with and denature a wide variety of digestive and non-digestive enzymes[71]. Chemically a plant "tannin" extract consists of a heterogeneous mixture of polymeric phenols of varying chain lengths. These polymers attach themselves to protein chains by hydrogen bonding between phenolic hydroxyls and peptide groups. It is the very heterogeneous and nonspecific nature of tannin action which may render herbivore digestive enzyme adaptations difficult. We have found that the proteolytic enzymes of the grasshopper, *Astroma quadrilobatum*, a creosote bush herbivore, are inhibited no less by creosote bush resin than are porcine enzymes, lending weight to the above hypothesis[75]. This argument, for general inhibitory activity as a counter to adaptations of digestive function in the herbivore, does not, of course, apply to specific digestive enzyme inhibitors.

Specific proteinase inhibitors exhibit properties which, we think, place these substances in a position that is intermediate for both cost and effectiveness between toxins and generalized digestibility-reducing substances (Fig. 5b). Like tannins, these substances act within the gut and probably must be synthesized *in situ*, in the tissues in which they occur. Like toxins, specific proteinase inhibitors are active and present in small quantities and should represent a minimal sequestration problem.

Ephemeral leaf tissues, then, are defined primarily by cheap divergent toxic chemical systems effective against generalized herbivores, but of low effectiveness against specialized herbivores (Figs. 4, 5a). Predictable tissues are defended, primarily, by more costly convergent

digestibility-reducing systems effective against both
specialist and generalist herbivores. The postulated scheme
for the distribution of chemical defenses among plants of
various predictability status is depicted in Fig. 5b.
Ephemeral plants utilize toxins and specific digestibility-
reducing systems in their ephemeral and more predictable
tissues respectively. The corresponding defenses of
predictable plants are toxins and generalized digestibility-
reducing systems. Annual plants are less predictable and
available, as a food resource, to herbivores than woody
perennials, mainly because of their short lifetime coupled
with their occurrence at a different physical location each
year. Other predictability gradients envisaged are: seral
stage, deciduous versus evergreen and rare versus common.
Thus in this scheme the most highly defended leaf tissue
should be the mature leaves of a long-lived, common, evergreen, climax woody perennial.

EFFECT OF PLANT DEFENSIVE CHEMISTRY ON TISSUE PREFERENCE AND
DIGESTIBILITY BY HERBIVORES

    The net influence of high toxicity and digestibility,
for ephemeral leaves, versus low toxicity and digestibility,
for predictable leaves, on grazing patterns by specialist
and generalist herbivores, should be as follows: Given
equal tissue availability, specialist herbivores should
prefer the most ephemeral leaves, since they are easier to
digest and usually contain higher nutrient concentrations[35]
than more predictable leaves. High levels of toxins, in
ephemeral leaves, should exert little repulsive effect or
perhaps even an attractive effect on specialists adapted to
the plant. On the other hand, generalists should be
repulsed by both toxins and the digestibility-reducing system.
This will result, in many cases, in a preference by generalist
herbivores for the more mature tissues, which, though they
are more difficult to digest, are lower in toxins.

    We have tested this hypothesis by determining the relative
tissue preference (RTP) of specialist and generalist insect
larvae for young, intermediate and mature leaves of their
host plants in the deserts of Arizona and Argentina (Table 6).
All insects involved in this study were lepidopteran larvae,
except for a single sawfly larva, a woody perennial specialist. RTP values were determined by dividing the number of

Table 6. Relative tissue preference (RTP) of specialist and generalist insect larvae for young, intermediate and mature leaves of desert plants.

	YOUNG		INTERMEDIATE		MATURE		Number of Herbivore Species
	RTP	N	RTP	N	RTP	N	
Annuals							
Specialist (one host plant)	100	(447)	36	(107)	11	(88)	2
Generalist (more than one host plant	17	(103)	60	(219)	100	(345)	6
Woody perennials							
Specialist (one host plant)	100	(787)	12	(268)	3	(140)	8
Generalist (3 host plants)	100	(294)	9	(88)	2	(53)	1
Generalist (6 host plants)	3	(18)	15	(328)	100	(803)	2

RTP = Relative Tissue Preference; N = Number of grazing observations.

field grazing observations (N) on a particular tissue type, by the relative abundance of the tissue, to give a preference value, and normalizing, so that the most highly preferred tissue is assigned a value of 100. For a more detailed description of this method and methods for determination of food utilization and assimilation parameters see reference 17

    In agreement with the hypothesis outlined above, all specialist insects (one host plant) studied, on both annuals and perennials, preferred the youngest leaves of their host plants, whereas generalists (more than one host plant) preferred the mature leaves, in all cases except one (Table 6). Significantly, the exceptional generalist is the least generalized, of the generalist insects on woody perennials, utilizing three host plants, on which it prefers the young leaves, in all cases.

These results contrast with a previous generalization that has arisen in the literature[78] which states that all herbivores should prefer the youngest leaves of their host plants. It seems, that this generalization is particularly applicable to specialist herbivores but less so to generalists which, we think, will in many cases be found to prefer the more mature leaves.

It is important to remember here that the level of grazing experienced by ephemeral and predictable leaves in the field from specialist and generalist herbivores will be determined, not only by herbivore tissue preference, but also by tissue availability. For instance, though specialized herbivores should prefer the most ephemeral leaves, we have postulated that such leaves will often escape in space and time from specialist grazing.

If ephemeral plants utilize a less effective digestibility-reducing system than do predictable plants (Fig. 5b), ephemeral plants should be more digestible to their herbivores than should predictable plants.

We have tested this hypothesis, by comparing nitrogen assimilation and mass utilization data for desert insects feeding on preferred tissue of their annual and woody perennial host plants (Table 7). All insects in this study were lepidopteran larvae except for a single beetle, a woody perennial specialist. Insects feeding on annuals assimilated a significantly higher proportion of total leaf nitrogen (NAE) and assimilated nitrogen at a significantly higher rate (NAR), on average (overall $\bar{x}$), than did insects feeding on woody perennials. On the other hand, mass utilization parameters (MUE and MUR) showed no significant difference, on average, for insects feeding on annuals or woody perennials. Thus for nitrogen assimilation, but not for mass utilization, the results are consistent with the hypothesis that woody perennials are less digestible to their herbivores than are annuals.

COEVOLUTION BETWEEN PLANTS AND HERBIVORES

Coevolution is a cyclic process (Fig. 4c). Characteristics of one member of a coevolving pair will influence characteristics evolving in the second member of the pair. These

Table 7.   Nitrogen assimilation and mass utilization from preferred leaf tissues by desert insects.

Herbivore type	Amount of Leaf eaten(g)	NAE	MUE	NAR	MUR	Number of Herbivore Species
Annuals						
Specialist ($\bar{X}$)	0.201	57	41	5.0	83	2
(one host plant)	+0.03	+11	+0	+0.8	+11	
Generalist ($\bar{X}$)	0.209	52	35	4.0	72	6
(more than one host plant)	+0.08	+16	+7	+1.6	+37	
Overall $\bar{X}$	0.208	53*	36	4.1*	74	
	+0.07	+15	+7	+1.5	+34	
Woody Perennials						
Specialist ($\bar{X}$)	0.184	35	30	2.9	57	9
(one host plant)	+0.07	+9	+9	+1.6	+29	
Generalist ($\bar{X}$)	0.209	43	49	4.4	97	1
(3 host plants)	+0.06	+7	+6	+0.4	+44	
Generalist ($\bar{X}$)	0.143	24	31	1.9	48	1
(6 host plants)	+0.05	+8	+8	+1.5	+22	
Overall $\bar{X}$	0.178	33*	33	2.8*	61	
	+0.06	+10	+11	+1.6	+33	

NAE = % nitrogen assimilated of that consumed
MUE = % mass utilized of that consumed
NAR = mg nitrogen assimilated 13 hrs 1.26-1.5 g fresh wt larvae
MUR = mg mass (total nutrients) utilized 13 hrs 1.26-1.5 g fresh wt larvae
*$p < 0.05$, Student's t-test, for the vertical comparisons

evolving characteristics will, in turn, feed back to affect characteristics evolving in the first member, and so on. We will attempt to consider the coevolution between plants and herbivores (Fig. 4, Cycle I), largely ignoring coevolution between herbivores and their predators and parasites (Cycle II) and also ignoring other cycles, e.g., between

plants and pathogens, not depicted; the assumption being
that it is useful to consider an isolated cycle.

In order to consider a coevolving system, we believe
that it is important to first elaborate "independent variables"
which are major determinants of some of the observed patterns
in the coevolving system, but which are not subject, or are
minimally subject, to alteration by feed-back effects from
the coevolving system. The independent variables, by
producing initial patterns, can thus be considered to be the
driving force for the coevolving cycle and the initial
patterns produced by the independent variables provide an
entry into the cycle. In this way circular reasoning can
be avoided.

In the present case, we assume that the important independent variables are spatial and temporal heterogeneity in
the physical environment, together with competitive plant/
plant interactions (Fig. 4, 1), and that the initial patterns
produced by these variables are the existence of ephemeral
and predictable plants and tissues (Fig. 4, 2). With respect
to predictable and ephemeral tissues, the most important
variable is probably temporal environmental heterogeneity,
since plant growth and maturation are so closely tied to
the seasons. With respect to the existence of predictable
and ephemeral plants, the important independent variables are
probably temporal and spatial environmental heterogeneity.
Spatially and temporally heterogeneous environments, by
providing an opportunity for resource partitioning, coupled
to competitive plant/plant interactions, have probably given
rise to plants of various growth form and seral stage.

As described previously, ephemerality and predictability
of food resource should select for generalism and specialism,
respectively (Fig. 4, 3) in the herbivores. Generalized
grazing should then select for divergent defense in ephemeral
resources and specialized grazing should select for convergent defense, in predictable resources (Fig. 4, 4).

It is at this stage that the "leveling effect of natural
selection" becomes operative. Convergent defenses should
select for generalism (Fig. 4, 5) in the herbivore, since if
the resource defenses are all very similar, it should not
be adaptive for the herbivore to expend time and energy
seeking a particular sub-group of the resources. Conversely,

divergent defenses should select for specialism in the
herbivores. Since all the resource defenses are very
different, the ability of a generalist to accommodate any
one of the defenses should be significantly less than of a
specialist herbivore - jack-of-all-trades, master-of-none -
and selective advantage should accrue to specialists. Therefore, though from predictability considerations alone, we
would expect ephemeral plants to experience mainly grazing
by generalists and predictable plants to experience mainly
grazing by specialists, the evolved defenses of the two plant
groups will tend to level out this effect.

A further leveling may be expected from the influence
of predators, pathogens and parasites on the herbivores
(Cycle II). Specialist herbivores, particularly those
specializing on predictable plants, are more predictable
to their predators and parasites than are generalists and thi
will select for generalism in the herbivores, to a greater
extent in the case of predictable plants than for ephemeral
plants. The combined effects of resource predictability,
evolved defenses and Cycle II may, therefore, result in the
relative levels of grazing pressure by specialist and generalist herbivores being quite similar for predictable and
ephemeral resources. However, for the above arguments to
hold true, it seems that the ratio of generalist to specialis
grazing pressure on ephemeral resources must be at least
slightly higher than the corresponding ratio for predictable
resources.

$$\frac{\text{generalist pressure}}{\text{specialist pressure}} > \frac{\text{generalist pressure}}{\text{specialist pressure}}$$

(ephemeral resources)   (predictable resources)

No predictions can be made concerning the numbers of
species of generalist and specialist herbivores utilizing
ephemeral and predictable resources, since it is grazing
level and not species number that is important to the
evolution of plant defenses. For instance though the
species richness of specialist insect herbivores is higher
than that of generalist insect herbivores on desert woody
perennials (Table 6), average abundance of individuals for
generalist species is higher than that for specialist
species[17]. A few common species of herbivores can exert
grazing pressure equal to or exceeding that of many rare

species. A further complication concerns large vertebrate
grazers, such as deer, many of which tend to be very
generalized in feeding habits, and to utilize mainly
ephemeral resources. In many environments, the abundance
of vertebrate grazers has been drastically and recently
altered by the activities of man. Plants in these environments may be adapted to grazing conditions which no longer
exist.

SUMMARY

We have highlighted what we consider to be an important
factor in the evolution of plant defenses against herbivores;
namely, predictability and availability of the plant or plant
tissue as a food resource to herbivores. Assuming that plant
defenses are costly to the time and energy budget of plants,
the observed distribution of toxic and digestibility-reducing
defensive systems, both between leaves of different stages
of maturity and between plant species, can be explained in
terms of greater investment in chemical defense for predictable plants and tissues than for ephemeral plants and tissues.
Since escape, particularly escape from specialist herbivores,
is high for ephemeral plants and ephemeral leaf tissues
(usually young leaves), they are defended by a cheap, divergent,
toxic chemical defense affording some protection against
generalist herbivores. Escape is low for predictable plants
and predictable plant tissues (usually mature leaves) which
thus utilize a more costly convergent digestibility-reducing
chemical defense, effective against both specialist and
generalist herbivores. Predictable plants utilize toxins
in their ephemeral leaf tissues and generalized digestibility-
reducing systems, particularly tannins, in their predictable
leaf tissues. Ephemeral plants utilize toxins in their
ephemeral tissues and are postulated to utilize specific
digestive enzyme inhibitors, in their mature leaves.

Fixed nitrogen and energy are both absorbed, by herbivores, in the form of small monomeric molecules. Since both
nitrogen and energy are largely present in plants in the form
of polymers, plants have evolved systems to reduce the
depolymerizability (digestibility) of their macromolecules.
Plant defensive systems that reduce the availability of
nutrients other than nitrogen and energy are likely. For
instance, thiaminase present in bracken fern[48] and the

"anti-vitaminic" factors of legume seeds[62] may be examples of such systems.

   Much of what we have said here is very similar to, and lends support for, Feeny's (this volume) concept of "quantitative" and "qualitative" defenses for "apparent" and "less apparent" plants, respectively. Most qualitative plant defenses, so far known, appear to be toxic (as previously defined) systems, whereas most quantitative defenses appear to be designed to reduce the availability of plant nutrients.

APPENDIX

Partition of Creosote Resin Between Diethyl Ether and Aqueous Solutions. Fresh frozen leaves and terminal stem portions (*ca.* 250 g) of *Larrea tridentata* (tetraploid), consisting of pooled material from five bushes, obtained from the Avra Valley, near Tucson, Arizona, were soaked in Et_2O (500 ml) for 12 hours. The ether layer was decanted, filtered and stored at $5°C$ in a brown glass bottle. One ml of this stock solution (24.45 mg resin/ml; by evaporation, $80°C$, 29" Hg vac., one min.) was then shaken by hand in a 25 ml volumetric flask (2 min.) with 5 ml aqueous solution to allow complex formation (if any), followed by addition of 14 ml Et_2O and further shaking (2 min.). After the layers had separated (2 min.), a 10 ml aliquot of the ether layer was evaporated. No corrections were made for the slight volume changes which occur on equilibration of anhydrous ether with aqueous solutions. In the case of the ether/1.5% gelatin partition (Table 2) it was shown that an "equilibrium" partition had been established during the final 2 minutes shaking.

   The interaction between resin and gelatin, is a highly non-ideal process (as is the interaction between tannin and protein). In the above method, concentrated resin solution (24.45 mg in 1 ml of ether) was shaken with the gelatin solution (5 ml, 1.5% in pH 7.0, 0.2 m sodium phosphate buffer) after which the bulk of the ether was added for the partitioning process and only 27% of the resin was recovered in the ether layer. However, when 24.45 mg of resin in 15 ml of ether was shaken with 5 ml of the gelatin solution for periods of up to 18 hrs., 97% of the resin was recovered in the ether layer! This non-ideal behavior is strongly suggestive of cooperative action between resin components so that stable complexes are formed only under conditions of high resin concentration.

ACKNOWLEDGEMENTS

Many of the ideas and syntheses presented here originated from discussions with Gordon H. Orians and John C. Schultz, to whom we are indebted. In addition, many ideas and data have been culled from the literature and we apologize to those authors who have not been specifically cited as originators. Special thanks are extended to Mary L. Paulson for the illustrations.

REFERENCES

1. Akahori, A., F. Yasuda, M. Togami, K. Kagawa and T. Okahishi. 1969. Variation in isodiotigenin and diosgenin content in aerial parts of *Dioscorea tokoro*. *Phytochem.* 8:2213-2217.

2. Anderson, A. B., R. Riffer and A. Wong. 1969. Monoterpenes, fatty and resin acids of *Pinus contorta* and *Pinus attenuata*. *Phytochem.* 8:2401-2403.

3. Anderson, J. D., P. Andrews and L. Hough. 1957. Occurrence of 2-0-methyl-L-fucose as a constituent of plum leaf polysaccharides. *Chem. and Ind.* 1453.

4. Andrews, P. and L. Hough. 1956. The isolation of 2-0-methyl-d-xylose from plum leaf hemicellulose. *Chem.and Ind.* 1278.

5. Aspinall, G. O. and A. Canas-Rodriguez. 1958. Sisal pectic acid. *J. Chem. Soc.* (London) 810:4020-4027.

6. _____ and R. S. Fanshaw. 1961. Pectic substances from Lucerne (*Medicago sativa*). Part I. Pectic Acid. *J. Chem. Soc.* (London) 822:4215-4225.

7. Bacon, J. S. D. and M. V. Cheshire. 1971. Apiose and mono-o-methyl sugars as minor constituents of the leaves of deciduous trees and various other species. *Biochem J.* 124:555-562.

8. Bate-Smith, E. C. and C. R. Metcalf. 1957. Leucoanthocyanins 3. The nature and systematic distribution of tannins in dicotyledenous plants. *J. Linn. Soc. Bot.* 55:669-705.

9. Bate-Smith, E. C. and T. Swain. 1967. New leucoanthocyanins in grasses. *Nature 213:*1033-1034.
10. Beck, S. D. 1965. Resistance of plants to insects.
    *Ann. Rev. Ent. 10:*207-232.
11. Bell, E. A. 1972. Toxic amino acids in the leguminosae.
    In Harborne, 1972:163-174.
12. Blohm, H. 1962. Poisonous plants of Venezuela. Harvard
    University Press. Cambridge, Mass. 136pp.
13. Brodie, B. 1961. Comparative biochemistry of drug
    metabolism. Proc. Int. Pharmacol. meeting. First
    Int. Cong. on Pharmacol. 6:299 pp.
14. Bonner, J. and J. E. Varner. 1965. Plant biochemistry.
    Academic Press, New York. 1054 pp.
15. Bull, L. B., C. C. J. Culvenor and A. T. Dick. 1968.
    The pyrrolizidine alkaloids. Elsevier, N. Y. 293 pp.
16. Burnett, W. C., S. B. Jones, T. J. Mabry and W. G.
    Padolina. Sesquiterpene lactones - insect feeding
    deterrents in *Vernonia.* (in press)
17. Cates, R. G. and D. F. Rhoades. *Prosopis* as a leaf
    resource. *In* A tree in perspective; mesquite
    (*Prosopis* spp.) in desert scrub ecosystems. B.
    Simpson (ed.) (in press).
18. Chestnut, V. K. and E. V. Wilcox. 1901. Stock poisoning
    plants of Montana. U.S.D.A. Div. of Bot. Bull. 26.
19. Clark, B. 1962. Balanced polymorphism and the diversity
    of sympatric species. System. Assoc. Publ. 4, Taxonomy
    and Geography: 47-70.
20. Clayton, R. B. 1970. The chemistry of nonhormonal
    interactions. Terpenoid compounds in ecology. p.
    235-275 *In* Sondheimer and Simeone. 1970.
21. del Moral, R. 1972. On the variability of chlorogenic
    acid concentration. *Oecologia 9:*289-300.
22. Dement, W. A. and H. A. Mooney. 1974. Seasonal changes
    in the production of tannins and cyanogenic glycosides
    in the chaparral shrub, *Heteromeles arbutifolia.*
    *Oecologia 13:*62-76.
23. Difeo, D. R., M. Sakakibara and T. J. Mabry. 1975.
    Flavonoid aglycones of two disjunct species of *Larrea.*
    Proc. 15th Annual Meeting of the Phytochem. Soc. of
    N.A.
24. Duffy, S. S. and G. G. E. Scudder. 1972. Cardiac
    glycosides in North American asclepiadaceae. A basis
    for unpalatability in brightly colored Hemiptera and
    Coleoptera. *J. Insect Physiol. 18:*63-70.

25. Ehrlich, P. R. and P. H. Raven. 1965. Butterflies and Plants. A study in coevolution. *Evolution* 18:586-608.
26. Eidt, D. C. and C. H. A. Little. 1970. Insect control through induced host-insect asynchrony. A progress report. *J. Econ. Entomol.* 63:1966-1968.
27. Eloff, J. N. and R. N. Grobbelaa. 1969. Isolation and characterization of N-methyl-L-serine from *Dichapetalum cymosum*. *Phytochem.* 8:2201-2204.
28. Erdtman, H. and T. Norin. 1966. The chemistry of the order Cupressales, *In* Progress in the chemistry of organic natural products. L. Zechmeister (ed.). Springer-Verlag, N.Y. 475 pp.
29. Erickson, J. M. and P. P. Feeny. 1974. Sinigrin: A chemical barrier to larvae of the black swallowtail butterfly, *Papilio polyxenes*. *Ecol.* 55:103-111.
30. Esau, K. 1953. Plant anatomy. John Wiley and Sons. N.Y. 735 pp.
31. Ettlinger, M. G., and A. Kjaer. 1968. Sulphur compounds in plants. *In* Recent Advances in Phytochemistry, 1:59-144.
32. Feeny, P. P. 1968. Seasonal changes in the tannin content of oak leaves. *Phytochem.* 7:871-880.
33. Feeny, P. P. 1970. Seasonal changes in oakleaf tannins and nutrients as a cause of spring feeding by winter-moth caterpillars. *Ecol.* 51:656-681.
34. Feeny, P. P. 1975. Biochemical coevolution between plants and their insect herbivores. *In* Coevolution of Animals and Plants. L. E. Gilbert and P. H. Ravens (eds.). Univ. of Texas Press. Austin (in press).
35. Fraenkel, G. 1953. The nutritional value of green plants for insects. Symp. of the 9th Int. Congr. Ent. 90-100.
36. Hanover, J. W. 1975. Physiology of tree resistance to insects. *Ann. Rev. Ent.* 20:75-95.
37. Harborne, J. B. (ed.). 1972. Phytochemical Ecology. Academic Press, N.Y. 272 pp.
38. Henderson, R. 1971. Catalytic Activity of α-chymotrypsin in which histidine-57 has been methylated. *Biochem. J.* 124:13-18.
39. Hill, R. and R. van Heyninger. 1951. Ranunculin: The precursor of the vesicant substance of the buttercup. *Biochem. J.* 49:332-335.
40. Hoch, J. H. 1961. A survey of cardiac glycosides and genins. Univ. of South Carolina Press, Charleston. 93 pp.

41. Horber, E. 1972. Alfalfa saponins significant in
     resistance to insects. *In* Rodriguez. 1972:611-627.

42. Ishaaya, I. and Y. Birk. 1965. Soybean saponins IV.
     The effect of proteins on the inhibitory activity of
     soybean saponins on certain enzymes. *J. Food Sci.*
     *30:*118-120.

43. Jacobson, J. and D. G. Crosby. 1971. (eds.). Naturally
     Occurring Insecticides. Dekker, N.Y. 585 pp.

44. James, W. D. 1950. *In* The alkaloids VI. R.H.F. Manske
     and H. L. Holms (eds.). Academic Press, N.Y. 525 pp.

45. Jones, D. A. 1962. Selective eating of the acyanogenic
     form of the plant *Lotus corniculatus* L. by various
     animals. *Nature 193:*1109-1110.

46. Jones, D. A. 1972. Cyanogenic glycosides and their
     function. *In* Harborne, 1972. p. 103-122.

47. Joseffson, E. 1967. Distribution of thioglucosides in
     different parts of *Brassica* plants. *Phytochem. 6:*
     1617-1627.

48. Kingsbury, J. M. 1964. Poisonous plants of the United
     States and Canada. Prentice-Hall, N.Y. 626 pp.

49. Kircher, H. W., W. B. Heed, J. S. Russell and J. Grove.
     1967. Senita cactus alkaloids and sonoran desert
     *Drosophila. J. Insect Physiol. 13:*1869-1871.

50. Krieger, R., P. P. Feeny, and C. Wilkinson. 1971.
     Detoxification enzymes in the guts of caterpillars:
     an evolutionary answer to plant defenses? *Science*
     *172:*579-580.

51. La Du, B. N., H. G. Mandel, and E. L. Way. 1971.
     Fundamentals of drug metabolism and drug disposition.
     The Williams and Wilkins Co., Baltimore. 615 pp.

52. Langenheim, J. H. 1973. Leguminous resin-producing
     trees in Africa and South America. *In* Tropical
     forest ecosystems in Africa and South America. A
     comparative review: 89-104. Smithsonian, N.Y.

53. Laskowski, M. 1970. *In* Structure - Function relation-
     ships of proteolytic enzymes. P. Desnuelle, H.
     Neurath and M. Ottesen (eds.). Academic Press, N.Y.
     309 pp.

54. Lawton, J. H. The structure of the arthropod community
     on bracken (*Pteridium aquilinium*) (L.) (Kuhn). (in
     press).

55. Loomis, W. D. and J. Battaile. 1966. Plant phenolics
     and the isolation of plant enzymes. *Phytochem. 5:*
     423-438.

56. Levin, D. A. 1971. Plant phenolics: an ecological perspective. *Am. Nat.* *105:*157-181.
57. _____. 1973. The role of plant trichomes in plant defense. *Quart. Rev. Biol.* *48:*3-15.
58. _____. 1975. Alkaloid-bearing plants: an ecographic perspective. Am. Nat. (in press).
59. McKey, D. 1974. Adaptive patterns in alkaloid physiology. *Am. Nat.* *108:*305-320.
60. Merck Index of Chemicals and Drugs VIIth Ed. 1960. Merck, Rathway, N.J. 1642 pp.
61. Metcalf, R. H. 1967. Mode of action of insecticide synergists. *Ann. Rev. Entom.* *12:*229-256.
62. Milner, M. 1975. (ed.) Nutritional improvement of food legumes by breeding. Proceedings of a Symposium, Rome, July 1972. Wiley Interscience, N.Y. 400 pp.
63. Muller, C. H. 1969. The "co" in coevolution. *Science*, *164:*197-198.
64. _____. 1970. Phytotoxins as plant habitat variables. *Rec. Adv. Phytochem.* *3:*106-121.
65. Munakata, K. 1970. Insect antifeedants in plants. *In* Wood. 1970:179-181.
66. Nash, J. Undated. Poisonous Plants. Etchells and Macdonald, London. 85 pp.
67. Nowacki, E. 1963. Inheritance and biosynthesis of alkaloids of Lupin. *Genetica Polonica 4:*161-202.
68. Oliver, B. F., F. G. Maxwell and J. N. Jenkins. 1971. Growth of the bollworm on glanded and glandless cotton. *J. Econ. Entomol.* *64:*396-398.
69. Pammell, L. H. 1911. A manual of poisonous plants. The Torch Press, Cedar Rapids, Iowa. 977 pp.
70. Paulson, D. R. 1973. Predator polymorphism and apostatic selection. *Evolution 27:*269-277.
71. Pridham, J. B. 1963. Enzyme Chemistry of Phenolic Compounds. Macmillan and Co., N.Y. 142 pp.
72. Rees, C. J. C. 1969. Chemoreceptor specificity associated with choice of feeding site by the beetle *Chrysolina brunsvicensis* on its food-plant *Hypericum hirsutum.* *Ent. Exp. Appl.* *12:*565-583.
73. Rehr, S. S., D. H. Janzen and P. P. Feeny. 1973(a). L-Dopa in legume seeds. A chemical barrier to insect attack. *Science 181:*81-82.
74. Rehr, S. S., P. P. Feeny, and D. H. Janzen. 1973(b). Chemical defense in Central American non-ant-acacias. *J. Anim. Ecol.* *42:*405-416.

75. Rhoades, D. F. The anti-herbivore defenses of *Larrea*.
     In The biology and chemistry of the creosotebush
     (*Larrea*) in the new world deserts. T. J. Mabry,
     J. Hunziker and D. R. DiFeo, Jr. (eds.). (in press).
76. Rice, E. L. 1974. Allelopathy. Academic Press, N.Y.
     354 pp.
77. Ribereau-Gayon, P. 1972. Plant phenolics. University
     Reviews in Botany, 3. V. H. Heywood (ed.). Oliver
     and Boyd, Edinburgh, 234 pp.
78. Rockwood, L. L. 1974. Seasonal changes in the suscep-
     tibility of *Crescentia alata* leaves to the flea beetle
     Oedionychus sp. *Ecology 55:*142-148.
79. Rodriguez, J. G. 1972. Insect and mite nutrition.
     North-Holland, Amsterdam, 702 pp.
80. Rothschild, M. 1973. Secondary plant substances and
     warning coloration in insects. *In* H. F. van Emden
     (ed.). 1973.
81. Sakai, W. S., M. Hanson and R. C. Jones. 1972. Raphides
     with barbs and qrooves in *Xanthasoma ragittifolium*.
     *Science 178:*314-315.
82. Sakakibara, M., D. DiFeo, Jr., N. Nakatani, B. Timmermann
     and T. J. Mabry. Flavonoid methyl ethers on the
     external leaf surfaces of *Larrea tridentata* and *L.
     divaricata* (Zygophyllaceae). *Phytochem.*, (in press).
83. Schmeltz, I. 1971. Nicotine and other tobacco alkaloids
     In Jacobson and Crosby. 1971.
84. Schoonhoven, L. M. 1972. Secondary plant substances
     and insects. *In* Recent Advances in Phytochemistry,
     V. C. Runeckles and T. C. Tso (eds.). pp. 197-224.
     Vol. 5. Academic Press, N.Y. 350 pp.
85. Schultz, J. C., D. Otte and F. Enders. *Larrea* as a
     habitat component for desert arthropods. *In* The
     biology and chemistry of the creosotebush (*Larrea*)
     in the new world deserts. T. J. Mabry, J. Hunsiker
     and D. R. DiFeo (eds.). (in press).
86. Schwerdtfeger, F. 1956. Is the density of animal
     populations governed by chance? Tenth Int. Congr.
     Ent. 4:115-122. E. C. Becker, (ed.).
87. Self, L., F. Guthrie, and E. Hodgson. 1964. Adaptations
     of tobacco hornworms to the ingestion of nicotine.
     *J. Insect Physiol. 10:*907-909.
88. Smith, B. D. 1966. Effect of the plant alkaloid spartei
     on the distribution of the aphid *Acrythosiphon spartii*
     *Nature 212:*213-214.

89. Smith, J. N. 1962. Detoxification mechanisms. *Ann. Rev. Ent.* 7:465-480.

90. Smith, R. H. 1961. The fumigant toxicity of three pine resins to *Dendroctonus brevicomis* and *D. jeffreyi*. *J. Econ. Entomol.* 54(2):359-365.

91. _____ and L. E. Green. 1972. Xylem resin in the resistance of the Pinaceae to bark beetles. U.S.D.A. For. Serv. Gen. Tech. Rep. PSW-1, 7 pp.

92. Sondheimer, E. J. and J. B. Simeone (eds.). 1970. Chemical Ecology. Academic Press, N.Y. 336 pp.

93. Stein, W. D. 1967. The Movement of Molecules Across Cell Membranes. Academic Press, N.Y. 369 pp.

94. Treub, M. 1896. Nouvelles Researches sur le role de l'acide cyanhyrique dans les plants vertes. *Ann. Jard. Bot. Buitzenbor,* Ser. II, 6:79-106.

95. van Emden, H. F. 1972. Aphids as phytochemists. *In* Harborne. 1972:25-43.

96. _____. 1973. Insect/Plant Relationships. (ed.) John Wiley and Sons, N.Y. 215 pp.

97. _____ and M. A. Bashford. 1969. A comparison of the reproduction of *Brevicoryne brassicae* and *Myzus persicae* in relation to soluble nitrogen concentration and leaf age (leaf position) in the brussels sprout plant. *Ent. Exp. Appl.* 12:351-364.

98. Vogel, R., I. Trautschold, and E. Werle. 1968. Natural Proteinase Inhibitors. Academic Press, N.Y. 159 pp.

99. Williams, C. M. 1970. Hormonal interactions between plants and insects. *In* Sondheimer and Simeone. 1970: 103-132.

100. Williams, R. J. and E. M. Lansford. 1967. The Encyclo-pedia of Biochemistry. Reinhold, N.Y. 876 pp.

101. Wood, D. L. 1973. Selection and colonization of Ponderosa Pine by Bark Beetles. *In* van Emden, 1973: 101-117.

102. _____, R. M. Silverstein and M. Nakajima. 1970. Control of insect behavior by natural products. Academic Press, N. Y. 345 pp.

Chapter Five

BIOCHEMICAL PARALLELISMS OF REPELLENTS AND ATTRACTANTS

IN HIGHER PLANTS AND ARTHROPODS

ELOY RODRIGUEZ

Department of Botany

University of British Columbia

Vancouver, B. C., Canada

DONALD A. LEVIN

Department of Botany

University of Texas

Austin, Texas

INTRODUCTION

In reviewing the distribution of micromolecular constituents (secondary compounds) that are known to function as defensive repellents* in arthropods, it becomes evident that a large number of chemical communicatory substances are also common metabolites of higher plants. This remarkable parallelism in inter and intra-specific chemical communication has been noted by numerous biologists, with Eisner (1970) commenting that "the very fact that plants possess the same materials that in other organisms are known to be defensive, may in itself be considered to be circumstantial evidence in support of the latter view."

A similar parallel is evident in secondary products that serve as insect sex pheromones or attractants (long-range and/or aphrodisiac) and floral pollination allurements or attractants. In many cases it has been demonstrated that some insects sequester sex pheromones directly from the host plants that they feed on, or they collect the floral fragrances (e.g. Euglossine male bees and orchid interaction).[28]

In this chapter we review the distribution of some representative secondary metabolites which are common to arthropods and higher plants, and which are known to be repellents and/or attractants. Plant products are discussed only with regard to arthropods. Comprehensive reviews on arthropod defensive secretions have been published[9,36,101]. and literature on arthropod sex attractants has been dealt with in great detail[6,8,37,61,62,81]. Information on the distribution of plant repellents and attractants has also been documented[23,27,27,75,114] as has the known distribution of phytochemical constituents in higher plants[42,56,89].

PRESENTATION OF REPELLENTS AND ATTRACTANTS BY PLANTS AND ARTHROPODS

Eisner (1970, p.159) in his review of defensive secretions in arthropods points out that there are two major types of

*Poisonous venoms (macromolecular substances) from spiders, ants, bees, scorpions and vertebrates are not considered in this review (See Karlson, 1973).

defensive substances produced: those synthesized in special
exocrine glands (glandular structures) and those contained
in the blood or elsewhere in the body and not of glandular
origin. In the former, the repellents are believed to be
produced *de novo*, while in the latter, they are either
sequestered from toxic plant and/or arthropod hosts.
Figure 1 is a schematic presentation of the more common
means of displaying repellents by arthropods. Numerous
secondary constituents, which are known arthropod secretions,
are also present in glandular hairs (trichomes) of higher
plants and function primarily in defense (Figure 2) [75].
It should be noted that although similar secretory structures
and glands are present in arthropods and plants, their chemical exudates can differ greatly. The following summarizes
the major means by which plants and arthropods display repellents and attractants:

PLANTS   ARTHROPODS

I. SECRETORY STRUCTURES   I. OOZE GLANDS
   A. Sunken secretory   A. External secretion
      glands (desert   (Millipedes; qui-
      plants; phenolics)   nones)
   B. Resin canals   B. Ooze droplets
      (*Pinus*;monoterpenes   (Dung beetles)
      etc.)

II. GLANDULAR TRICHOMES   II. "HORN" GLANDS
    A. Unicellular trichomes   A. Cornicles (trigly-
       (Numerous plants)   cerides) (Aphids)
    B. Capitate-glandular   B. Osmeteria (simple
       hairs (*Solanum*;   carboxylic acids)
       phenolics)   (Butterfly larvae

III. VOLATILE-CONTAINING   III. SPRAY GLANDS
     GLANDS   A. Ejection
     A. Rupture-released   (*Eleodes*; quinone
        (*Dyssodia*; monoter-   termites; monoter
        penes)   penes)

IV. URTICATING GLANDS   IV. BARB GLANDS
    A. Injection   A. Injection
       (*Urtica*; amines)   (Moth; amines)

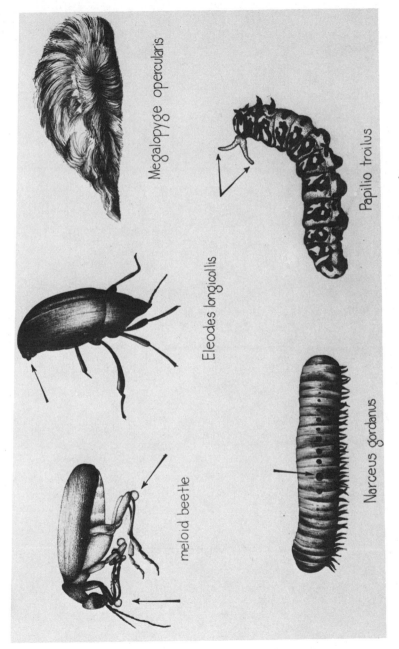

Figure 1. Defensive mechanisms of arthropods.

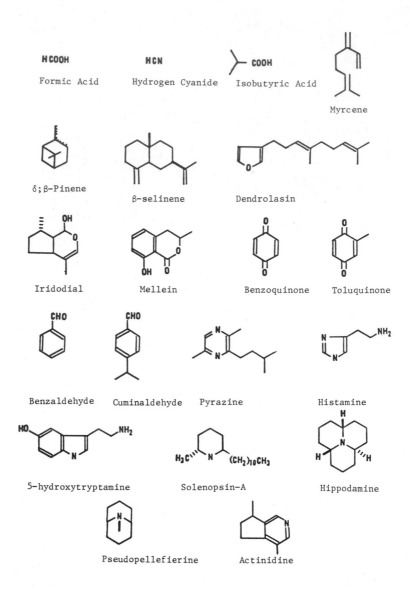

Figure 2.   Repellents found in arthropods and plants.

NON–GLANDULAR

PLANTS	ARTHROPODS
V. LATEX COMPONENTS (Euphorbiaceae; diterpenes)	V. BLOOD COMPONENTS (Meloidae beetles; sesquiterpenes)

VI. SEQUESTERED PLANT TOXINS

(Lepidopteran larvae; cardiac
glycosides) (Parasitic plants;
cardiac glycosides)

Quinones As Repellents In Plants And Arthropods. Benzoquinones are the most prevalent quinoid constituents of
arthropod repellent secretions. More than 100 species from
at least seven orders belonging to the classes Arachnida
(spiders), Diplopoda (millipedes), and Insecta (termites to
beetles) elaborate benzoquinones and anthraquinones as
defensive irritants [36,112,117,136,154]. Tables 1 and 2 list
some representative taxa of arthropods and higher plants
that are known to produce identical or structurally similar
quinones. The simplest quinone, p–benzoquinone, has been
found as the major repellent constituent in arthropod
secretions. In addition these secretions generally contain
aliphatic hydrocarbons, carbonyl compounds, carboxylic acids
and terpenes. Benzoquinones have been identified in numerous
species of millipedes (orders: Cambalida, Julida, Spirobolida,
Spirosteptida)[155] such as *Sprostraptus castaneus*[4] and *Oriulus
delus*[69], in the termite, *Odontotermes badius* [161], the grasshopper, *Romalea microptera* [35] and in at least ten genera of
Coleoptera, in particular tenebrionid beetles[141], and in the
earwig, *Forficula auricularia* (Hymenoptera)[115] Methylated
and methoxylated benzoquinones and simple phenols (m-cresol,
o–cresol) are also found in secretions of many arthropods,
with species of the millipedes, *Bollmaniulus* sp., *Eurhinocri-
cus* spp., *Leptogoniulus nasresi*, *Orthoporus ornatus*, *Rhinocri-
cus holomelanus*, *Sailus* spp. and *Trigoniulus lumbricinus*,
containing benzoquinone, toluquinone and 2–methyl-3-methoxy-
benzoquinone[30]. The bombardier beetles of the genus *Brachinus*
(Carabidae) synthesize toluquinone and benzoquinone in the
lumen of the gland reservoirs at the very moment of discharge,

TABLE 1

BENZOQUINONE IN ARTHROPODS AND P—HYDROQUINONE IN PLANTS

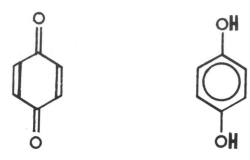

ARTHROPODS	PLANTS

Diplopoda (millipedes)
  Spirostreptus sp.
  Metiche sp.
 	
Isoptera (termites)
  Odontotermes badius*
  Mastotermes sp.
 	
Orthoptera (grasshoppers)
  Romalea microptera
 	
Coleoptera (20 genera)
  Blaps
  Eleodes
  Brachinus
  Tenebrio
  Lomechusa
	Ericaceae
	  Rhodendron sp.
	
	Rosaceae**
	  Pyrus sp.
	
	Proteaceae
	  Protea sp.
	
	Compositae
	  Xanthium sp.

Fungi also produce benzoquinones (Thomson, 1971)
* Secretion mixture of quinones and proteins.
** In some higher plants p—hydroquinone occurs as the glycosylated form
  (e.g. arbutin in Pyrus communis, Grisdale and Towers, 1960).

TABLE 2

QUINONES COMMON TO PLANTS AND ARTHROPODS

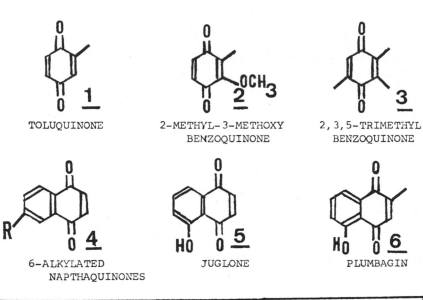

| TOLUQUINONE | 2-METHYL-3-METHOXY BENZOQUINONE | 2,3,5-TRIMETHYL BENZOQUINONE |

6-ALKYLATED NAPTHAQUINONES   JUGLONE   PLUMBAGIN

ARTHROPODS	QUINONES	PLANTS

Diplopoda (Millipedes)
  Orthoporus ornatus
  Rhinocricus holomelanus
  Bollmaniulus spp.
 	
Coleoptera (Beetles)
  Agroporis sp.
	1-2 **5**
	
	6
	4
	6
		Juglandaceae
		  Juglans nigra
		  Carya orata
		
		Droseraceae (sundews)
		  Drosera rotundifolia
		
		Plumbaginaceae
		  Plumbago pulchella

Napthaquinones also common to Hyphomycetes and Basidiomycetes (Thomson, 1971).

the respective hydroquinones are oxidized by hydrogen peroxide in an enzyme-catalyzed, explosive reaction.[1]

Approximately 20 families contain plants which produce benzoquinones [136]. These compounds have been found in all plant parts. Benzoquinones generally occur as p-hydroxy quinones which upon destruction of the leaf cells are readily oxidized to the more reactive benzoquinones. Para-hydroxy quinones occur in at least four plant families (Table 1).

Insect damage to plants may be accompanied by the oxidation of phenolics to quinones. The salivary glands of most phytophagous Hemiptera secrete a polyphenol oxidase[87,88]; whether this enzyme or peroxidase is of plant or animal origin is unknown. Glandular hairs of *Primula obconica* contain copious amounts of primin (2-methoxy-6-n-pentyl p-benzoquinone) which repels potential predators[76] and causes contact dermatitis in humans[90]. It is clear, however, that quinones (be they constitutive or induced) are insect feeding inhibitors or toxins.

Naphthaquinones have also been implicated as repellents in numerous insects and plants. In a comparative study of over 147 species of tenebionid beetles (e.g. *Tenbrio, Eleodes,* and *Cratidus*) from 55 genera and 16 tribes, Tschinkel (1975) found that benzoquinones were relatively rare, but toluquinones, ethyl quinones and 6-alkyl-1,4-naphthaquinones were present in most species. The 6-alkyl-1,4-naphthaquinones were noted to be the major constituents of the western American tenebrionid beetle, *Agroporis alutacea*.

The naphthaquinones are present in 20 angiosperm families the following being the most important: Juglandaceae, Plumbaginaceae, Boraginaceae, Lythraceae, Ebenaceae and Droseraceae[136]. Naphthaquinones occur in leaves, flowers, wood, bark and fruit. These compounds seem to occur most often as individual compounds rather than in mixtures; this is also the case for the benzoquinones. The former do not occur as glycosides.

Gilbert and co-workers (1967) have demonstrated that juglone (Table 2),produced in *Carya ovata,* is a feeding deterrent to the bark beetle, *Scolytus multistriatus.* Other 1,4-naphthaquinones have a similar property[96]. However,

juglone is not an effective deterrent to feeding by all
organisms which use *Carya ovata* as a host, as evidenced
by the fact that *Scolytus quadrispinosus* is not inhibited
by the compound[45]. The inhibitory effect of 1,4-naphtha-
quinones on feeding also has been demonstrated in
Periplaneta americana[3,96].

One well-known dianthrone is hypericin, which is a
broad spectrum-feeding inhibitor. This compound is present
in several species of *Hypericum*. This genus constitutes a
largely unexploited food source, except for several species
of the beetle *Chrysolina* which have evolved a mechanism to
detoxify hypericin. Moreover, *Chrysolina brunsvicensis* uses
hypericin as a feeding stimulant, the tarsel sensilla being
highly sensitive to hypericin at levels which are met in
the food plant[106]. This beetle does not eat the hypericin-
lacking *Hypericum androsaemum*, even though the plant sometimes occurs among *Hypericum perforatum* plants upon which
beetles are feeding. *Chrysolina brunsvicensis* accumulates
hypericin elsewhere than in the gut, and Rees[106] suggests
that the compound might render the beetle distasteful to
predatory animals. Thus, the beetle not only has "turned"
the plants' defense compound upon the plant, it may also be
using the plants' defense compound to insure its survival
and continued predation of the plant.

In summary, benzoquinones appear to be the major
constituent of many arthropod defenses. It is not surprising
that quinones are so widespread in arthropods, since in
insects the exoskeleton is produced by complexing special
proteins with polymeric aromatic or quinoid material derived
from tyrosine. Furthermore, sclerotization occurs between
orthoquinones (most defensive secretions contain paraquinones)
and free amino groups of the proteins[138]. Plants usually
contain dihydroquinones which are converted to quinones.
Therefore, quinones can be considered as "general repellents"
which are very reactive. They have been shown to complex
proteins and intercellular amino acids, interfere with co-
factor and enzyme synthesis, alter cellular redox potentials,
and inhibit specific enzyme systems[76]. Besides their toxicity to insects, quinones are also reported to effect bacteria[97,112].

Phenolic Aglycones. At least nine phenolic aglycones
and some aromatic aldehydes (which are common constituents of
higher and lower plants) have been identified in the repellent

secretions of Diplopoda Orthoptera, Heteroptera and
Coleoptera (Tables 3 and 4). As noted by Duffield and
co-workers (1974), the phenolic compounds, p-cresol and
o-cresol, are major components of the secretion of the
chordeumoid millipede, *Abacion magnum*, and the parajulid
millipede, *Oriulus deles*, which produces 2-methoxy-3-
methyl-1,4-benzoquinone as the major compound. Guaiacol
has been identified as a minor constituent of the
benzaldehyde-rich repellent of the millipedes *Euryrus
leachi*, *E. australis* (Family Euryuridae), *Othomorpha
coarcatat* and *Oxidus gracilis*[11,31]. Furthermore, the
same research groups demonstrated that the millipede
secretions effectively repelled the ant, *Solenopsis
richeteri*. The grasshopper *Romalea microptera* produces
a defensive froth containing a mixture of phenols,
terpenoids, benzoquinones (which includes guaiacol) and
2,5-dichlorophenol, a herbicide derivative ingested by
the grasshopper and found to be repellent to ants[35].

Extensive comparative biochemical investigations by
Schildknecht and associates[115,116,117] on land and water
beetles (Dyticidae and Colymbetinae) has led to the
identification of benzoic and p-hydroxy benzoic acids and
p-hydroxy benzaldehyde (Table 5). Schildknecht (1971)
noted that "injurious microorganisms are not only killed
by the phenolic compounds in the secretion, but at the
same time became embedded in a glycoprotein network which
is formed on contact with air from the large cysteine
cysteine component."

A similar mode of "immobilization" of an insect by
a plant phenolic exudate has been demonstrated by Gibson
(1971). He noted that glandular hairs of *Solanum
polyadenium*, *S. berthaultii* and *S. tarijense* discharge
an exudate when the cell walls are ruptured only by
contact with an aphid (either *Myzus persicae* or
Macrosiphum euphorbeae). The water soluble material,
which is released from the trichomes, changes to a dark
insoluble substance on contact with atmospheric oxygen and
precipitates on the aphids' legs. In *Populus deltoides*,
insect feeding was greater on newly expanded leaves in
which the resin had dried rather than on buds which are
covered with fresh liquid resin[25]. In laboratory tests,
larvae of the cottonwood leaf beetle, given leaves in
which only the margin was resinous, fed normally, pupated

TABLE 3

SOME PHENOLIC AGLYCONES COMMON TO ARTHROPODS AND PLANTS

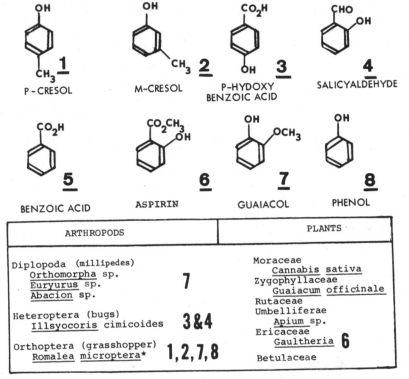

ARTHROPODS		PLANTS

Diplopoda (millipedes)
  Orthomorpha sp.
  Euryurus sp.
  Abacion sp.
	7	Moraceae
		  Cannabis sativa
		Zygophyllaceae
		  Guaiacum officinale
		Rutaceae

Heteroptera (bugs)
  Illsyocoris cimicoides
	3 & 4	Umbelliferae
		  Apium sp.
		Ericaceae

Orthoptera (grasshopper) Romalea microptera*	**1, 2, 7, 8**	Gaultheria **6** Betulaceae

*2,5-dichlorophenol also present in R. microptera (Eisner et al.,1971).

TABLE 4

CYANOGENIC COMPOUNDS COMMON TO SOME PLANTS AND INSECTS

ARTHROPODS	PLANTS

Dipliopoda (7 genera)
  Polydesmus (3 spp.)
  Nannaria sp.
  Oxidus gracilis
  Harpaphe haydeniana
Chilopoda (centipedes)
  Pachymerium ferrugineum
Insecta
  Procoris geryon (Lepidoptera)
  Zygaena (3 spp.)
Paropsis atomaria (Coleoptera)
	Occur in 80 families (800 spp.)
	  Rosaceae*
	  Leguminosae
	  Caprifoliaceae
	  Sapotaceae
	  Rutaceae
	
	  Sapindaceae**
	  Boraginaceae

*These families produce primarily cyanogenic glycosides (structure 3).
**Contain primarily cyanogenic lipids (structure 4) (Seigler, 1974).

TABLE 5

PHENOLIC CONSTITUENTS OF THE PYGIDIAL BLADDERS OF VARIOUS WATER BEETLES (ADOPTED FROM SCHILDKNECHT, 1971)

BEETLES	$COOH$ (benzoic acid)	$COOH$, OH	$COOCH_3$, OH	CHO, OH	$COOCH_3$, OH, OH	$COOC_2H$, OH, OH	OH, OH	CH_2COOCH_3, OH, OH
Dytiscinae								
Dytiscus marginalis	+		+	+	+	+		+
Cybister lateralimarginalis	+		++	++				
Cybister tripunctatus	++		++	++				
Acilius sulcatus								
Graphoderus cinereus	+	+						
Colymbetinae								
Rhantus exsoletus	++	++	+++	+++				
Colymbetes fuscus			++	++	+		++	
Copelatus ruficollis	+	+	+	+				
Agabus sturmi					+		+	
Agabus bipustulatus								
Ilybius fenestratus	+							

and emerged as normal adults. Larvae given sticky resin-covered leaves failed to eat them, and most died. Those that pupated did not complete their life cycle. Rhoades (this volume) demonstrated that the resinous (containing NDGA and methylated flavonoids) coating on the leaves and terminal stems of *Larrea divaricata* is a deterrent to feeding by species of *Cibolacris, Bootettix, Ligurotettix, Astroma,* and *Geometrida*; the youngest leaves, which have 2-3 times the resin content of the older ones, experienced the least damage. Leaf palatability is not a simple function of resin content alone, but is dependent upon the catechol content of the resin. He suggests that resins may disrupt digestion by non-specific binding to plant proteins and herbivore digestive enzymes.

Todd and associates (1971) reared biotype B greenbugs, *Schizaphis graminum*, on chemically defined diets containing commercially available phenolic and flavonoid compounds and found that compounds having ortho-hydroxyl groups, such as catechol, tannic acid, quercetin, chlorogenic acid, and protocatechuic acid,were detrimental to greenbug growth; the number and survival of progeny were drastically reduced when these compounds were included in the diet. Cis-caffeic acid caused drastic reduction in growth with no reproduction of greenbugs feeding on it; whereas, trans-caffeic acid caused some reduction in weight gain but had no effect on reproduction.

Another group of compounds were benzoic or cinnamic acid derivatives, having a para-hydroxyl group. Although less toxic, these compounds effectively reduced survival of the progeny. Less than 20% of the greenbugs survived feeding on diets containing vanillic, sinapic, syringic, gentisic or ferulic acids. They concluded that since many of these compounds are constituents of barley leaves, it is likely that at least part of the resistance to greenbugs of some barley varieties is attributable to the presence of these phenolic and flavonoid substances in quantities sufficient to retard the insect's growth and reproduction.

Hydrogen Cyanide. The utilization of hydrogen cyanide (HCN) and an aromatic aldehyde as defensive repellents in arthropods appears to be limited to millipedes (Diplopoda), centipedes (Chilopoda) and moths (Lepidoptera)[36,117]. Biosynthetic studies by Towers and associates (1972) on hydrogen cyanide formation in the polydesmid millipede, *Oxidus gracilis* Koch, demonstrated that dietary phenylalanine is converted into benzaldehyde and hydrogen cyanide. Duffey (1974) later showed that mandelonitrile is a precursor of HCN and benzaldehyde in the millipedes *O. gracilis* and *Harpaphehaydenia,* as previously known for higher plants. Earlier studies by Pollares (1946) on the millipede *Polydesmus vicinus,* showed the presence of the glucoside of p-isopropyl mandelonitrile which on enzymatic hydrolysis yields HCN and cuminaldehyde (Table 4).

Hydrogen cyanide in plants is stored in the form of cyanogenic lipids and/or glycosides. Seigler (1974) in a review of the chemistry and distribution of cyanogenic compounds notes that at least 800 species representing 70 to 80 plant genera are cyanogenic (Table 4). The wound sap of higher plants, which contain cyanogenic glucosides, usually contain β-glucosidase, which converts the glucosides to glucose and cyanohydrin and which readily breaks down (via oxynitrile) to HCN and aglycone.

The storage of cyanogenic glycosides is a very effective defense against phytophagous insects[47]. Greshoff (1967) reported that HCN evolved by *Arum maculatum* L. killed any insects which entered the plant after pollination. The resistance of some varieties of *Sorghum vulgare* to insects has been attributed to the presence of dhurrin (p-hydroxymandelonitrile-glucoside) in high concentrations. Perhaps the best demonstration that cyanide is toxic to insects is the use of cyanide jars for killing insects. However, no defense is sacrosanct. Parsons and Rothschild (1964) report that the larvae of *Polyommatus icarus* (Common Blue Butterfly) and *Hypera plantaginis*, a weevil, are capable of detoxifying the cyanogenic compounds of *Lotus corniculatus*. Cyanide resistance was observed in larvae of the moth *Malacosoma neustria* which feeds upon *Prunus laurocerasus*[99,108] and in the southern army worm[107]. As we have seen before, tolerance of a toxic compound may lead in evolutionary time to that compound being used as a feeding cue or for the animal's defense. The cyanogenic glucosides phaseolunatin and lotaustrin are feeding incitants for the Mexican bean beetle on *Phaseolus* spp. The lepidopterans *Zygaena* and *Acraea* release HCN from body tissues whether they are reared on cyanogenic plants or not; however, both genera feed on the cyanogenic plant, *Lotus*

corniculatus. The Heliconid butterflies ostensibly secrete
HCN, and may have evolved this characteristic in stepwise
fashion with their food plant, *Passiflora*[46].

   <u>*Terpenoids As Repellents*</u>. It is well-known that
arthropods require a dietary source of sterols, since they
are unable to synthesize the sterol nucleus from acetate
and mevalonate[24,65]. It is therefore not too surprising
to find that diterpenic acids, steroids, triterpenes,
cardenolides and cardiac glycosides,utilized as defensive
compounds by insects,are sequestered from numerous toxic
plants. The distribution of secondary constituents
sequestered by arthropods has been thoroughly reviewed by
Rothschild (1972), and therefore these repellents will
not be discussed in detail.

   Monoterpenes. Herout (1970), Karlson (1970) and
Clayton (1970),in their reviews of terpenoids in insects
and plants,note that volatile monoterpenes are the most
common repellents encountered in insects, particularly
in the alarm-defense systems of ants. Acyclic monoterpenes,
such as, geraniol, citronellal, citral and limonene have
been identified as alarm-defense substances (Table 6) in
numerous ant species of the subfamilies Formicinae,
Myrmicinae and Dolichoderinae[154,160]. Using radiotracer
incorporation techniques, Happ and Meinwald (1965) established that the walking stick, *Anisomorpha buprestoides*,
and the ant, *Acanthomyops claviger*, synthesize monoterpenes
similar to higher plants.

   In the family Termitidae, soldiers produce cephalic
secretions,containing α-pinene, β-pinene, limonene and
terpineole,which are ejected in a sticky resinous mixture
containing diterpenes and unsaturated hydrocarbons[93].
The flightless grasshopper, *Romalea microptera*,releases
a defensive froth containing phenols, quinones, and volatile
terpenes, including the monoterpene verbenone[35].

   Essential oils isolated from plants consisting of
acyclic and moncyclic monoterpenes, similar to those in
arthropod secretions, have been shown to be effective
repellents against insects (Table 7). Leaf oils from
Backhousia myrtifolia, *Malaeuca bracteata* and *Zieria
smithii* repel mosquitos, March flies and sand flies[51,52].
Limonene and myrcene are known feeding repellents that

TABLE 6

ACYCLIC MONTERPENES OF ALARM-DEFENSE SYSTEMS OF ANTS
AND TERMITES AND KNOWN REPELLENTS IN PLANTS

ANTHROPOD	MONOTERPENES*	PLANT
Hymenoptera (ants) 　Formicinae (2 genera) 　　Lasius sp.	Citronellal (1)	Pinaceae** 　Pinus Myrtaceae 　Eucalyptus sp.
Myrmicinae (3 genera) 　Alta sexdens	Citronellol (2) Geraniol (3)	Verbenaceae 　Lippia sp.
Dolichoderinae 　Conomyrma sp.	Citral (5)	Rutaceae Labiatae Gramineae
Isoptera (Termites) 　Amitermes vitiosus	Myrcene (4)	Canellaceae 　Canella alba

*Although some plants produce one major monoterpene, a plant or insect repellent secretion generally consists of numerous constituents acting synergistically.

**Zavarin (1975) has summarized the distribution of biologically active monoterpenoids in Pinaceae.

TABLE 7

MONOTERPENES WHICH ARE KNOWN REPELLENTS IN ARTHROPODS
AND PLANTS

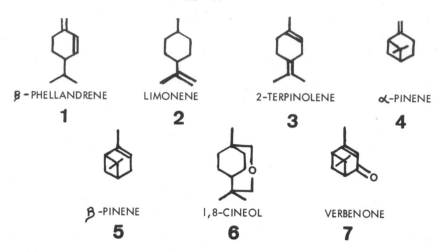

β-PHELLANDRENE   LIMONENE   2-TERPINOLENE   α-PINENE
1   **2**   **3**   **4**

β-PINENE   1,8-CINEOL   VERBENONE
5   **6**   **7**

ARTHROPODS	MONOTERPENES	PLANTS
Isoptera (termites) Nasutitermes longipennis Tumulitermes pastinator Amitermes vitiosus	**1 - 5** **4 - 5** **2 - 5**	Pinaceae Pinus sp. Picea sp. Compositae* Tagetes sp.
Coleoptera (beetle) Stenus bipunctatus	**6**[**]	Magnoliaceae Rutaceae
Orthoptera (grasshopper) Romalea microptera	**7**	Umbelliferae Labiatae

*
For distribution of volatile terpenes in leaf glands of Tagetes
and related genera see Rodriguez and Mabry (1975).
**
In some cases, different concentrations of monoterpenoids function
in attracting insects for oviposition and feeding (Hedin et al.,19

cause mortality in western pine beetles feeding upon
Pinus ponderosa. Repellency and toxicity of a given
species of bark beetles to pure resin vapors varies
with the source of resins. Bark beetles (*Dendroctonus*)
are more tolerant of the vapors of their particular
host pines than other species of pines, indicating
counteradaptation by these insects[19].

Menthol, a component of trichomes in genera of
the Labiatae is known to be a repellent to the silkworm
(c.f. Levin, 1973). Saturated hydrocarbons from *Pinus
densiflora* with straight-chains of C_5 to C_{10} are active
against the beetle, *Monochamus alternatus*[132]. These
same workers also established that ethane, the major
volatile gas in pine needles of *P. densiflora*, repelled
M. alternatus. The presence of saturated hydrocarbons,
such as, n-pentane, n-nonane and n-heptane, in mixtures
of repellent secretions has been reported for numerous
arthropods[154], and it would seem that gymnosperms which
contain similar saturated hydrocarbons are effective
"repellent carriers" when co-occurring in mixtures of
phenols and quinones.

As noted earlier, many arthropods either secrete or
eject chemical repellents through special glands or
sequester plant toxins in their blood and release (reflex
bleeding) the chemical when molested. Beetles of the
family Meloidae are good examples of "reflex bleeders"
which contain the highly oxygenated monoterpene, cantharidin. Contact by humans with the cantharidin-containing
blood of the "spanish fly" beetle results in blisters[91].
Recent studies by Carrel and Eisner (1974) on cantharidin
have shown that predaceous ants and carabid beetles are
repelled at concentrations of 10^{-5}M. Anemonin, a monoterpene ($C_{10}H_8O_4$) related to cantharidin and isolated from
Anemone (Ranunculaceae), produces subepidermal blistering
of the skin and more than likely is a feeding deterrent
to insects[91].

Perillen, a furanomonoterpene isolated from the
alarm-defense secretion of the ant *Dendrolasius*[160] was first
isolated from the mint, *Perilla citriodora*.

Iridoids. An interesting group of terpenoids which are common to ants and higher plants are the cyclopentane monoterpenes or iridoids. Iridoidal aglycones have been identified in the alarm-defense secretions of the ant genera *Conomyrma, Dolichoderus, Tapinoma* and *Iridomyrmex* (Table 8). Similar iridoids are also constituents of the Labiatae and Actinidiaceae. Nepetalactone and metatabilactone, identified from *Nepeta cataria* and *Actinia polygama* respectively, are known to repel insects, particularly nepetalactone[36]. In a recent chemosystematic review of iridoids in angiosperms, Jensen and associates (1975) note that iridoids have been detected in 44 species from 36 families, with iridoid glucosides found in Reziaceae, Dipsacaceae, Calyceraceae, Roridulaceae and Goodeniaceae.

Sesquiterpenes. Sesquiterpenes (Table 9), unlike monoterpenes and simple hydrocarbons, are not common repellents of arthropods, but are widespread in the plant kingdom. The eudesmane, β-selinene, first isolated from *Apium graveolus* (Umbelliferae) and members of the Compositae [104], was recently isolated as a defensive substance from the osmeterial secretion of the caterpillar, *Battus polydamas*[33]. Dendrolasin, a furanosesquiterpene, isolated from *Torreya nucifera* (Taxaceae) and *Ipomea batata* (Convolvulaceae), is also a constituent of the alarm-defense secretion of *Dendrolasius fuliginosus*[59]. Recently, dendrolasin was detected in the sponge, *Oligoceras hemorrhages*[148].

The aphid, *Myzus persicae*, is repelled by the terpenoid odor of droplets released from the cornicles and squashed aphids of the same species[68]. An allenic sesquiterpene similar to β-ionone from *Boronia megastigma* (Rutaceae) was isolated from the defensive froth of *Romalea microptera*; the sesquiterpene probably originated from the degradation of the carotenoids, fucoxanthin and neoxanthin[86].

Other Terpenoids. Neocembrene-A, a diterpene known from *Pinus*[164], has been isolated from sticky repellent of the termite, *Natsutitermes*, and is believed to also function as a trail pheromone. Eisner et al. (1974) have shown that the sawfly, *Neodipron sertifer* discharges an oral effluent consisting of α-pinene and β-pinene and various diterpenic acids identical to those produced by the host plant *Pinus sylvestris*. The fluid, which is stored in compressible diverticular pouches in the foregut is an effective

TABLE 8

IRIDOIDS COMMON TO PLANTS AND ARTHROPODS

ARTHROPODS	IRIDIODS		PLANTS
Hymenoptera Conomyrma sp. Iridomyrmex (6 spp.)	IRIDODIAL	NEPETALACTONE	Labiatae Nepeta cataria
Dolichoderus scabridus Iridomyrmex nitidus Tapinoma nigerrimum (IRIDIOMYRMECIN)	A	B	Actinidiaceae Actinia polygama
Dolichoderus clarki Iridomyrmex rufoniger Anisomorpha buprestoides	C	D	Eucommiaceae Eucommia ulmoides
			Stychnaceae[**] Strychnos nuxyomica Cornaceae Cornus sp.

[*]Compounds-A (iridomyrmecin), -B (matatabilactone),-C (dolichodial) and
-D (eucommiol) (Bate-Smith and Swain, 1966).

[**]Iridoid glycosides are not known to be present in arthropod repellent
secretions (for distribution in higher plants see Jensen et al., 1975).

TABLE 9

SESQUITERPENES AND DITERPENES WHICH ARE KNOWN REPELLENTS IN PLANTS AND ARTHROPODS

ARTHROPOD	SESQUI- AND DITERPENES	PLANT
Lepidoptera _Battus polydamas_ (larvae)	β-SELINENE*	Umbelliferae _Apium graveolus_ _Libanotis_ sp.
Hymenoptera _Dendrolasius fulig-_ _inosus_	DENDROLASIN	Taxaceae _Torreya nucifera_ Convolvulaceae _Ipomea batatas_ _Oligocera hemorrhages_ (Sponge)
Hymenoptera _Dendrolasius_ sp.	PERILLIN	Labiatae _Perilla citriodora_
Orthoptera _Romalea microptera_	1 2**	Rutaceae _Boronia megastigma_
Isoptera _Nasutitermes_ (3 spp.)	NEOCEMBRENE-A	Pinaceae _Pinus obovata_

*   Also found in the osmerterial secretion of _Battus polydamas_ was the
      sesquiterpene, selin-11-en-4a,ol[83].

**   β-ionone (Geissman and Crout, 1969).

repellent. Here we have another case (e.g. *Hypericum* X *Chrysolina* beetle) where the defensive chemistry of a plant has been breached and used to protect the herbivore.

One of the most intensively studied terpenoid repellents is gossypol, a product of cotton produced from subepidermal glands. This compound is toxic to larvae of bollworm and tobacco budworm which are major cotton pests[127]. In one of the more important experiments involving this plant-herbivore relationship, Shaver et. al. (1970) demonstrated that an experimental line of Upland cotton containing 1.7% gossypol was utilized less efficiently by the bollworm that was a standard line containing 0.5% gossypol. Also when gossypol was incorporated into the normal bollworm diet, it depressed feeding in budworm and bollworm larvae. Correlatively, in replicated field plants with cage experiments, high concentrations of gossypol bore significantly smaller budworm populations than those with standard concentrations[77]. On the other hand, this terpene stimulates feeding in the boll weevil[54].

In recent years, it has become apparent that the plant kingdom is a rich source of steroidal substances that possess molting hormone activity in insects[55]. Over 30 different active compounds have been identified from plants, and in some instances their concentration is higher in the plant than in the insect. The phytoecdysones bear a striking similarity to insect ecdysones. The physiological function of these substances in plants is unknown; however it also appears that they do not play a current role when administered orally to insects[156]. Whether this is an unusual case of parallel evolution, or whether these plant products might have had a detrimental effect to primitive insects, which perhaps lacked a means of degrading these compounds, is a matter of conjecture. The latter is suggested by the fact that natural (or synthetic) analogues of phytoecdysone may cause severe derangement of growth and metamorphosis.

Schildknecht (1970) has reviewed the secondary chemistry of water beetles (*Dytisus*) which sequester steroidal compounds from their host plants and use them as repellents.

Saponins, cardiac glycosides and triterpenoid glycosides occur as repellents in fruits and seeds[67,123].

Fruits rich in compounds of this type are present in such
genera as *Agrostemma*, *Strophanthus*, *Aleurites*, *Actaea* and
Phytolacca. In other genera (e.g. *Daphne*, *Croton* and
Jatropha) resins in the fruit are the active deterrents
to feeding. The majority of the terpenoid toxins occur
in, seeds as well as fruits, and are gastrointestinal
irritants,with the exception of the cardiac glycosides.

Nitrogen-Containing Repellents. The alkaloids and
amines are the most prominent constituents in lists of
insect feeding repellents and deterrents[53]. In spite of
a rather limited taxonomic base, alkaloids seem to repel
or poison representatives of all major groups of herbivores[23]. Among arthropods, only a handful of insects are
capable of producing alkaloid-like constituents. In most
cases alkaloids are sesquestered from plants. The pyrrol-
izidine-type alkaloids are well-represented in lists of
plant toxins sequestered by insects for their defense[114].
For example, pyrrolizidine alkaloids are stored by:
Amphicallia, which feeds upon *Crotalaria* (Leguminosae);
Utethesia, which feeds upon *Heliotropium*, and *Tournfortia*
(Boraginaceae); *Tyria*, which feeds upon *Senecio* (Compositae);
and ctenuchid moths when they feed upon *Pentzia* and *Mikania*
(Compositae) and *Lolium* and *Thelopogon* (Gramineae). The
products are not necessarily sequestered in the proportions
present in the host plant, nor in the same structural form.
The cinnabar moth caterpillar (*T. jacobaea*) sequesters and
stores all six alkaloids present in *Senecio jacobaea*.
However, 40%-60% of these alkaloids are jacobine, jacozine
and jacoline, whereas they constitute 10%-15% of those
present in the adult moth. In contrast, 60%-70% of those
stored by the moth are senecionine,seneciphylline and
integerrmine, which constitute 40%-60% of these in the
plant.

The fire ant, *Solenopsis saevissima*, produces
piperidine alkaloids (e.g. solenopsin-A; Table 10) which
exhibit necrotoxic, phytotoxic, antibiotic and insecticidal
properties[80]. Similar alkylated piperidines are known from
Cassia excelsa[110]. Actinidine, an iridoidal-alkaloid,
present in the defensive secretion of rove beetles[5] is
known from *Actinia polygama* (Table 10).

Volatile repellent secretions of the cosmopolitan
Argentine ant, *Iridomyrmex humilis* (Mayr),were found to

TABLE 10

PYRAZINE AND ALKALOID DERIVATIVES COMMON TO INSECTS AND PLANTS

ARTHROPOD	COMPOUNDS	PLANT
Hymenoptera Camponotus herculeanus (MELLEIN)	(MELLEIN)*	Hydrangeaceae Hydrangea macrophylla
Hymenoptera Iridomyrmex humilis Odontotomachus O. hastatus O. clavus O. brunneus	 PYRAZINES**	Rubiaceae Caffea sp. Sterculiaceae Theobroma cacao
Coleoptera Hesperus semirufus Philonthus politus	 ACTINIDINE	Actinidiaceae Actinidia polygama
Lepidoptera Arctia caja (moth) Phragmatobia fuliginosa Tyria jacobaeae	 HISTAMINE [a] 5-HYDROXYTRYPTAMINE [a] ACETYL CHOLINE[a]	Urticaceae Urtica dioica U. urens U. passiflora Giradinia heterophylla Leguminosae Mucuna pruriens

* Mellein is not known from higher plants, but occurs in numerous
   fungi, such as Aspergillus melleus (Brand et al., 1973; Thomson, 1964).
** For reference on pyrazine-containing plants see Takken et al.(1975)
[a] Found in barbs of moths and trichomes (stinging hairs) of plants.

contain the alkylated pyrazines, 2-isobutyl-2-methoxy,
3-s-butyl-2-methoxy and **3**-isopropoyl-2-methoxy pyrazine[22].
Alkylpyrazines (e.g. 3-isobutyl-2-methoxy pyrazine) have
been identified in species of *Caffea* (Rubiaceae) and
Theobroma cacao (Sterculiaceae). Mandular secretions of
the ponerine ants *Odontomachus hastatus*, *O. clavus*, and
O. brunneus contain alkypyrazines that function as powerful releasers of alarm behavior for *Odontomachus* workers
and are also utilized as defensive compounds.

The stinging hairs of Urticaceae, which includes
Laportea, *Urera* and *Urtica*, contain the active substances,
acetylcholine, histamine, and 5-hydroxytryptamine (Table 11).
These same compounds are also known from the barbs of the
moths *Arctia caja*, *Phagmatobia fulginosa* and *Tyria
jacobaeae*[113].

In recent years bicyclic and tetracyclic alkaloids
(coccinelline-type) have been identified from the repellent
secretions of the ladybugs of the Coccinellidae[143,144].
Almost identical alkaloids have been found in *Poranthera
corymbosa* (**compound** porantherine) and *Euphorbia atoto*
(9-aza-1 methyl bicyclo- 3,3,1-nonan-3-one), both of the
Euphorbiaceae. The bitter alkaloids present in the
coccinelid beetles are effective against ants and quail[100].
The similarity of alkaloid structures in the euphorbs and
ladybugs would suggest that the ladybugs are sequestering
the alkaloids from aphids that are feeding on the toxic
plants. As of yet, no experimental evidence is available
to support this suggestion.

Alkaloids are the most prominent elements in lists of
insect feeding deterrents and poisons[53,120,121]. The compounds
tested thus far are primarily from the Solanaceae,
Leguminosae and Papaveraceae. In spite of their rather
limited taxonomic base, they seem to inhibit or **poison**
representatives of all the major groups of herbivorous
insects. The hundreds of alkaloids present in and beyond
the aforementioned families may be expected to have similar
effects upon some insects (Levin, in press). Alkaloids
rarely are feeding stimulants. The alkaloid, α-tomatine
of *Solanum* and *Lycopersicon*, interferes with growth and

TABLE 11

ALKALOIDS IN ARTHROPODS AND PLANTS

ARTHROPODS	ALKALOIDS	PLANTS
Hymenoptera <u>Solenopsis saevissima</u>	 SOLENOPSIN-A	Caricaceae <u>Carica papaya</u> * Leguminosae <u>Cassia excelsa</u> *
Coleoptera <u>Propylaea</u> sp. Coccinella septempunctata	 PROPYLEIN PORANTHERINE	Euphorbiaceae <u>Poranthera corymbosa</u>
Coleoptera <u>Adalia bipunctata</u>		Euphorbiaceae <u>Euphorbia atoto</u>

* Solenopsin A is not known from plants, but similar piperidine
  alkaloids are known (Rice and Coke, 1965).

development of nymphs of *Melanoplus bivittatus* and larvae
of *Aedes aegyptii*[52]. Potato leaves containing α-tomatine
repel the Colorado potato beetle, *Leptinotarsa decemlineata*[131], the potato leafhopper, *Empoasca fabae*[26],and
Manduca sexta[121]. Moreover, this alkaloid deters inhibition in 2 species of *Dysdercus*[122]. Tomatine is inhibitory
to feeding of *Pieris brassicae*, but in the presence of
sinigrin it is a phagostimulant, and higher concentrations
of the alkaloid are necessary to reduce feeding[79]. The
effect of nicotine and related compounds has been studied
more than other alkaloids for obvious reasons. This
alkaloid is toxic to most of the insect which
were subjected to it in one form or another[118]. Self
et. al. (1964) examined tobacco-feeding insects which
manage to survive the large quantities of nicotine in the
diets. They found that in some species most of the
nicotine had been metabolized to the nontoxic conlinine.
Other species failed to metabolize nicotine but rather
rid themselves of the alkaloid through efficient excretory
systems. Whether, and to what degree, the ability to cope
with nicotine offers a species immunity from the toxic
effect of other alkaloids remains to be determined.

It is evident that all alkaloids are not equally toxic
or repellent. Consider the adverse effects of alkaloids
on the Colorado potato beetle larvae (*Leptinotarsa
decemlineata* Say)[16]. Nicotine, colchicine, veratrine,
aconitine and delphinine cause sudden death soon after
ingestion. Alkaloids retarding growth and development,or
which are toxic at high concentrations,include atropine,
scopolamine, theobromine and morphine. Tomatine, strychnine, cocaine, chinchonine, papaverine, berberine, and
sanguinarine affect larvae by acting as feeding repellents;
they are detrimental if ingested. A disparity in the
repelling effect of different alkaloids also is apparent
in the neurophysiological responses of *Pieris brassicae*
larvae. This species is most sensitive to quinine and
strychnine, less sensitive to brucine, nicotine, berberine
and atropine, and weakly sensitive to salicine colchicine
and caffeine[78].

Miscellaneous Repellents. As noted earlier, an
arthropod or plant repellent secretion generally consists
of a mixture of micromolecular constituents, such as,
simple carboxylic acids (isobutyric acid, tiglic acid,

valeric acid, caprylic acid), alkanes (n-undecane), alkanols, alkanones (2-hexenal) and aldehydes; many of these function as carriers[154]. Eisner (1970) has suggested that lipophylic constituents (e.g. n-tridecane) which are associated with polar irritants and/or repellents aid in the penetration of insect cuticles and animal skin. Blum and Brand (1972) have also suggested that alkenes (C_9-C_{13}) may competitively inhibit the chemoreceptive sites on insect antennae. In a similar manner, Wright (1975) proposes that mosquito chemical repellents function by blocking the pores in the cuticle of sensory hairs.

Glucosinolates (mustard oil glycosides), common constituents of the Capparidaceae and Cruciferae, have been demonstrated to be insect repellents (e.g. sinigrin)[121]. They are sequestered by the lepidopteran *Pieris brassicae* and *P. rapae* and used in defense[114].

Polyacetylenic compounds, which are widespread in the family Compositae and some Umbelliferae[12], are also known in arthropods, with the beetle *Chanliognathus lecontei* producing dihydromatricaria (both *cis* and *trans* dihydromatricaria are known from plants)[86].

Biosynthesis of repellents. In most arthropods that have been studied, the production of simple repellents (HCN, acyclic monoterpenes and simple alkanes) parallels the biosynthetic pathways of higher plants and some fungi. Duffey (1974) demonstrated that HCN production, *via* mandelonitrile, was similar in millipedes and higher plants. *Lasius fuliginosus* incorporated exogenous acetate-1-^{14}C, mevalonate-2-^{14}C and glucose -^{14}C into the furanoterpenoid, dendrolasin[152]. Meinwald et. al. (1966) showed that the monoterpene, anisomorphal, was synthesized from acetate and mevalonate in the phasmid, *Anisomorpha buprestoides*, and in the plant, *Nepeta* (catmint).

DISTRIBUTION OF SIMILAR ATTRACTANTS IN
PLANTS AND ARTHROPODS

Insect Sex Pheromones (Attractants) and Floral Scent Attractants. Sensitivity to volatile chemical constituents, either of insect or plant origin, plays an important role in the sexual and feeding lives of arthropods. Floral scent allurements provide the olfactory stimuli for

flower-visitations and the sex attractants (either
long-range or aphrodisiac) involved in the courtship
of moths, butterflies, beetles, flies, bugs and bees.

Chemical communications is a paramount mode of
communication among insects and vertebrates[159] and it
is becoming evident that it is an important means of
attracting potential pollinators by plants. Floral
allurements are an assemblage of visual, olfactory and
gustatory substances which, through their composition
and temporal pattern, provide a diagnostic and alluring
signal complex to one or more insects.

In this section, the biochemical and structural
parallelism of chemical communicatory substances, in
this case olfactory constituents, released by both
insects and plants is discussed. Emphasis is placed
on the parallel evolution of secondary constituents
used in the chemical communication systems of plants
and arthropods.

*Insect sex attractants (long-range, aphrodisiac
and aggregators).* Sex pheromones are widely, if not
universally, distributed in the Insecta and appear to
have reached their evolutionary peak in the Lepidoptera,
where they have been demonstrated for more than 170
species[81]. Most sex attractants have molecular weights
between 200 and 300[157]. Often these attractants occur
in mixtures of two or more synergistic compounds;
individual compounds of a mixture may be inactive by
themselves. The threshold value of olfactory sex
attractants are very small[17], and male insects are
allured to the female scent even when the source is many
tens of meters away.

The chemistry of insect attractants,
has in the last five years been the
subject of detailed reviews[6,8,37,61,62,81,111]. The
reader is referred to these reviews for pertinent
information on the chemistry, physiological action
and production of insect sex attractants.

According to Evans and Green (1973), insect sex
attractants fall into two broad categories:
volatile compounds which produce a positive response
for mating, aggregation and foraging and those compounds

of plant origin which are key chemical cues for locating
food resources (host plants) and for oviposition. In this
chapter, only mating sex attractants will be considered.

The more commonly occurring sex attractants in the
Lepidoptera (Families: Arctiidae, Bombycidae, Lymantriidae,
Noctuidae, Pyralidae and Tortidicidae) are straight-chain
fatty alcohols, esters and aldehydes of C_{12} to C_{24} chain
length (monoene and diene) that operate over long distances
and persist in time[37,128]. Among the Lepidoptera, moths
(superfamily Noctuiodea) produce the more diverse structural
types, ranging from the unsaturated fatty alcohols and acetates to a saturated hydrocarbon and epoxide (refer to
Table 12 for representative long-range sex attractants).
Most moth sex attractants are produced by females and are
potent. For example the male gypsy moth (*Portheria diapar*)
is attracted to a female from a distance of a fourth to a
half mile, while the male of *Bombyx mori* (silkworm) can
"detect" a female which is two and half miles away[150,151].

In addition to the Lepidoptera, attractants have been
identified in members (male and female) of the Coleoptera,
Hymenoptera, and Diptera.[61] Straight-chain hydrocarbons
and carboxylic acids are known from beetles (*Limonius
californicus, Trogoderma* and *Attagenus* species) and bees
(Bombus and *Apis)*. A phenol was identified as the sex
attractant in *Costelytra zealandica* (grass grub beetle)
and is believed to be produced by symbiotic bacteria within
the glands of the female[37]. The sesquiterpene alcohol,
farnesol[107] has been identified as a sex attractant in the
mite, *Tetranychus uriticae*, and in species of *Bombus*[130]
and *Macropsis*[71] bees.

Aggregation (population pheromones) attractants of
terpenoid origin (acyclic and bicyclic monoterpenes) are
known primarily from taxa of the family Scolytidae
(Coleoptera) and in some cases are derived from their
infested host trees. Table 12 includes some pheromones
isolated and identified from the bark beetles, *Dendroctonus*
and *Ips* [13,74].

The Nasonov gland of worker honeybees produces a
perfume which is important in transmitting information
necessary for successful foraging. The gland secretion
is composed primarily of the acyclic terpenes, citral and

Table 12. Representative Insect Sex Attractants*

INSECT SPECIES	ATTRACTANT

MOTHS — Long-range

Argyrotaenia velutina
Choristoneura rosaceana   } .....
Ostrinia nubilis
Protheria dispar   ....
Arctiidae (7 species)   ....

BEETLES

Limonius californicus   ....
Costelytra zealandica   ....

Aggregator

Anthomus grandis   ....
Dendroctonus frontalis   ....
Ips confusus   ....

BEES

Apis mellifera   ....
Bombus sp.**   ....

*From Evans and Green (1973)
**From Kullenberg and Bergstrom (1975).

geraniol; geranic and nerolic acids are minor components[18].

Unlike the large molecular weight attractants that are common to female moths and operate for considerable distances, many butterflies and moths produce aphrodisiatic pheromones (short-range or arresting pheromones) which prepare the opposite sex (in most cases the female) for copulation after they are brought together by olfactory attractants or visual and auditory means[8,128]. Table 13 lists representative aphrodisiacs isolated from male andeoconia, scent-organs, hair-pencils, hair-brushes and other secretory organs of moths and butterflies. These compounds range from simple carboxylic acids (e.g. iso-valeric acid), aromatic phenolic alcohols and aldehydes to simple monoterpenes[8]. Meinwald et. al. (1971), Schneider and co-workers,1975 and Edgar and Culvenor(1974) have identified numerous dihydropyrrolizidine alkaloids, aliphatic acids and sesquiterpenoids from the hair-pencils of males in the closely related nymphalid subfamilies Ithomiinae and Danaunae.

Origin Of Some Insect Sex Attractants. Hendry and associates (1975 and this volume) have reported that male oak leaf roller moths (*Archip semiferanus*) become sexually active whenever near damaged oak leaves and attempt to copulate with host leaves. Analysis of the leaves demonstrated the presence of many compounds identified as components of sex pheromones of female moths. Edgar and Culvenor (1974) and Pliske (1975a,b) have shown that hair-pencil dessiminated pheromones are dihydropyrrolizidines of plant origin. Neotropical male euglossine bees are known to collect orchid fragrances consisting of terpenoids[27]. Tumlinson et. al. (1970)have suggested that the production of the sex attractant of the male boll weevil (*Anthonomus grandis*) may be diet-related, with the weevil converting a monoterpenoid constituent into the sexually attractive compounds.

Silberglied (1975), in an excellent review on chemical communcation in Lepidoptera, summarizes the latest findings on lepidopteran sex attractants and notes that "it would certainly be ironic if Mullers' (1888) suggestion, that floral odors attract lepidopterans due to a similarity to sexual attractants were to be reversed with the implication that the order of the opposite sex would excite a moth or butterfly through the remembrance of food fragrances past."

Table 13.   Aphrodisiacs (Short-range) From Insects*

NOCTUIDAE MOTHS

Leucania impura
L. conigera
L. pallens

Pseudaletia unipuncta

Apamea monoglypha

Phlogophora meticulosa

DANAIINAE BUTTERFLIES

Danaus gilippus

D. plexippus

Lycorea ceres

*From Birch, 1974

Floral Scent Attractants (Allurements). "Tropical
nights are filled, almost beyond belief with the fragrances
of beetle-pollinated blossoms...", so state Faegri and van
der Pijl (1971). This observation and the fact that plants,
such as cycads, emit pervasive odors led them to suggest that
odor is an older attractant than color. Whatever the sequence,
it is clear that floral perfumes are very important in the
attraction of bees, flies, carrion beetles, hawkmoths, moths,
butterflies and bats. Unlike insect sex attractants, which
have received considerable attention from chemists and biologists, there are few chemical studies available on the odorants of flowers, particularly on the role of floral attractants in flower-pollinator coevolutionary interactions. In
this section, the distribution of floral odorants, albeit
meager, is discussed, with structural similarities to insect
sex attractants noted.

Information on floral attractants comes from the analysis of species whose scents are of commercial importance.
These studies are reviewed by Guenther (1948-1952) in his
series on the essential oils and from Nicolas (1973) on
miscellaneous volatile constituents from higher plants.
Only recently have evolutionists considered this character,
which is so important, in the ethological isolation of species.

The scent of flowers is quite diverse and consists of
straight-chain hydrocarbons derived from fatty acids,
monoterpene, sesquiterpene and phenolic alcohols, aldehydes,
ketones, simple carboxylic acids, pyrrolizidine and indole
alkaloids, amines and glucosinolates. Figure 3 includes
some representative floral attractants and the plant source,
while significant compounds are enumerated for several
species in Table 14. Most species contain a few to several
compounds often representing more than one class of compounds. Among alcohols, eugenol, linalool geraniol,
benzyl alcohol and cineole, are the most frequent contribu-
tors to fragrances. In *Majorica holtensis* and several
species of *Catasetum* cineol constitutes over 60% of the
volatle oil from flowers. Benzyl-acetate is the most
important ester in flower fragrances in terms of species
occurrence and percentage of volatile oils. It accounts
for 65% of the oil of *Jasmium officinale*. The chemical
analysis of the fragrances of flowers has been restricted
to those which are primarily bee pollinated. We know
little about the identity of the odorants responsible for

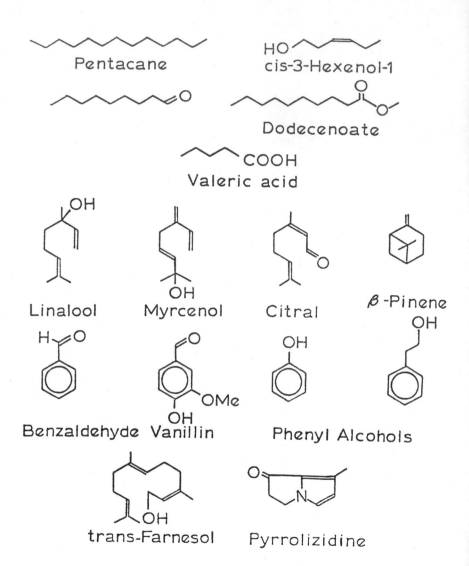

Figure 3.   Floral Attractants.

Table 14. Important odorants of flowers

Species	Compound Class			
	Alcohols	Esters	Acids	Aldehydes Phenols and Ethers
Rosa damescana	phenylethyl alcohol citronellol geraniol nerol			ethyl phenols
Acacia farnesiana	benzyl alcohol linalool cresol mixture		benzoic acid butyric acid	
Cananga odorata	benzyl alcohol nerol farnesol phenylethyl alcohol eugenol linalool geraniol	benzyl acetate methyl benzoate methyl salicylate p-cresol acetate	benzoic acid salicylic acid	
Jasminum officiniale	linalool benzyl alcohol	benzyl acetate linalyl acetate		jasmone
Polyanthus tuberosa	geraniol nerol farnesol benzyl alcohol eugenol	benzyl benzoate methyl salicylate	butyric acid	
Narcissus taxetta	eugenol benzyl alcohol cinnamyl alcohol		benzoic acid	benzaldehyde
Gardenia spp.	terpineol linalool	benzyl acetate styrolyl acetate linalyl acetate methyl anthranilate		

the fruity, fermenting, spermatic or aminoid fragrances
of flowers pollinated by beetles, the carrion fragrances
of fly-pollinated flowers, the musty and sour fragrances
of bat-pollinated flowers or the pungent and sweet
fragrances of moth-pollinated flowers[39].

The only substantial studies on floral scents conducted within the framework of pollination ecology have
been on orchids[28,29,59,72,73,149]. The work by Dodson[28,29,
59] and his associates is the most illuminating and will be
summarized here. They found approximately 60 different
compounds in a survey of the fragrances of 150 species from
16 genera. Most species produce seven to ten compounds,
but the range is from 2 to 18. Almost 60% of the species
produced 1,8-cineole, although the amount varied from as
much as 90% of the odor to 7% depending on the species.
Other important compounds are eugenol, benzyl acetate,
methyl cinnamate, and methyl salicylate. Of particular
importance is the distribution of these and other compounds
in congeneric orchid species, and the percentage of the
total fragrance complement they comprise. This can best
be seen in the data on *Catasetum* (Table 15). The differences in both parameters are striking and explain why
most species have rather distinctive odors.

Williams and Dodson (1972) ingeniously demonstrated
the role of orchids in the attraction of euglossine bees.
Pads saturated with pure compounds were attached to trees
to determine the number and species of euglossine bees
which they attracted. Certain compounds were more effective than others in attracting the pollinators, and the
preference of the pollinator varies *inter se*. However,
when pads contain mixtures of the compounds the number of
visitors is reduced and the differential preferences of
the bee species is emphasized. Even a 2% adulteration
of a pure compound with a second elicits a pronounced
response. Thus, it is clear that the role of orchid
odorants is to attract euglossine bees and that the
nature and proportion of each compound in the ensemble
determines in large measure which species will be
attracted. Ethological isolation is the consequence of
acute scent perception by the pollinators.

Dodson (1974) continuing his studies on the relationship of euglossine bees and neotropical orchids has noted

Table 15.   Percentages of fragrance compounds in Catasetum (after Hills et al., 1972)

Species	Compound													
	a-pinene	β-pinene	myrcene	cineole	ocimene	linalool	methyl benzoate	benzyl acetate	d-carvone	methyl salicylate	2-phenyl ethyl acetate	nerol	2-phenyl ethanol	methyl cinnamate
C. roseum	32.1	1.7	--	52.9	6.2	--	0.7	--	--	--	--	--	--	1.1
C. russellianum	14.1	0.6	--	77.9	--	--	--	--	--	--	--	--	--	--
C. thylaciochilum	--	--	--	--	--	--	--	--	--	--	91.3	--	7.8	--
C. werczewitzii	1.1	--	--	3.5	--	1.3	--	--	--	--	--	5.7	--	--
C. dilectum	1.4	--	0.2	98.3	--	--	--	--	--	--	--	--	--	--
C. discolor	70.1	1.1	--	4.4	--	--	0.4	1.4	1.4	--	--	--	--	--
C. longifolium	59.4	0.8	--	7.0	--	--	--	5.9	2.3	--	8.2	--	--	--
C. maculatum	67.2	1.1	--	5.8	--	--	0.2	3.2	3.0	--	--	--	--	--
C. collare	0.3	--	--	2.8	--	--	64.2	--	--	32.3	--	--	--	--
C. gnomus	5.5	--	--	9.9	--	--	9.3	--	--	59.1	--	--	--	--
C. luridum	10.5	--	--	83.9	--	--	--	--	--	--	--	--	--	--
C. ochraceum	--	--	--	--	13.9	--	--	--	--	--	--	--	--	--
C. callosum	1.1	--	--	5.3	83.7	--	--	--	--	--	--	--	--	--
C. cirrhaeoides	35.4	--	--	3.1	--	--	--	--	--	--	--	--	--	--
C. atratum	2.41	0.2	89.5	--	--	--	--	--	--	--	--	--	--	--

that only male bees visit orchids while the females visit
other (non-orchid) flowers. During the time that the male
bees are visiting the flowers, they scrape off the waxy
cuticle which contains floral fragrances (e.g. cineole,
geraniol and other monoterpenes) and pack the fragrances
into specialized pouches on the hind tibia. Males,which
obtain the orchid attractants, act as centers about which
swarms of males gather ("lek" formation), thereby attracting
females by the communal buzzing and visual displays.

A system which parallels the bee-orchid-terpene
interaction is the relationship of pyrrolizidine-contain-
ing plants and butterflies. Pliske (1975a,b) has found
that live plants of *Heliotropium indicum* and *Eupatorium
xestolepis* (Compositae) are attractive to numerous
Lepidoptera (Danainae, Ithomiinae, Ctenuchidae, Arctiidae)
which congregate on dead shoots and infloresences to feed.
Pliske also established that the male Ithomiinae and
Danainae ingest alkaloidal precursors for sex attractants
(aphrodisiacs). Furthermore, the fragrance of *E.
xestolepis* attract male Ithomiinae and Ctenuchidae; these
constitute nearly 100% of the pollinators.

Von Frisch (1919) and Ribbands (1955) demonstrated
scent perception in honeybees by showing that they could
be trained to differentiate between different floral
odorants. Bees trained to 1% phenyl ethyl alcohol could
distinguish this scent from a 1% mixture containing 119
parts phenyl ethyl alcohol and 1 part geraniol; similarly
1% benzyl acetate could be distinguished from 119:1
mixture of benayl acetate and linalool in the same concentration. Honeybee thresholds for perceiving floral
odorants vary from 1/10 to 1/100 of the thresholds for
humans. They can recognize the presence of phenyl ethyl
alcohol, benzyl acetate and methyl benzoate in concentrations of $1:10^8$ and linalool in $1:5 \times 10^6$. Using electrophysiological techniques and analyzing single olfactory
cells of antennal plate organs Vareschi (1971) demonstrated
that honeybees could distinguish among floral fragrances
(e.g. amyl acetate, geraniol, linalool, benzyl alcohol).
In contrast, not all odorants can be distinguished by
these organisms. Alfalfa flowers have four primary
volatile components, ocimene, myrcene, limonene and
linalool. Honeybees could not distinguish among the
first three odorants, but could distinguish between
linalool and the other three[153].

Not only can bees discriminate between floral odorants, species of *Apis* display species-and race-specific odor preferences. Koltermann (1973) demonstrated such with three races of *Apis mellifera* and *A. cerana*. In laboratory trials, the differences in preference were related to the flora in which a taxon forages, as evident in the fact that scents eminating from flowers native to their own native area are preferred. Since the foraging preferences of bees are genetically determined[40] as well as determined by the availability of different food sources, genetic differentiation among related bees for response to floral signals is to be expected if the regions in which they forage have different floral composition and dominance. Koltermann (1973) also demonstrated that the rapidity with which an unattractive scent is chosen is also specific to different races and species of bees. *Apis cerana*, indigenous to India, are the quickest to respond to the unattractive scent, whereas the races of *Apis mellifera* from Egypt, Italy and central Europe were most hesitant in that order. These differences may be adaptations to differences in the reliability to major food sources, the less reliable the sources the more plastic the behavior of the pollinator. Experiments of this type have not been conducted on other classes of pollinators, but it seems likely that if odor were an important cue the results would be similar. Indeed this should be a very rewarding area of pollinator-plant coevolution.

The olfactory sensitivities of beetles, flies and moths are poorly understood. We have only anecdoctal or circumstantial evidence that they have acute scent perception. Proctor and Yeo (1972) provide a good account of the material.

Olfaction ostensibly assists nectivorous bats in their search for food, but the extent to which it is normally relied upon is not clear[133]. It is noteworthy that nectivorous and frugivorous bats have larger olfactory bulbs than do insectivarous genera. The unpleasant odor of bat flowers (butyric acid) is similar to that produced by glandular secretion by bats and used for communication. It may well be that the exploitation of bats by flowering plants has involved mimicking the smell of bats[38].

A similar situation may prevail in moth visited flowers which are known to produce odorants which are

detected by night-moths from distances comparable to those
noted in moths detecting long-range sex attractants. It
therefore seems plausible that long-range and some short-
range floral attractants are mimicking insect sex attractants, with the compounds being in accord with the
theoretical constraints on molecular size in insect chemical communication systems[14, 158].

CONCLUSIONS

No species of arthropod nor plant can survive without
communicating information for attracting mates, or pollinators and seed dispersers, or without being able to avoid
enemies, or to find food as in the case of arthropods. To
accomplish these ends, most species rely on chemical
signals, which in plants and arthropods (especially insects)
show a striking parallelism. The employment of similar
signals in part owes to the intimate contact and reciprocal
selection between arthropods and plants, and to the vulnerability of both groups of organisms to invertebrate and
vertebrate predators.

The defensive substances of plants are similar or
identical to those used by arthropods. Selection in plants
for effective inhibitors of herbivory has resulted in the
sequestering or exudation of low molecular weight compounds
and potent olfactants which afford plants a type of defense
used by arthropods against arthropods or other predators.
The extraordinary diversity in natural products utilized
by plants and arthropods underscores the biochemical
versatility of these groups. Most defensive secretions
in arthropods are highly volatile, and as a rule are
highly odoriferous; many are irritants or bitter tasting[36,121].
The same is true of glandular secretions in plants and to
a lesser extent of substances stored beneath the surface
of leaves. Most plant and many arthropod defensive
secretions are mixtures, rather than single compounds.

Although it seems likely that the chemical defenses
of plants to predation have evolved in tandem with those
of arthropods, we do not mean to imply that all defensive
substances used by arthropods are synthesized by arthropods.
Indeed there are many well known examples of arthropods
sequestering specialized plant products and using them
for their own defense.

Many arthropods with chemical defenses, especially those conferring unpalatability, are elements in mimetic assemplages of one kind or another and have warning coloration[36]. The extent to which molecular mimicry has developed in plants is poorly understood. The best examples involve the mimicking of fleshy fruits by seeds for the purpose of attracting seed dispersers[145]. However, fruits or seeds do have distinctive colors or odors. Accordingly mimicry of a poisonous or unpalatable fruit or seed by a nonpoisonous one could reduce the level of fruit or seed predation. Similarly, mimicry of distinctive volatile substances of poisonous or unpalatable plants by non-poisonous ones could reduce the level of foliage loss.

To a greater extent than is true for defensive substances, compounds serving as sex attractants in some phytophagous insects occur in their host plants. Indeed, diversification of insect species could be primarily due to the potential sex attractants available during the evolution of host plants rather than to the evolution of biosynthetic pathways independent of changes in certain dietary chemicals[57]. If this were true, it would help explain why the odorants of flowers and sex pheromones of their pollinators (especially lepidopterans and hymenopterans) tend to be variations on the same theme. It might also permit us to predict the general nature of the floral odorants in species whose flowers have not been studied from knowledge of the sex attractants of their pollinators. Conversely, knowledge of floral odorants might also permit predications about the sex attractants of pollinators. These predications may be valid even if the host was not the source of the sex attractant. A similarity between sex attractants and flower scent still would be expected since natural selection is likely to favor plants which fall within the sensory realm of a pollinator and which the pollinator finds attractive.

The design-features of chemical communication between plants and pollinators parallels that employed within insect species. Communication beyond a few meters is primarily by olfactory signals, especially terpenes. Floral scents tend to be comprised of compounds which have molecular weights that are usually between 200 and 300, and have low olfactory thresholds. Signal systems often are

characterizes by a high specificity of the chemical
signal, which is realized either by a highly specific
structure, or by a characteristic combination of quite
simple components[9,72]. Floral scent exhibits temporal
patterning in quantity, especially when pollination
is at dawn and dusk[2,38]. Qualitative aspects of temporal
patterning are poorly understood.

Finally, it is important to recognize that different
species of plants and arthropods frequently utilize the
same compound for entirely different functions, a situation which reflects the biological peculiarities of each
species. For example, citral is a component of floral
fragrance, it may be a feeding repellent, it is the
primary component of the Nasonov gland secretions in
honeybees, and it is a component of alarm and trail lay-
ing phermones in a number of insects[10]. Several other
terpenes play multiple roles in communities. The
multiple function of secondary compounds has fostered
the development of communicative versatility in plants
and arthropods and has enabled both groups to undergo
extensive radiation and to develop the elaborate
coevolutionary adaptations in spite of biosynthetic
constraints.

Acknowledgements: The authors wish to thank Drs. G.H.N.
Towers, J.C. Mitchell, C.K. Wat and C. Stanley (Department
of Botany, University of British Columbia, Vancouver) for
helpful comments on the manuscript and F. Bond for her
illustration of Figure 1. ER thanks the Canadian Medical
Research Council (Grant MT 2164 to JCM) for financial
assistance.

1. Aneshanseley, D., T. Eisner, J. M. Widom and B. Widom.
    1969. Biochemistry at 100 C: the explosive dis-
    charge of bombardier beetles (Brachinus). *Science*
    165:61.

2. Baker, H. G. and P. D. Hurd. 1968. Intrafloral ecology.
    Ann. Rev. Entomol. 13:385-414.

3. Baker, J. E. and D. N. Norris. 1972. Effects of feed-
    ing-inhibitory quinones on the nervous system of
    Periplaneta. Experimentia 28:31-32.

4. Barbier, M. and E. Lederer. 1957. Sur les benzoquinones
    du venum de troi especies de myriopodes. *Biolimlea*
    22:169.

5. Bellas, T. E., W. V. Brown and B. P. Moore. 1974. The
    alkaloid actinidine and plausible precursors in
    defensive secretions of rove beetles. *J. Insect.*
    Physiol. 20:277.

6. Beroza, M. 1971. Insect Sex Attractants. *Amer.*
    Scientist. 59:320.

7. Birch, A. J., W. V. Brown, J. E. T. Corrie and B. P.
    Moore. 1972. Neocembrene-A, a termite trial
    pheromone. *J. C. S. Perkin Trans. 2653*.

8. Birch, M. C. 1974. Aphrodisiac pheromones in insects.
    In "Pheromones". (M. C. Birch, ed.) p. 115. North-
    Holland, Amsterdam.

9. Blum, M. S. 1971. Dimensions of chemical sociality,
    pp. 147-159. *In* Chemical Releases In Insects
    (A. S. Tahori, ed.). Gordon and Breach Sci. Publ.,
    New York.

10. Blum, M. S. and J. M. Brand. 1972. Social insect pher-
    omones: Their Chemistry and function. *Amer. Zool.*
    12:533.

11. Blum, M. S., J. G. MacDonnell, J. M. Brand, R. M.
    Duffield and H. M. Fales. 1973. Phenol and benzal-
    dehyde in the defensive secretion of a millipede.
    Ann. Ent. Soc. 66:235.

12. Bohlmann, F., T. Burkhardt, and C. Zedro. 1973.
    Naturally Occurring Acetylenes. Academic Press, New
    York.

13. Borden, J. H. 1974. Aggregation pheromones in the
    Scotylidae. *In* "Pheromones", (M. C. Birch, ed.)
    p. 134. North-Holland, Amsterdam.

14. Bossert, W. H. and E. O. Wilson. 1963. The analysis
    of olfactory communication among animals. *J.*
    Theoretical Bio. 5:443.

15. Brand, J. M., H. M. Fales, E. A. Sokoloski, J. G.
    MacDonnell, M. S. Blum and R. M. Duffield. 1973.
    Identification of mellein in the mandibular gland
    secretions of carpenter ants. *Life Sciences 13*:201.
16. Buhr, H., R. Toball and K. Schreiber. 1958. Die
    Wirkung von einigen pflanzenlichen Sonderstoffen,
    inbesondere von Alkaloiden, auf die Entwicklung
    der Larven des Kartoffelkafers. (*Leptinotarsa
    decemlineata* Say). *Entomol. Exp. & Appl. 1*:209-224.
17. Butler, C. G. 1967. Insect pheromones. *Biol. Rev.
    42*:42-87.
18. Butler, C. G. and D. H. Calam. 1969. Pheromones of
    the honey bee--the secretions of the Nassanoff
    gland of the worker. *J. Insect Physiol. 15*:237-244.
19. Callaham, R. Z. 1966. Nature of resistance of pines
    to bark beetles, p. 197-201. *In* Breeding pest resis-
    tant trees (H. D. Gerhold,ed.). Pergamon Press, New
    York.
20. Carrel, J. E. and T. Eisner. 1974. Cantharidin:
    Potent feeding deterrent to insects. *Science
    183*:755.
21. Cavill, G. W. K. and D. V. Clark. 1971. Ant secretions
    and cantharidin. *In* "Naturally Occurring Insecti-
    cides". (M. Jacobson and D. G. Crosby, eds.) p. 271.
22. Cavill, G. W. K. and E. Houghton. 1974. Some pyrazines
    derivatives from Argentine ant, Iridomyrmex humilis.
    Austr. J. Chem. 48:228.
23. Chapman, R. F. 1974. The chemical inhibition of feed-
    ing by phytophagus insects: A review. *Bull Ent.
    Res. 64*:339.
24. Clayton, R. B. 1970. The chemistry of nonhormonal
    interactions. Terpenoid compounds in plants. *In*
    "Chemical Ecology". (E. Sondheimer and J. B.
    Simeone, eds.) Academic Press, New York.
25. Curtis, J. D. and N. R. Lersten. 1974. Morphology,
    seasonal variation, and function of resin glands
    on buds and leaves of *Populus deltoides*. (Salicaceae).
    Amer. J. Bot. 61:835.
26. Dahlman, D. and E. T. Hibbs. 1967. Response of
    Empoasca fabae to tomatine, solanine, leptine I,
    tomatidine, solanidine and demissidine. *Ann.
    Entomol. Soc. Amer. 60*:732-740.
27. Dodson, C. H. 1974. *In* "Coevolution of Animals and
    Plants. (L. E. Gilbert and P. Raven eds.). Univ.
    of Texas, Press, Austin.

28. Dodson, C. H., R. L. Dresser, H. G. Hills, R. M. Adams, and N. H. Williams. 1969. Biologically active compounds in orchid fragrances. *Science*. *164*:1243.

29. Dodson, C. H. 1970. The role of chemical attractants in orchid pollination. *In* "Biochemical Coevolution". (K. L. Chambers, ed.) Oregon State Univ., Corvalis.

30. Duffey, S. S. 1974. The Biosynthesis of Defensive Chemicals by Millipedes: Parallelism with Plant Biosynthetic Pathways. Unpublished Ph.D. thesis, University of British Columbia, Vancouver.

31. Duffield, R. M., M. S. Blum and J. M. Brand. 1974. Guaicol in the defensive secretions of polydesmid millipedes. *Ann. Ent. Soc. Am. 67*:821.

32. Edgar, J. A. and C. C. J. Culvenor. 1974. Pyrrolizidine ester alkaloids in danaid butterflies. *Nature* (Lond.) *248:*614.

33. Eisner, T., J. S. Johnessee, J. Carrel, L. B. Hendry and J. Meinwald. 1974. Defensive use by an insect of plant resin. *Science 184*:996.

34. Eisner, T., L. B. Hendry, D. B. Peakall, and J. Meinwald. 1971. 2,5-dichlorophenol (from ingested herbicide) in defensive secretion of grasshopper. *Science 172*:177.

35. Eisner, T., A. F. Kluge, M. I. Ikeda, Y. C. Meinwald and J. Meinwald. 1971. Sesquiterpenes in the osmeterial secretion of a papilionid butterfly, **Battus polydamas**. *Insect Physiol. 17:*245.

36. Eisner, T. 1970. Chemical defense against predators in arthropods. *In* "Chemical Ecology". (E. Sondheimer and J. B. Simeone, eds.). Academic Press, New York.

37. Evans, D. A. and C. L. Green. 1973. Insect attractants of natural origin. *Chemical Reviews 12*:75-97.

38. Faegri, K. and van der Pijl. 1971. Principles of Pollination Ecology. Pergamon Press, New York.

39. Faegri, K. and L. van der Pijl. 1973. The principles of pollination ecology. 2nd ed. Pergamon Press, New York.

40. Free, J. B. and I. H. Williams. 1973. Genetic determination of honeybee (*Apis mellifera L.*) foraging preferences. *Ann. Appl. Bio. 73:*137.

41. Geissman, T. A. and D. H. G. Crout. 1969. Organic Chemistry of Secondary Plant Metabolism. Freeman, Cooper and Co., San Francisco.

42. Gibbs, R. D. 1974. Chemotaxonomy of Flowering Plants. Vols. I-III. McGill-Queen's Univ. Press, Montreal.

43. Gibson, R. W. 1971. Glandular hairs providing resis-
     tance to aphids in certain wild potato species.
     *Ann. Appl. Biol. 68:*113.

44. Gilbert, B. C., J. E. Baker and D. M. Morris. 1967.
     Juglone (5-hydroxy-1,4-napthaquinone) from *Carya
     ovata*, a deterrent to feeding by *Scolytus
     multistriatus*. *J. Insect Physiol. 13:*453.

45. Gilbert, B. C., and D. M. Harris. 1968. A chemical
     basis for bark beetle distinction between hosts
     and non-host. *J. Insect Physiol. 14:*106.

46. Gilbert, L. E. 1971. Butterfly-plant coevolution:
     Has *Passiflora adenopoda* won the selectional race
     with heliconine butterflies? *Science 172:*585.

47. Gilmour, D. 1961. The Biochemistry of Insects.
     Academic Press, New York.

48. Grisdale, S. K. and G. H. N. Towers. 1960. Bio-
     synthesis of arbutin from some phenylpropanoid
     compounds in *Pyrus communis*. *Nature* (Lond.)
     *18:*1130.

49. Guenther, E. 1952-58. The Essential Oils. 6 Vols.
     Van Norstrand Reinhold, New York.

50. Happ, G. M. and J. Meinwald. 1965. Biosynthesis of
     arthropod secretions. I. Monoterpene synthesis in
     an ant. *J. Amer. Chem. Soc. 87:*2507.

51. Harley, K. L. 1967. A note on the influence of plant
     chemicals on the growth and survival of *Aedes
     aegypti* L. larvae. *Canad. J. Zool. 45:*1297-1300.

52. Harley, K. L. and A. J. Thorsteinson. 1967. The
     influence of plant chemicals on the feeding behavior,
     development, and survival of the two-striped grass-
     hopper, *Melanoplus bivittatus* (Say), Acarididae.
     *Canad. J. Zool. 45:*315-319.

53. Hedin, P. A., F. G. Maxwell and J. N. Jenkins. 1974.
     Insect plant attractants, feeding stimulants,
     repellents, deterrents and other related factors
     affecting insect behavior. *In* "Proc. Summer
     Institute on Biological Control of Plant-Insects
     and Disease". (F. G. Maxwell and F. A. Harris, eds.)
     Univ. of Mississippi Press, Jackson.

54. Hedin, P. A., L. R. Miles, A. C. Thompson and J. P.
     Minyard. 1968. Constituents of the cotton bud.
     Formulation of a boll weevil feeding stimulant
     mixture. *J. Agric. Food Chem. 16:*505-513.

55. Heftman, E. 1974. The biological significance of
     steroids in plants. *Phytochem. Bull. of the Bot.
     Soc. Am. 7:*55.

56. Hegnauer, R. 1962–1966. Chemotaxanomie der Pflanzen. Vols. I–IV, Birkhauser, Basel.
57. Hendry, L. B., J. K. Wichman, D. M. Hindenlang, R. O. Mumma, and M.E. Anderson. 1975. Evidence for the origin of insect sex pheromones present in food plants. *Science 187*:608.
58. Herout, V. 1970. Some relations between plants, insects and their isoprenoids. *In* "Progress in Phytochemistry, Vol. 2 (L. Reinhold and Y. Liwschitz, eds.). Interscience Publish., New York.
59. Hills, H.G., N.H. Williams and C.H. Dodson. 1968. Identification of some orchid fragrance components. *Amer. Orchid Soc. Bull. 37*:967.
60. Hills, N.G., N.H. Williams and C.H. Dodson. 1972. Floral fragrances and isolating mechanisms in the genus *Catasetum* (Orchidaceae). *Biotropica 4*:61.
61. Jacobson, M. 1974. Insect Pheromones. *In* "Physiology of Insects". (M. Rockstein, ed.) Vol. 3. Academic Press, New York.
62. Jacobson, M. 1972. Insect Sex Pheromones. Academic Press, New York.
63. Jensen, S.R., B.J. Nielsen and R. Dahgren. 1975. Iridoid compounds their occurrence and systematic importance in the angiosperms. *Botaniska Notiser. 128*:148.
64. Jones, D.A., J. Parsons and M. Rothschild. 1962. Release of hydrocyanic acid from crushed tissues of all stages in the life–cycle of species of Zygaeninae Lepidoptera. *Nature* (Lond.) *193*:52.
65. Karlson, P. 1970. Terpenoids in insects. *In* "Natural Substances Formed Biologically from Mevalonic Acid. (T. W. Goodwin, ed.). Academic Press, New York.
66. Karlsson, E. 1973. Chemistry of some potent animal toxins. *Experientia 29:* 1319.
67. Kingbury, J.M. 1964. Poisonous Plants of the United States and Canada. Prentice–Hall, Englewood Cliffs, New Jersey.
68. Kislow, C.J. and L.J. Edwards. 1972. Repellent odours in aphids. *Nature* (Lond.) *235:* 108.
69. Kluge, A.F. and T. Eisner. 1971. Defense mechanisms of arthropods. XXVII. A quinone and a phenol in the defensive secretion of a parajulid millipede. *Ann. Ent. Soc. Amer. 64:*314.
70. Koltermann, R. 1973. Rassen-bsq. artesezifische Duftbewertung bei der Honigbiene und okologische adaptation. *J. Comp. Physiol. 85:* 327.

71. Kullenberg, B. 1953. Some observations on scents among bees and wasps (Hymenoptera). *Entomol. 74:* 1-7.

72. Kullenberg, B. and G. Bergstrom. 1975. Chemical Communication between living organisms. *Endeavor 122:* 59.

73. Kullenberg, B. and G. Bergstrom. 1973. The pollination of *Ophrys* orchids. *In* "Nobel Symposium 25: Chemistry in Botanical Classification". Academic Press, New York.

74. Lanier, G.N. and W.E. Burkholder. 1974. Pheromones in speciation of Coleoptera. *In* "Pheromones". (M.C. Birch, ed.). North-Holland, Amsterdam.

75. Levin, D.A. 1973. The role of trichomes of plant defense. *Quart. Rev. Biol. 48:* 3.

76. Levin, D.A. 1971. Plant phenolics: An ecological perspective. *Amer. Nat. 105:*157.

77. Lukefahr, M.H. and J.E. Houghtaling. 1969. Resistance of cotton strains with high gossypol concentrations to *Heliotis* spp. *J. Econ. Entomol. 62:* 588-591.

78. Ma, Wei-chun. 1969. Some properties of gustation in the larvae of *Pieris brassicae*. *Entomol. Exp. and Appl. 12:* 584-590.

79. Ma, Wei-chun. 1972. Dynamics of feeding responses in *Pieris brassicae* as a function of chemosensory input: a behavioral, ultrastructural and electrophysiological study. *Meded. Landhoogegsch, Wagingen 72.* 192 p.

80. MacConnell, J.G., M.S. Blum and H.M. Fales. 1970. Alkaloid from fire ant venom: Identification and synthesis. *Science 168:* 840.

81. MacDonnell, J.G. and R.M. Silverstein. 1973. Recent results in pheromone chemistry. *Angew. Chem. Internat. Edit. 12:*643.

82. Meinwald, J., C.J. Boriack, C.J. Schneider, D. Boppre, M. Wood, and T. Eisner. 1974. Volatile ketones in the hair-pencil secretion of danaid butterflies *(Amauris* and *Danaus).* *Experientia* (Basel) *30:*721.

83. Meinwald, J.K. Erickson, M. Hartshorn, Y.C. Meinwald and T. Eisner. 1968. Defensive mechanisms of arthropods. XXIII. An allenic sesquiterpenoid from the grasshopper *Romalea microptera*. *Tet. Lett. 25:* 2959.

84. Meinwald, J., W.R. Thompson and T. Eisner, 1971. Pheronomes. VII. African Monarch: Major components of hair pencil secretion. *Tet. Lett. 38:*3485.

85. Meinwald, J., G.M. Happ, J. Lapows, and T. Eisner. 1966. Cyclopenanoid terpene biosynthesis in a phasmid insect and in catmint. *Science 151:* 79.

86. Meinwald, J., Y.C. Meinwald and A.M. Chalmers. 1968b. Dihydromatricaria acid: Acetylenic acid secreted by soldier beetle. *Science 160:*890.

87. Miles, P.W. 1968. Insect secretions in plants. *Ann. Rev. Phytopathol. 6:* 137.

88. Miles, P.W. 1969. Interaction of plant phenols and salivary phenolases in the relationship between plants and Hemiptera[a]. *Ent. Exp. and Appl. 12:*736.

89. Miller, L.P. 1973. Phytochemistry. Vols. I-III. Van Norstrand Reinhold Co., New York.

90. Mitchell, J.C. 1975a. Biochemical basis of geographical ecology. Part I.*Int J of Dermatology 14:* 239.

91. Mitchell, J.C. 1975b. Biochemical basis of geographical ecology. Part II.*Int J of Dermatology 14:* 301.

92. Moore, B.P. 1967. Hydrogen cyanide in the defensive secretions of larval *Paropsini* (Coleoptera: Chrysomelidae). *J. Austral. Entomol. 6:* 36.

93. Moore, B.P. 1968. Biochemical studies on termites. *In* "Biology of Termites". (K. Krishna and F.M. Weesner, eds.). Academic Press, New York.

94. Muller, H. 1888. The Fertilization of Flowers., trans. by D'Arcy W. Thompson. MacMillan and Co., London.

95. Nicholas, H.J. 1973. Miscellaneous volatile plant products. *In* "Phytochemistry", Vol. III. (L.P. Miller, ed.). Van Norstrand Reinhold Co., New York.

96. Norris, D.M. and H.M. Chu. 1974. Chemosensory mechanism in *Periplaneta americana:* electroantennogram comparisons of certain quinone feeding inhibitors. *J. Insect Physiol. 20:* 1687-1696.

97. Oku, H. 1959. Biochemical studies on *Cochliobolus miyabeanus* V. β-glucosidase activity of the fungus and the effect of pheno glucosidase on the mycelial growth. *Annu. Rep. Takamine Lab. 11:* 190 (cited in Levin, 1971).

98. Pollares, E.S. 1946. Notes on the poison produced by the *Polydesmus (Fontaria) vicinus* (L.). *Arch. Biochem. 9:* 105.

99. Parsons, J.A. and M. Rothschild. 1964. Rhodanese in the larva and pupa of the common blue butterfly *(Polyommatus icarus* Rott.) Lepidoptera. *Entomologists Gazette 15:* 58.

100. Pasteels, J.M., C. Deroe, B. Tursch, J.C. Braekman, D. Daloze, and C. Hootele. 1973. Distribution et activities des alcaloides defensifs de Coccinelidae. *J. Insect Physiol. 19:* 1771.

101. Pavan, M. and M.V. Dazzini. Toxicology and pharmacolog Arthropoda. *In* "Chemical Zoology". (M. Florkin, ed.). Academic Press, New York.

102. Pliske, T.E. 1975a. Attraction of Lepidoptera to plants containing pyrrolizidine alkaloids. *Environ. Ent. 4:* 455.

103. Pliske, T.E. 1975b. Pollination of pyrrolizidine alkaloids containing plants by male lepidoptera. *Environ. Ent. 4:*474.

104. Pliva, J., V. Herout, B. Schneider and F. Sorm. 1953. On terpenes XLIII. Infra red investigation of teroenes-IV. *Coll. Czech. Comm. 18:* 500.

105. Proctor, M. and P. Yeo. 1972. The Pollination of Flowers. Taplinger Publish. Co.

106. Rees, C.J.C. 1969. Chemoreceptor specificity associated with choice of feeding site by the beetle, *Chrysolina brunsvicensis* on its food plant, *Hypericum hirsutum. Ent. Exp. Appl. 12:* 565.

107. Regev, S. and W.W. Cone. 1975. Evidence of farnesol as a male sex attractant of the two-spotted spider mite, *Tetranychus urticae* Koch (Acarina: Tetranychidae). *Environ. Entomol. 4:* 307-311.

108. Rher, S.S., D.H. Janzen and P.O. Feeny. 1973. L-dopa in legume seeds: A chemical barrier to insect attack. *Science 181:* 81.

109. Ribbands, C.R. 1955. The scent perception of the honeybee. *Proc. Royal Soc. B. 143:* 367.

110. Rice, W.Y. and J.Y. Coke. 1965. Structure and configuration of alkaloids. II. Cassine. *J. Org. Chem. 31:* 1010.

111. Roelofs, W.L. and R.T. Carde. 1974. Sex pheromones in the reproductive isolation of Lepidopterous species. *In* "Pheromones". (M.C. Birch, ed.). North-Holland, Amsterdam.

112. Roth, L.M. and T. Eisner. 1962. Chemical defenses of arthropods. *In* "Annual Reviews of Entomology". (R.F. Smith and E.A. Steinhaus, ed.). Academic Press, New York.

113. Rothschild, M., R. Aplin, T. Reichstein, J. von Euw and R.R. Harman. 1970. Toxic lepidoptera. *Toxicon 8:* 293.

114. Rothschild, M. 1972. Secondary plant substances and
          warning colouration in insects. *In* "Insect-
          Plant Relationships". (H. F. van Enden ed.).
          Symposium of Royal Entomological Society of
          London.
115. Schildknecht, H.K. Holoubek, K. Weis and H. Krammer.
          1964. Defensive substances of arthropods: Their
          isolation and identification. *Angew. Chem. Inter.
          Edit. 3:* 73.
116. Schildknecht, H. 1970. The defensive chemistry of
          land and water beetles. *Angew. Chem. Internat.
          Edit. 9:* 1.
117. Schildknecht, H. 1971. Evolutionary peaks in the
          defensive chemistry of insects. *Endeavour 30:*136.
118. Schmeltz, I. 1971. Nicotine and other tobacco alka-
          loids, p. 99-136. *In* "Naturally occurring
          insecticides". (M. Jacobson and D.G. Crosby,
          eds.). Macell Dekker, New York.
119. Schneider, D., M. Boppre, H. Schneider, W.R. Thompson,
          C.J. Boriack, R.L. Petty and J. Meinwald. 1975.
          A pheromone precursor and its uptake in male
          Danaus butterflies. *J. Comp. Physiol. 97:* 245.
120. Schoonhoven, L.M. 1972. Secondary plant substances
          and insects. *In* "Recent Advances in Phyto-
          chemistry, Vol. 5. (J. B. Harborne, ed.)
          Academic Press, New York.
121. Schoonhoven, L.M. 1972. Secondary plant substances
          and insects, pp. 197-224. *In* "Recent Advances
          in Phytochemistry Vol. 5: Structural and
          functional aspects of phytochemistry". (V.C.
          Runeckles and T.C. Tso, eds.). Academic Press,
          New York.
122. Schoonhoven, L.M. and I. Derksen-Koppers. 1973.
          Effects of secondary plant substances on drink-
          ing behavior in some Heteroptera. *Entomol. Exp.
          and Appl. 16:* 141-145.
123. Schwartung, A.E. 1963. Poisonous seeds and fruits.
          Prog. Chem. Toxicol. 1: 385-401.
124. Seigler, D.S. 1975. Isolation and characterization
          of naturally occurring cyanogenic compounds.
          Phytochem. 20: 850.
125. Self, L., F. Guthrie and E. Hodgson. 1964. Metabolism
          of nicotine by tobacco-feeding insects.
          *Nature 204:*300-301.

126. Shaver, T.M., M.J. Lukefar and J.A. Garcia. 1970.
      Food utilization, ingestion, and growth of larvae
      of the bollworm and tobacco budworm on diets
      containing gossypol. *J. Econ. Entomol. 63:*
      1544-1546.

127. Shaver, T.N. 1974. Biochemical basis of resistance
      of plants to insect antibiotic factors other than
      phytohormones, pp. 485-493. *In* "Proc. Summer
      Inst. Biol. Control of Plant Insects and Diseases"
      (F.G. Maxwell and F.A. Harris, eds.). Univ.
      Mississippi Press, Jackson.

128. Silberglied, R.E. 1975. Communication in the
      Lepidoptera (in press).

129. Staddon, B.W. and J. Weatherston. 1967. Constituents
      of the stink gland of a water bug *Ilyocoris
      cimicoides* L. (Heteroptera:Naucoridae). *Tet.
      Lett. 46:* 4567.

130. Stein, G. 1963a, b. Cited in Regev and Cone.

131. Sturchow, B. and I. Low. 1961. Die Wirkung siniger
      Solanum-alkaloid-glykoside auf den Kartofellkafer,
      Leptinotarsa decemlineata Say. *Entomol. Exp.
      and Appl. 4:* 133-142.

132. Sumimoto, M., M. Shigara, and T. Kondo. 1975. Ethane
      in pine beetles preventing the feeding of the
      beetle, *Monochamus alternatus. J. Insect. Physiol
      21:* 213.

133. Suthers, R.A. 1970. Vision, olfaction, taste. *In*
      "Biology of Bats". (W.A. Wimsett, ed.). Vol. 2.
      Academic Press, New York.

134. Theigles, B.A. 1960. Altered polyphenol metabolism
      in the foliage of *Pinus sylvestris* associated with
      European pine sawfly attack. *Can. J. Bot. 46:* 726.

135. Thomson, R.H. 1964. Reactivity of phenolic compounds.
      In "Biochemistry of Phenolic compounds". (J.B.
      Harborne, ed.). Academic Press, New York.

136. Thomson, R.H. 1970. Naturally Occurring Quinones.
      Academic Press, New York.

137. Todd, G.W., A. Getchum and D.C. Cress. 1971. Resis-
      tance in barley to greenbug, *Schizaphis graminum*
      L. Toxicity of the phenolic and related com-
      pounds and related substances. *Ann. Entomol.
      Soc. Amer. 64:* 718-722.

138. Towers, G.H.N., and P.V. Subba Rao. 1972. Degradation
      metabolism of phenylalanine, tyrosine and dopa. *In*
      "Recent Advances in phytochemistry, (V. Runeckles,
      ed.). Vol. 4. Appleton-Century Crofts, New York.

139. Towers, G.H.N., S.S. Duffey, and S.M. Siegel. 1972. Defensive secretion: biosynthesis of hydrogen cyanide and benzaldehyde from phenylalanine by millipede. *Can. J. Zool. 50:* 1047.

140. Tumlinson, J.H., R.C. Guelder, D.D. Hardee, A.C. Thompson, P.A. Hedin and J.P. Minyard. 1970. The boll weevil sex attractant. *In* "Chemicals Controlling Insect Behavior". (M. Beroza, ed.). Academic Press, New York.

141. Tschinkel, W.R. 1975. A comparative study of the chemical defensive system of tenebrionid beetles: Chemistry of secretions. *J. Insect Physiol. 21:* 753.

142. Tschinkel, W.R. 1972. 6-alkyl-1,4-naphthaquinones from the defensive secretion of the tenebrionoid beetle, *Argoporis alutacea. J. Insect. Physiol. 18:* 711.

143. Tursch, B., J.C. Braekman, D. Daloze, C. Hootele, D. Losman, R. Karlsson, and J.M. Pasteels. 1973. Chemical ecology of arthropods. VI. Adaline, a novel alkaloid from *Adalia bipunctata* L. (Coleoptera, Coccinellidae). *Tet. Lett. 3:* 201.

144. Tursch, B., D. Daloze and C. Hootele. 1973. The alkaloid of *Propylaeae quatuordecimpunctata* L. (Coleoptera, Coccinellidae). *Chimia 25:* 74.

145. van der Pijl, L. 1969. Principles of Dispersal in Higher Plants. Springer-Verlag, New York.

146. van der Pijl, L. and C.H. Dodson. 1966. Orchid Flowers and Their Pollination and Evolution. Univ. of Miami Press, Coral Gables.

147. Vareschi, E. 1971. Duftunterscheidung bei der honigbiene-Einzelsell-Alleitungen und Verhaltenreaktionen. *Z. Vergl. Physiol. 75:* 143.

148. Vanderah, D.J. and F.J. Schmitz. 1975. Marine natural products: Isolation of dendrolasin from the sponge *Oligoceras hemorrhages. Lloydia 38:* 271.

149. Vogel, S. 1966. Parfumsammelinde bienen als bestauber von Ochidaceen und Gloxinia. *Osterr. Bot. Z. 113:* 302.

150. Vollmer, J.J. and S.A. Gordon. 1975. Chemical communication: Part II. Between Plants and insects and among social insects. *Chemistry 48:* 5.

151. Vollmer, J.J. and S.A. Gordon. 1974. Chemical communication. Part I. Non-social insects. *Chemistry 47:* 6.

152. Waldner, E.E. Ch. Schlatter and H. Schmid. 1969. Zur
      biosynthesis des Dendrolasins, eines, Inhalt-
      stoffes der Ameise Lasius fluginosus Latr.
      Helv. Chim. Acta 52: 15.
153. Waller, G.D., G.M.Loper and R.L. Berdel. 1974.
      Olfactory discrimination by honeybees of terpenes
      identified from volatiles of alfalfa flowers.
      J. Apiculture Res. 13: 191-197.
154. Weatherston, J. and J.E. Percy. 1970. Arthropod de-
      fensive secretions. *In* "Chemical Controlling
      Insect Behavior". (M. Beroza, ed.). Academic
      Press, New York.
155. Weatherston, J. 1971. Aspects of the chemical defense
      of millipedes. *In* "Chemical Releasers in
      Insects". Vol. III. (A.S. Tahori, ed.). Gordon
      and Breach Publish.
156. Williams, C.M. 1970. Hormonal interactions between
      plants and insects. *In* "Chemical Ecology".
      (E. Sondheimer and J.B. Simeone, eds.). Academic
      Press, New York.
157. Wilson, E.O. 1968. Chemical systems. *In* "Animal
      communication". (T. Sebeok ed.). Indiana Univ.
      Press, Bloomington.
158. Wilson, E.O. and W.H. Bossert. 1963. Chemical com-
      munication among animals. *Recent Progress in
      Hormone Research 19:* 673.
159. Wilson, E.O. 1970. Chemical communication within
      animal species. *In* "Chemical Ecology". (E.
      Sondheimer and J.B. Simeone, eds.). Academic
      Press, New York.
160. Wilson, E.O. and F.E. Regnier. 1971. The evolution
      of the alarm-defense system in the formicine ants.
      Amer. Nat. 105: 279.
161. Wood, W., W. Truckenbrodt and J. Meinwald. 1974.
      Chemistry of the defensive secretion from the
      African termite *Odontotermes badius Ann. Ent.
      Soc. Amer. 68:*359.
162. Wright, R.H. 1975. Why mosquito repellents repell.
      *Scien. Amer. 233:*104.
163. Yeo, P.R. 1973. Floral allurements for pollinating
      insects. *In* "Insect-Plant Relationships".
      (H.F. van Enden, ed.). Blackwell Scientific
      Publications, London.
164. Zavarin, E. 1975. The nature, variability and bio-
      logical metabolities from Pinaceae. *Phytochem.
      Bull. 8:*6.

Chapter Six

COTTON PLANT AND INSECT CONSTITUENTS THAT CONTROL BOLL

WEEVIL BEHAVIOR AND DEVELOPMENT

P.A. HEDIN, A.C. THOMPSON, and R.C. GUELDNER

Boll Weevil Research Laboratory, Agricultural Research Service, USDA, Mississippi State, Mississippi

INTRODUCTION

The boll weevil, *Anthonomus grandis* Boheman, was introduced from Mexico into the United States (near Brownsville, Texas) about 1892[71]. By 1922 the pest has spread into cotton-growing areas of the United States from the eastern two-thirds of Texas and Oklahoma to the Atlantic Ocean. Its recent extension into west Texas now threatens cotton in New Mexico, Mexico, Arizona and California. As early as 1895, recognition of the tremendous damage caused by this insect was noted, and Townsend (1895) suggested that cotton growing be terminated in the region then infested, and that the cotton free zone be maintained along the Rio Grande River bordering Mexico. Several other entomologists studied and suggested ways to control the boll weevil. Howard (1896) reported on the use of, and lack of response to, light traps. Malley (1901) studied the use of poisons for weevil control as well as the use of cotton as trap plants. An act was passed in 1903 offering $50,000 as a cash reward for a practical remedy for controlling the boll weevil. Sanderson (1904) reported that hand picking of infested squares had been tried and was meeting with little success.

Hunter (1904) made the following statement: "It is concluded that there is not even a remote possibility that the boll weevil will be eradicated." Since that time, numerous methods of control have been tested and reported, 65 years have elapsed, and the boll weevil still has not been eradicated. From 1917 until the late 1940's the most effective method of control was dusting with calcium arsen-ate[69,76]. During World War II, the development of DDT and other chlorinated hydrocarbons made available a completely new group of insecticides for controlling many insect pests[21]. However, in 1954 scientists in Louisiana reported that boll weevils were becoming resistant to the chlorinated hydro-carbons[168], and within two years resistance was widespread throughout the cotton belt.

The organophosphate insecticides have been successfully used, since then, to control boll weevils without any wide-spread resistance problems. However, even though $70 million is spent annually for boll weevil control, the pest still causes an estimated $300 million in crop losses each year[89,1?]. In recent years, these figures may have increased by 50% because of inflation, even with 10-20% less cotton grown.

More than three-fourths of all insect losses to cotton in the United States have been attributed to the boll weevil. It is generally agreed that cotton cannot be profitably grown in areas where it occurs without adequate control measures. Coker (1958) estimated that 64,877,000 bales of cotton and 27,917,000 tons of cottonseed valued at $7,680,000,000 were destroyed by the weevil from 1909 to 1954. Approximately one-third of the insecticide used for agriculture, in the United States, is required for boll weevil control[100].

In recent years, the cotton industry has been confronted with major competition from the synthetic fiber industry, which has penetrated markets that historically have belonged to American cotton producers. Competition is also felt from foreign cottons. In order to maintain cotton in a competi-tive position, the U.S. Congress has provided relief in two areas. (1) From the early 1950's until 1973, growers were subsidized at approximately $0.10 per pound, which allowed them to compete with foreign cotton which sold for

approximately $0.20 per pound until 1970-2. Congress no
longer provides price supports. The current domestic
price of about $0.45 per pound (June 1975) is also the world
price. (2) Cost reduction research was instigated
approximately 15 years ago to improve yields of cotton and
to decrease costs of production. A part of this program
was the effort to control or eradicate the boll weevil with
decreased dependence on insecticides. Funds were provided
to intensify research at existing facilities, and for the
construction and staffing of a larger central laboratory
at Mississippi State, Mississippi. This review article
describes the chemical and biochemical aspects of this
program.

HOST PLANT RESISTANCE

     The complex interactions between phytophagous insects
and their hosts are the results of a long and continuing
evolutionary process. Tendencies toward parallel evolution
are generally restricted to the development of defense
mechanisms, by the plant, and of counteracting adaptations
by the insect. Beck (1965) divides relationships between
insects and plants into two principal aspects: (1) host
selection, attractants or feeding stimulants and (2) resistance to the insect by the plant. These aspects are not
fully separable, and analysis of the causes of plant resistance must also include consideration of the behavioral
and physiological characteristics of the insect. Conversely,
studies of host selection must include an analysis of the
role of plant characteristics which tend to reduce the
ultimate suitability of a plant species or variety as host
for an insect species or biotype.

     Painter (1951, 1958) has defined plant resistance as
the relative amount of heritable qualities possessed by a
plant, which influence the ultimate degree of damage done
by the insect. In practical agriculture, it represents the
ability of a certain variety to produce a larger crop of
good quality, than do ordinary varieties, at the same level
of insect population. Painter divided plant resistance
mechanisms into 3 main categories: (1) preference and non-
preference, in which a plant displays a degree of resistance
by exerting an adverse effect on the insect's behavior:
(2) antibiosis, in which a plant is resistant by exerting

an adverse influence on the growth and survival of the
insect; and (3) tolerance, in which a plant is capable of
supporting an insect population without loss of vigor and
reduction of crop yield. Beck (1965) excludes tolerance
because it implies a biological relationship between
insects and plants that is quite different from resistance
in the strict sense.

   Host plant resistance was one of the areas selected for
major emphasis at the Boll Weevil Research Laboratory (BWRL).
One of the initial tasks was the accumulation and screening
of the world germ plasm of cotton. Techniques for measuring
preference and antibiosis mechanisms in cotton to the boll
weevil were also developed. The investigations have been
concentrated in the following areas:
   1) Oviposition suppression factor in cotton. Detection
in *Gossypium barbadense* L. and transference to *G. hirsutum*.
2) Seedling screening technique for boll weevil resistance.
3) Field plot studies with frego bract cottons. 4) Biologically active substances in cotton and certain alternate
hosts affecting boll weevil behavior. 5) Abscission factor
in boll weevil larvae.

   In 1963, Maxwell[113] found that S. I. Seaberry, a *G.
barbadense* cotton, suppressed boll weevil oviposition by
about 40%, compared with commercial uplands, when weevils
were confined on squares (cotton buds), in the laboratory.
Subsequent work was directed to determine if the 40% oviposition reduction in S. I. Seaberry was (1) genetic in effect,
(2) if genetic, to move the genes responsible into a plant
more like upland cotton in plant type and, (3) to test
developed lines in field plot tests to determine the effects
of the oviposition suppression factor under field plot
conditions. Subsequently, the oviposition suppression factor
was shown to have moved into upland cotton. Several lines
were selected, which have upland fruiting patterns and plant
type, that carry 25-40% reduction in oviposition as compared
to standard commercial type cottons. Working with Frego,
a bract mutant controlled by a single recessive gene which
Hunter and Pierce (1912) had shown to inhibit feeding and
oviposition, Buford et al. (1968) showed that the oviposition
suppression factor was under genetic control, and the low
oviposition character did not appear to behave as a simple
dominant to high oviposition.

In recent work, Jenkins et al. (1973) found that Frego
cotton suppressed the boll weevil population from 69 to 79%,
and weekly applications of insecticides could be delayed
4 weeks. This was additionally significant, because this
was through the period of peak activity of the bollworm
complex. The limited investigations to determine whether
there is a chemical basis for the Frego resistance have been
inconclusive. Holder et al. (1973-74) found that Frego and
Deltapine 16 strains have the same sugars with only minor
quantitative differences. Also, there were only minor
differences in the GLC profiles of the essential oils and
air space volatiles (Hedin, unpublished data). It has now
been postulated (Jenkins, private communication) that the
basis of resistance of Frego bract cotton is morphological,
that is, the flared bract does not provide the cover that
the boll weevil requires for ovipositional activities.

Since experimental lines of glandless cottons were
first developed by McMichael (1960), investigators[9,79,144]
have found the entomological ramifications of considerable
interest. Most of the early work consisted of studies of
the comparative preference, of both host-specific and nonhost-
specific insects, for glanded and glandless cottons. The
observed entomological variations were then explained[110] by
the presence of the $gl_2gl_2gl_3gl_3$ genes, that is, in chemical
terms, by the absence of the stored form of gossypol in the
plant. Maxwell et al. (1965a) noted that the striped blister
beetle, *Epicauta vittata* (F.), consumed the lower leaves of
both "Acala" 4-44-77 glandless and "Rex Smoothleaf" glandless
lines, without any apparent damage to the glanded counterparts.
In field studies, the bollworm, *Heliothis zea* (Boddie); the
grape colaspis, *Colaspis flavida* (Say); and *Gastrophysa cyanea*
(Melsheimer) also showed a marked preference for glandless
lines[79]. Maxwell et al. (1966) found significantly larger
boll weevils, *Anthonomus grandis* Boheman, were reared, in the
laboratory, on a powder diet of reconstituted lyophilized
buds (squares) from 3 glandless lines than from the glanded
parent lines, although adult weevils preferred the glanded
lines. However, Jenkins et al. (1967) observed, in field
studies, that the presence of the glandless genes of gl_2 and
gl_3 in 14 different genetic backgrounds neither increased
nor significantly decreased their susceptibility to boll
weevil attack, over that of their glanded counterparts. They
observed a significant increase in the size of weevils
emerging from 5 of the 14 glandless lines.

Lukefahr and Martin (1966) found that an increased concentration of gossypol in the diet of the bollworm and the tobacco budworm, *Heliothis virescens* (F.), resulted in increased time to pupation, smaller larval weight, and increased mortality of both species. They attributed these adverse effects to reduced food intake (starvation), metabolic inhibition, and/or both acting in concert. Moreover, Lukefahr et al. (1966) observed that the larval growth of both species was 48-61% greater when they were fed on buds from glandless lines. Thus, the dietary effects of gossypol were still in question because the weight gain of larvae, on a diet from glanded cotyledons, was 90% greater than the gain of those reared on the glanded bud, though the cotyledons contained as much as 3 times more gossypol than the bud. The cotyledons also contained a greater concentration of unsaturated fatty acids per gram fresh weight than that observed in the buds of glanded "Deltapine Smoothleaf" cotton [192].

The importance of dietary lipids and lipogenic factors, for optimum larval development and egg production by adults, is well documented[96,207]. These investigators concluded that their test insects were not capable of forming poly-unsaturated fatty acids from simpler metabolites. Thus, the presence of larger concentrations of these nutritionally necessary constituents would overshadow, to some extent, the toxic effects of gossypol. Such an effect appeared to occur with the boll weevil, since its larval development and adult egg production are enhanced on a diet of cotyledons which are richer in unsaturated fatty acids and gossypol than the cotton bud. Also, since low concentrations of the gossypol salt of acetic acid are known to produce some feeding activity for the boll weevil[53], its dietary presence may neither be toxic nor objectionable to the insect, at low concentrations.

Even though there did not appear to be a difference in the susceptibility of the glandless lines to the boll weevil, antibiosis tests showed slightly larger weevils on some glandless lines[80]. Therefore, a study was made of the lipids from Acala 4-42-77 and Rex Smoothleaf cotton and their isogenic lines, to establish if there existed a variation in the composition and chemical state of lipids and their fatty acids.

The total fatty acids in the glanded and glandless lines did not vary significantly, but the phospholipids and glycolipids of glanded lines contained 50% more fatty acids than did those of the glandless lines. This difference resulted primarily from differences in the amounts of linoleic acid ($C_{18:2}$) and linolenic acid ($C_{18:3}$). About 50% of the total fatty acids in the glandless lines were contained in a highly polar lipid complex that was obtained only after the silicic acid was extruded from the column and hydrolyzed with an acid solution[192].

A number of factors in cotton and other malvaceous plants has been shown to elicit various effects on boll weevil behavior. Keller et al. (1962, 1963) have prepared water extracts, from cotton, which stimulate feeding by boll weevils; they have also extracted the ice, obtained in the process of freeze dehydrating cotton squares, with chloroform to yield an oil that was attractive to the weevils. By fractional distillation of this oil, under vacuum, the same team showed that the pot residue was repellent to the insects[108].

Cross et al. (1975) reported that the principal plant hosts of the boll weevil are in 4 closely related genera of the Malvaceae: *Gossypium*, *Cienfuegosia*, *Thespesia*, and *Hampea*.

Boll weevils are known to feed on *Hibiscus syriacus*, Rose of Sharon, in times of stress, when cotton has matured out, or in situations where weevil population pressures are high and there is a short supply of available food. Under these conditions, Rose of Sharon plants, near cotton fields, have been observed to be infested. In observing the nature of feeding and oviposition on the bud of this plant, it was noticed that the insect always sought the buds where the calyx had split, prior to blooming. It was also observed that if the calyx was removed, weevils fed and oviposited on this host equally as well as on cotton[111]. Weevils would not feed through the calyx, except on rare instances. In some tests, the calyx was removed from the bud and replaced with a debracted cotton square. No weevils fed through the Rose of Sharon calyx into the cotton square underneath. The calyx material was extracted and found to contain a non-volatile, water soluble material which prevented feeding of the adult as well as inhibiting larval development. Boll weevil attractant activity was not detected in *H. syriacus*.

It was concluded that the non-preference shown, in nature
to *H. syriacus*, by the boll weevil results from a low level
of attractant, a high level of repellent, and a highly
active feeding deterrent, in the calyx, which effectively
masks feeding and oviposition stimulants present in the
H. syriacus bud.

We made an extensive effort to isolate the feeding
deterrent. Several hundred grams of calyx tissue were ground
with water and extracted from the water phase with butanol.
The butanol extract was chromatographed on a Sephadex LH-20
column with methanol. An early eluting yellow band was
subsequently chromatographed on a Florisil column, from
which the deterrent was eluted with ethyl ether. Mass
spectral examination indicated the presence of 2-3 compounds
of M^+ 240 - 280, probably alcohols. During subsequent work-
ups, difficulties with the bioassay were encountered;
ultimately the work was suspended.

Fryxell and Lukefahr (1967) found *Hampea* sp. infested
by boll weevils in Vera Cruz, Mexico. Closer examination
showed that in nature, the male tree of this dioecious
species is infested heavily, while the female tree is
apparently resistant[104]. Plant materials were harvested,
and brought back to the laboratory for studies to determine
basic causes for differences.

Buds of both sexes and female capsules were extracted
and analyzed for feeding stimulant and plant attractant
activity. Male and female buds and female capsules were
homogenized and lyophilized. The resulting powder was
reconstituted into diets, on which weevils were reared, to
detect possible antibiosis factors. Water obtained from
each sample during freeze drying, was retained and bioassayed
for attractiveness to the weevil. Female *Hampea* buds were
found to have less feeding stimulant activity than cotton
and male *Hampea* buds. Female buds also produced smaller
weevils than Deltapine Smoothleaf cotton and male *Hampea*
buds. Male *Hampea* buds and female *Hampea* capsules both
produced significantly smaller weevils than did cotton.
Days to emergence were significantly prolonged by male and
female *Hampea* buds, and particularly by female capsules.
Antibiosis was strongly indicated in female buds and capsules
by low percent emergence.

Extracts of female capsules and buds were significantly less attractive than cotton and male *Hampea* buds. Interestingly, male and female extracts, stored at 5°C for any length of time, became equally attractive, indicating the loss or degradation of the repellent material present in the female *Hampea* bud extract.

From these results, it has been concluded[114] that the resistance of the female *Hampea* tree results from three primary factors: (1) lack of attractiveness resulting from a highly repellent substance(s) in the female buds and capsules that masks the attractive materials present; (2) the presence of less feeding stimulant in the female bud than in the male bud, which results in significantly less feeding and oviposition, and (3) the high antibiosis to the boll weevil larvae in female buds of *Hampea* resulting in significantly reduced and prolonged emergence, and reduced adult weight. The physiology of antibiosis to boll weevil larvae by the cotton bud has been the subject of a series of studies by workers at this laboratory.

When boll weevils oviposit in cotton squares, the squares drop from the plant in 5-6 days. The physiological reactions of the plant, resulting in abscission of the square and the particular phase of the life cycle of the larvae of the boll weevil responsible for the abscission, have not been previously determined. Coakley et al. (1969) studied the influence of natural and artificial feeding on abscission. No significant differences occurred in any of the methods except larvae implanted and larval extracts injected. Data showed that injection of homogenates of 1st instar larvae (less than 6 hours after hatching) gave no difference in abscission from the control. Injection of water homogenates of 2nd instar larvae (36 hours), however, gave a 3-fold increase and homogenates of 3rd instar larvae (144 hours), a 4-fold increase in abscission within 60 and 48 hours, respectively.

Results obtained with implantation of larvae and with injection of water homogenates of 2nd and 3rd instar larvae demonstrated that only these two larval instars caused the reaction, in the plant, that resulted in abscission of the squares.

King and Lane (1969) initially characterized the active component as protein. Additional work[87] has established that several insect enzymes are implicated in the abscission. They are one, or more, cellulases, an endo-and exo-polygalacturonase, an endo- and exo-pectinlyase, a pectin methylesterase and a protease. Since it is extremely unlikely that the enzymes of the insect are translocated from the anthers to the abscission zone, it follows that the enzymes induce metabolic changes, which in turn induce either a growth regulator imbalance or the production of a small molecule(s) which is translocated and promotes abscission. Insect control could conceivably be achieved through the use of a systemic agent, which either hastened square shed, or delayed it, so that larvae would be exposed to parasites, such as wasps, for a longer period.

FEEDING STIMULANTS

As discussed in the previous section on host plant resistance, Painter (1951) divided insect resistance into 3 bases or mechanisms: antibiosis, tolerance, and preference or non-preference. Mechanisms such as feeding stimulants, plant attractants, repellents, and arrestants should be included in this latter category. Until 1950, chemicals which affected the behavior of insects were termed either attractive or repellent. However, insect behaviorists began to realize that these 2 terms could not be applied to all insect reactions. Deithier et al. (1960) distinguished between the positive long-range response, which they defined "attractant", and the short range responses which they defined as "arrestant" and "feeding stimulant".

Keller et al. (1962) demonstrated a boll weevil arrestant and/or feeding stimulant response with cotton bud water extracts. The bioassay for this response has been described[7-8].

Employing the same bioassay several years later, Daum et al. (1969) investigated the ingestion of Calco Oil by the boll weevil and found that insects may make a large number of feeding punctures, but do not necessarily ingest the same quantity of dye with each puncture. In one test where the number of punctures per weevil per hour remained constant, the amount ingested per puncture decreased by half, with a change in formulation. The bioassay also does not measure "arrestant" potency since Deithier et al. (1960)

defined an arrestant as a chemical which causes insects to aggregate when in contact with it, the mechanism of aggregation geing kinetic or having a kinetic component. In practice, we have tested fractions which elicited an arrestant response, but little or no puncturing. However, from the standpoint of insect control, compounds which elicit puncturing, nearly always elicit a substantial amount of ingestion, and thus could be expected to stimulate the ingestion of some biological agent. Our evaluation of cotton plant fractions and components, plus a large number of commercially available compounds and preparations, has been with respect to their ability to stimulate the insect to puncture rolled agar plugs. With the adoption of any laboratory bioassay, certain calculated risks are assumed.

In our initial study[51], chloroform extracts of squares and also of freeze-dehydrated square powder were compared for their feeding stimulant potency with the water extract of Keller et al. (1962). The extracts were of comparable potency. To determine whether the polar and non-polar solvents were extracting the same principle, the residue from the chloroform extraction process was extracted with boiling water and vice-versa. In each case, high activity was obtained which gave presumptive evidence that at least 2 active principles were present. Subsequent investigations were to implicate a broad spectrum of compounds with some stimulant activity. A feeding threshold study was conducted with both water and chloroform extracts. A plot of activity vs log concentration gave a straight line threshold intersect of approximately 15 µg. Since the 15 µg was evenly applied to a 37 x 37 mm filter paper square, the actual perceptive ability of the insect must be several orders of magnitude greater. A cold water extract lost activity in 3-5 days, even when stored at -20°C; cold acid and base treatments had no effect on activity, but all activity was lost upon refluxing in 1N acid or alkali. A number of commercially available compounds were inactive or only slightly active; they included sucrose, glucose, ATP, ADP, AMP, choline, quercetrin, quercetin, morin, rutin, ascorbic acid, pH 1 and pH 12 phosphate buffers. Gossypol, with and without phosphate buffer and glucose, elicited a partial feeding response.

In further studies, on the polar components from the cotton bud[184], lyophylized bud powder was extractd for 24 hrs, in a Soxhlet, with petroleum ether, benzene,

chloroform, ethanol, methanol, and water. Essentially full
activity was obtained in each extract except that with benzene
The high activity of the water extract was quite remarkable
since extraction with the other solvents should have removed
such important chemical classes as the hydrocarbons, terpenes,
carotenoids, porphyrins, and flavonoids. The results from
successive Soxhlet extractions demonstrate the significant
thermal stability of the major polar feeding stimulant.
This extraction series clearly demonstrates the presence of
a highly polar feeding stimulant component or components
in cotton buds which is distinct from active feeding, nonpolar components in the same plant material.

The active principle in the water extract was not
adsorbed on a weak cation exchange resin (as are flavonoids)
but was precipitated with lead ions. Studies by gel
permeation suggested a molecular weight of <2,000 for the
active constituent(s) and thus eliminated proteins, large
peptides, and large polysaccharides. TLC showed that
flavonoids were not present.

In preliminary studies, flavonoids had been detected
in the highly active methanol extract. Procedures were
devised to remove the flavonoids before bioassay of the
remainder. The extract was streaked on preparative PC and
chromatographed with 22% isopropanol in water, which left
the flavonoid monoglycosides in the slowest migrating
fraction. Isoquercitrin and some diglycosides were found
in the median fraction, and some di- and triglycosides
plus other highly polar compounds in the 3rd fraction, which
also gave the highest feeding activity. Upon incubation
with emulsin, under conditions which hydrolyzed model
flavonoid glycosides quantitatively, the activity was retained
The activity was also retained after refluxing this fraction
in 1N HCl. Extraction of the methanol extract with sodium
bicarbonate showed that the activity cannot be attributed to
free or bound phenolic acids present in the extract. Thus
the highly active alcoholic and aqueous extracts of buds
were found to contain large amounts of flavonoids, but
subsequent hydrolytic and chromatographic results showed that
the major portion of the activity was retained when the
flavonoids were removed. However, since appreciable activity
could be attributed to these compounds, the major components
were isolated for bioassay. All except the anthocyanin[52],
had been previously reported[125,177]. These isolates were

bioassayed with and without 0.1M phosphate buffer, pH 7.0, and rhamnose, which, on occasion, had fortified or stabilized the activity of the isolates, presumably because they ameliorated the bitter flavonoid taste. Rhamnose was used in preference to glucose, fructose, or sucrose, all of which elicit similar feeding activities individually, because it appeared to provide better fortification.

Quercetin, quercetin-7-glucoside (quercimeritrin), and quercetin-3'-glucoside were moderately active; however, rhamnose and phosphate buffer did not appreciably increase activity. Quercetin-3-glucoside (isoquercitrin) and kaempferol were inactive. Cyanidin-3-glucoside was not active alone, but was appreciably synergistic with rhamnose and phosphate. The activity of these compounds has some added significance, because they were the first pure compounds, isolated from bud extracts, that elicited activity[53].

Two weakly acidic components present in a methanol extract of cotton buds were isolated by an ether-bicarbonate extraction technique and were shown to possess moderate feeding stimulant activity. UV analysis indicated that the 2 components were flavonoid. However, the most active fraction (I_c) obtained in this study was more polar and may be similar to that obtained in aqueous extracts.

The possibility exists that a combination of several components is necessary to produce a full feeding stimulus response. Although the water extract, obtained from the successive Soxhlet-extraction experiments, elicited the best feeding response, the other extracts were also active. Furthermore, in some of our experiments, the activity of complex mixtures tended to decrease or even disappear with further fractionation. Degradation during isolation has not been excluded as being responsible for the occasional dissipation of activity, although the thermal stability of the polar principle suggests that chemical instability is, perhaps, not the chief cause of activity loss.

The nonpolar components were also investigated[188]. The petroleum ether extracts of fresh and lypophylized buds gave intermediate feeding activities. TLC revealed the presence of triglycerides, fatty acids, fatty alcohols, and carotenoids, but few, if any, mono- and diglycerides or phospholipids. Attempts to isolate active fractions by column chromatography

were essentially unsuccessful. Fractionation of the chloroform extracts, on silicic acid, place the major activity in
the methanol eluate which contained pheophytins and other
pigments.

The acetone extract of fresh buds was partitioned
between 90% aqueous methanol and diethyl ether-petroleum
ether (1:4) and yielded an active, green-black oil upon
evaporation of the nonpolar phase. A yellow-brown oil, of
slighlty lower activity, was obtained by chloroform extraction of the polar phase. Again, several TLC- and column-
chromatographic procedures were used in attempts to concentrate the activity from both oils. Best resolution and
maximum activity was achieved on Mallinckrodt TLC-7^R silica
gel plates developed in CHCl$_3$; the bands at R_f4-0.6(T/S 50-
60) giving a 10-fold increase in concentration. Rechromatography of the active bands resulted in loss of activity.
Because of the similar behavior of the active components on
TLC, they are probably similar or identical. Also, neither
was extracted quantitatively into either solvent layer by
the partitioning. Pheophytin a was associated, in varying
quantities, with most of these active bands (see later
discussion).

Fresh cotton buds were also extracted 3 times with 200-
ml portions of chloroform-methanol (2:1 v/v) and the extract
taken to dryness. redissolved in chloroform and chromatograph
on a silicic acid column prepared in the same solvent. The
column was thoroughly washed with chloroform to remove
nonpolar materials and was then developed with chloroform
containing increasing amounts of methanol. The elution
pattern of the feeding-stimulant materials indicated 3 major
groups with at least 8 components varying in polarity.

TLC and spectral data of the combined (30:1) chloroform-
methanol fractions indicated the presence of phosphatidic acid
and pheophytin. This fraction, obtained either from dark-
grown seedlings (a source free of chlorophylls), light-grown
seedlings, or frozen buds, caused feeding activity that varied
directly with the content of phosphorus and nitrogen. After
elution from silica gel, TLC of the polar lipids yielded no
fractions with feeding activity. Recombination of the lipid
fractions, from TLC, did not regenerate a feeding activity
which was equal to the original fraction eluted during column
chromatography.

Further efforts to define the chemical nature of the feeding-stimulant components were made by partitioning the fresh bud extract (chloroform-methanol, 2:1) between 95% methanol and heptane. The methanol fraction was evaporated to dryness, and the residue dissolved in a minimal amount of benzene; treatment of this solution with 100 volumes of acetone, at 0°C for 24 hr, yielded a precipitate. The acetone solution, after filtration, was evaporated to dryness and both this fraction and the acetone insoluble one were chromatographed on silicic acid, in chloroform and methanol. Both fractions contained 3 major groups of feeding stimulants and the acetone-insoluble fraction contained an additional, more polar group of feeding-stimulants. However, repeated precipitation of the acetone-insoluble fraction, with acetone, did not yield a complete separation of these 3 less polar groups from the fourth more polar group. Exposure of the active fractions to silica gel or polyamide TLC resulted in the same loss of feeding activity described above. Both fractions were then chromatographed on silicic acid using petroleum ether-diethyl ether as the developing solvent. Most of the activity was eluted from the column with 5% diethyl ether.

Extraction of the bud or bud powder with petroleum ether, chloroform, or acetone all yielded feeding-stimulant, green-black residues. Because the visible spectra of these residues has a shoulder at 655 mμ and a peak at 667 mμ, the presence of pheophytins a and b was suggested. However, the pheophytins are formed by the removal of magnesium from the chlorophylls, and this conversion is favored in acidic systems; thus, their presence in the intact plant was not certain. Presumptive evidence that the pheophytins were not entirely artifacts was obtained in the following manner: Acetone extracts, with and without triethylamine, were partitioned between petroleum ether and water; the residue from each organic phase was analyzed on silica gel TLC in chloroform-ether, 19:1. Zones for chlorophyll a and pheophytin a were at R_f 0.15 and 0.50, respectively. If the extracts were made with acetone only, the ratio of pheophytin a to chlorophyll a was high; if the extracts were made with acetone-triethylamine, the ratio was low. As expected, the addition of dilute acid to the extract from acetone-triethylamine converted chlorophyll a to pheophytin a. The presence of small quantities of pheophytin a in extracts prepared in alkaline solutions, suggests its natural presence in the plant.

The best yields of the pheophytins were obtained by extraction of frozen, homogenized cotton squares with hot chloroform-methanol, 2:1. The chloroform-soluble fraction was applied to a silica gel H^R (Brinkmann) column and developed with chloroform. A green band containing a mixture of the pheophytins was eluted, banded on silica gel H plates and developed with chloroform to separate pheophytins a and b.

The samples of pheophytin showed moderate and consistent activity in a number of weevil assays. Pheophytin a had a T/S value between 0 and 90, with an average of 47, for 13 assays. Pheophytin b had a value between 3 and 88 with an average of 42, for 12 assays. To ascertain whether the activity remained with the pheophytins, a mixture containing pheophytins a and b was distributed for 25 transfers in a 50-cell Craig counter double-current distribution apparatus, in cyclohexane-ether-methanol-water (9:6:10:5). The pheophytins were found in 2 active fractions of the organic phase.

The mass spectrum of the fraction containing, pheophytin a, recorded at 250°C, showed no impurities. Thus the feeding stimulant activity of this sample must arise either from a nonvolatile compound, present as an impurity, or from the cotton pheophytin itself. In the mass spectrum of pheophytin b, peaks extended out to m/e 440. The most prominent of the unexplained peaks occurred at m/e 218 and 239.

Pheophytin a was treated with HCl to obtain pheophorbide a, which was found to be inactive. The pheophorbide a was esterified with methanol-sulfuric acid to obtain methyl pheophorbide a, and with phytolphosgene to obtain pheophytin a. Both products were identified by TLC and their visible spectra. However, the synthesized pheophytin a was not obtained in sufficient quantity to permit a rigorous proof of its structure. Both were inactive in the boll weevil assa'

In another approach, impure chlorophylls a and b were obtained by extracting spinach with acetone, chloroform-methanol (2:1), and with petroleum ether-methanol. Both chlorophylls had little, if any, activity. Acid hydrolysis of the chlorophyll mixture produced pheophytins a and b, which were purified on silica gel plates. Spinach pheophytin a had a modest, but consistent, activity, whereas spinach pheophytin b was inactive. The phytyl group of spinach

pheophytin a was removed to obtain pheophorbide a, which
was also inactive.

Since the existence of chlorophylls other than standard
a and b has been proposed[120], perhaps slight structural
differences in cotton and spinach pheophytins would yield
the reduced activity observed with spinach pheophytins.

Maxwell et al. 1963b, reported that boll weevils preferred feeding on cotton flowers and buds, but also fed
well on whole seedlings. Jenkins et al. 1963, reported
that water extracts of the flowers were a slightly stronger
feeding stimulant than water extracts of buds. We prepared
extracts of the flowers and buds. These were then fractionated,
as described above, and examined for feeding stimulation, a
methanol-soluble, active substance was found in both tissues
and another highly active fraction, not present in buds, was
recovered from flowers by re-extraction with methanol.
Analysis of these fractions showed that those from buds were
more stimulating than those of flowers[185].

Later Jenkins and Parrott (personal communication),
reported a definite feeding preference for the epicotyl of
the seedling, over that for the cotyledon, hypocotyl and
radical. Hanny et al. (1973) showed that primarily males
were attracted to the epicotyls, where they fed preferentially
and were strongly deterred from feeding on the hypocotyls.
Females were attracted to and fed on both the epicotyls
and hypocotyls. Although 36 compounds were identified from
the epicotyls, none could be specifically implicated as a
feeding stimulant.

Reports on insect feeding stimulants are limited. No
studies, thus far, have provided clear evidence that a single
compound acts independently. Furthermore, when a glycoside,
sugar, or purine has been implicated, it has often been
questionable whether the activity would have been maintained
after rigorous purification. When isolated compounds have
been demonstrated to give activity[42,167], it is possible
that other compounds may have been present as well which were
active alone or synergistic with the isolated compounds.
The concept of host-plant specificity implies that insects
can discriminate flavors. However, even though insects almost
certainly do respond to a single, dominant compound, when it
is present, there is probably a much larger number of

situations where no dominant compound exists. Consequently, an adequate response more likely requires a complicated profile of components.

Feeding-stimulant components from cotton buds, extractable with petroleum ether, chloroform, acetone, and chloroform methanol appear to belong to several major groups which cause feeding activity in the boll weevil. The physical and chemical properties of these active fractions clearly differentiate between polar and nonpolar constituents. Attempts to purify the major groups frequently result in a dissipation of feeding activity which cannot be fully regenerated by recombination, indicating a breakdown in the structure of some components. This inability, to regenerate full feeding activity by recombination, is especially true for many of the TLC systems studied. The labile nature of some components is further indicated by the gradual loss of the feeding stimulus activity of compounds stored in various solvent systems. For example, an extract will show only low levels of feeding stimulus activity after 12 hr in diethyl ether, but full activity can be retained for 2-3 weeks in either chloroform or methanol.

It is significant that the activity of lipolytic enzymes, in diethyl ether, is high. Even though no significant feeding stimulant activity can be attributed to the lipid components derived from cotton buds, the increase in the concentration of hydrogen ions caused by lipid degradation may have an adverse effect on the feeding-stimulant components. These considerations suggest that lability of some nonfeeding-stimulant components may occasionally contribute to the observed loss of activity during procedures of fractionation. More importantly, however, other studies indicate that a multicomponent system is necessary for optimum boll weevil feeding[184,185,188].

Since recombination or fortification, of fractions by sugars and buffers, often rejuvenated part of the activity, efforts were directed toward formulating an active feeding mixture from known cotton constituents, common metabolites, and compounds inducing primary mammalian sensations of taste and odor[53]. Of 286 compounds bioassayed, 52 elicited substantial activity, and of these 14 had previously been reported in cotton. They include gossypol, α-ketoglutaric acid, malonic acid, vanillin, formic acid, lactic acid,

1-malic acid, quercetin, β-sitosterol, succinic acid, valine, quercimeritrin, quercetin-3'-glucoside, and cyanidin-3-glucoside. The boll weevil was found to express preference for sweet, sour, and cooling taste properties, but odor preferences were difficult to establish. Sixteen carboxylic acids, 8 alcohols, 8 carbonyls, 8 phenols, and 10 amides or amines were among the most stimulatory. When the active components were grouped on the basis of taste and molecular size, it became apparent that sweet substances, having molecular weights above 200, were consistently well accepted. Since most of these compounds were di- or triterpenoids or steroids, hydroxylated, and much less sweet than the sugars, their activity may be associated with their predictable low rate of desorption. The most favored MW for sour and salty compounds was below 150. Bitter compounds of 100-200 MW and pungent compounds of 150-200 MW were most deterrent.

Formulation of mixtures was initiated on an empirical basis. More than 330 mixtures were tested with limited success. Those which gave the best feeding results included 17 different compounds, 6 of which were known to exist in cotton. The 17 were formulated into 4 to 8 component mixtures with some consideration given to taste class and intensity. Subsequently, several 8-component mixtures were found which gave feeding responses equal to cotton bud water extracts, and also elicited slightly more insect punctures. The best was a mixture of β-sitosterol, 15-pentadecanolide, cineole, N,N-dimethylaniline, vanillin, mannitol, rhamnose and phospate buffer pH 7.0[53].

For several reasons there has been no extensive effort to evaluate these mixtures in field trials. Cottonseed oil "foots" and cottonseed meal contain most of the components of the water or organic extracts, and are a cheaper and more convenient source for bait (boll weevil attractant) preparation. Secondly, the baits are not competitive during the period of greatest plant growth. Finally, the insects are overwhelmingly attracted to male boll weevils or the synthetic pheromone in the presence of baits or growing cotton.

McKibben et al. (1971a) investigated eight acids including: adipic, fumaric, malic, phosphoric, succinic, citric, lactic, and tartaric acids, as food acidulants. When added to cottonseed oil baits, they caused a 41-67% increase in the amount ingested but there were no significant differences among the acids tested.

Hedin et al. (1974a) made an exhaustive review of biologically active substances, in host plants, affecting insect behavior. The known feeding stimulants, classified by chemical structure, included: 20 glycosides, 15 acids, 8 flavonoid aglycones, 6 carbonyls, 5 phospholipids and terpenoids, and 17 miscellaneous compounds. When these feeding stimulant classes were further subdivided according to preferences shown by insects from various orders, somewhat more specific relationships were suggested. Among the Lepidoptera, 8 glycosidic feeding stimulants were found, along with 3 aglycones which could undergo cosylation there were responses from 9 miscellaneous compounds. Among the Coleoptera, 10 acids, 6 glycosides, 4 flavonoid aglycones, and 4 terpenes were found. There were also 8 miscellaneous classes. In the order Homoptera, both compounds reported were glycosides. In the order Orthoptera, acids, their esters and their salts, were most prevalent. These compounds accounted for 8 of the 14 listed responses.

Based on human responses, the tastes of the insect stimulatory substances appeared to be sour, cooling, semisweet, and salty, and the stimulatory odors appeared to be floral, musky, pepperminty, and camphoraceous. Repellent substances appeared to have the bitter tastes and strongly flavored (pungent?) characteristics afforded by the lactones. The similarity of these results to those which we observed with the boll weevil is striking.

PLANT ATTRACTANTS

Fraenkel (1959) has discussed the role of secondary plant substances in the development of host plant-insect relationships. Compounds produced by plants as waste, metabolic intermediates, defense mechanisms, or insect attractants or stimulants, influence the establishment of highly specific plant-insect ecologies. He suggested that the secondary plant substances, known to exist in many plants without having any apparent role in plant function, may have developed from past mutations and natural selections. The development of these substances would afford the plant protection against increased damage by insects or other

enemies. Certain insects, less susceptible to the undesirable
effects of the compounds, may then have adapted to the secondary substances as a stimulus for feeding or propagation.

Very little is known about the stimuli which attract
insects to their food supply. There are over 350,000 described plant-feeding insect species, and limited information
is available for only a few.

When the known attractants were classified by chemical
structure[59] there were: 21 terpenes, 7 alcohols, 4 esters,
4 acids, 4 sulfur containing, and 2 phenolics. There
were 7 miscellaneous substances including 3 which possessed
little or no volatility and may have been misclassified as
attractants. When these attractants were further subdivided
according to attractiveness to insects from various orders
somewhat more specific preferences were suggested. Among
the Coleoptera, 17 terpenes, 3 ketones, 2 alcohols, 2 acids,
and 4 miscellaneous substances were reported. A parallel
with the compounds reported as sex attractants for insects
of this order, is therefore suggested. These data are
predictable since a number of tree-infesting beetles are in
the group. Among the Lepidoptera, 5 esters, 2 glycosides
(probably improperly identified as attractants), 2 acids,
and 4 alcohols were reported. A parallel is also obvious
with the sex attractants of this insect order; esters, their
alcohols and their acids were most frequently reported.
Terpenes were also reported as sex attractants in the orders
Hymenoptera, Diptera, and Isoptera.

Viehoever et al. (1918) first demonstrated that the
volatile oil, obtained by steam distillation of Upland cotton
leaves, was attractive to the boll weevil. Its attractancy
was subsequently confirmed[33,44,84,147]. The oil was pale
brown and distilled mainly between 200 and 300°C. It was
different from that obtained earlier[161] from the root bark
of *Gossypium herbaceum*. The latter was pale-yellow and
distilled between 120 and 135°C under the same conditions.

Parrott et al. (1969b) evaluated the lyophylized bud
powder from several varieties and species of Malvaceae, for
plant attractancy to the boll weevil. The response, in each
case, was less than that for the extract from fresh cotton
bud squares. This difference may be, in part, attributed to

the loss of the more volatile components (up to C_{10}) by lyophylization.

McIndoo (1926) did extensive work in an attempt to explain how plants attract insects by smell. His assumption was that plants, like animals, emit odors and that insects in searching for food, either for themselves or their progeny, are guided by these odors. He suggested that the attractant might be put in traps and serve as a control measure for many insects including the boll weevil.

The principle of using trap plantings (larger cotton plants which fruit earlier than planted cotton) in controlling boll weevils was advanced shortly after the boll weevil had migrated into the U.S. However, the value of this approach was not initially recognized by entomologists. Howard (1892) advised the trapping of late weevils in the fall and overwintered weevils in the spring. Malley (1901) recommended destroying all the cotton in a field, except for a few rows to which boll weevils would be attracted, and which would then be poisoned frequently or grazed down. Single plants or groups of plants have been used by researchers for several years as a means of early detection of overwintered populations[154,182].

The question of whether the boll weevil is attracted to cotton, or whether it finds cotton by chance, has been debated for many years. In referring to fall migrations, when boll weevil population densities are often great and food supplies short, Hunter and Pierce (1912) were of the opinion that the "...weevil has no sense by which to locate cotton,...but the general aimless flight of thousands of individuals seems sufficient to account for the infestation of all fields in new territory...". Hunter and Coad (1923) reported, "The idea of attracting weevils to a few early plants or trap rows has frequently been advanced. Practical experience, however, has shown that the only possibility of success in such a procedure lies in the use of entire fields adjoining hibernation quarters, the fields to be poisoned later. The use of only a few rows as a trap crop has been found to be absolutely valueless." Whether or not they are attracted to cotton by an odor, the boll weevils have little or no difficulty reaching cotton because they usually hibernate near old cotton fields that will be replanted the following spring[130].

Isely (1932) first suggested the value of spot dusting, especially in areas of older or taller cotton, where the weevils congregated. Later he was of the opinion that an early cotton variety, particularly if it germinated well and was relatively cold hardy in the seedling stage, could be used as a trap crop in concentrating infestation of overwintered weevils[75]. The value of trap plants for weevil control was further demonstrated by Isely 1950[76].

Mistric and Mitchell (1966) and Mitchell and Taft (1966) have shown that migrating field boll weevils are attracted, in considerable numbers, to isolated groups of caged and uncaged fruiting cotton plants. Without alighting, the weevils were able to distinguish between fruiting and mature plants and only a few were seen to alight on the latter. Removal of all or part of the buds or squares did not lessen the attractiveness of the plants. The tests indicated that the primary attractant of cotton, in the field, was odor rather than plant shape or color, thus lending support to the results obtained with the laboratory tests of the essential oil.

Within the past 10 years, our laboratory has systematially isolated and identified compounds from the essential oil of the cotton bud of *Gossypium hirsutum* var. DPSL (see ref. 43, 54-56, 62, 63, 122-124, 126, 127). The purpose of these investigations has been to gain fundamental knowledge about the plant volatiles and to identify component(s) which are attractive to the boll weevil. The bioassays of oil fractions and pure compounds were performed in the laboratory in a glass unit developed by Hardee et al. (1967a). In the bioassays, the attractancy of different compounds and fractions was compared to that of a standard hot water extract of squares. The total essential oils from cotton buds was about 50% as attractive at 200 ppm as the standard. Removal of the hydrocarbon fraction by silica gel chromatography increased the attractiveness. Except for α-pinene, limonene, and $(-)$-β-caryophyllene, which were moderately to very attractive at levels approximately equal to their natural abundance in cotton bud oil, the hydrocarbon fraction was totally unattractive. The hydrocarbon free oil was then treated with the carbonyl reagent, Girard T. After regeneration, the carbonyls were found to possess only about 1/3 potency, but the remainder of the polar oil possessed full activity. By column and thin layer chromatography, several

isolates were obtained from this polar oil fraction,
identified, and bioassayed for attractancy.

The most abundant polar compound (5.6% of the oil) had
never been reported, although it has now been found in
several other malvaceous plants; its structure was elucidated
in this laboratory[126]. It is the tertiary sesquiterpene
alcohol β-bisabolol ($C_{15}H_{26}O$). Its activity as an attractant
is only about 50% of the standard but it appears to enhance
the activity of other oil constituents. Recently, a second
tertiary sesquiterpene alcohol, α-bisabolol[54] has been
identified. It is present in much lower concentration (0.6%)
and gives an attractancy of 64% at 2 ppm in water. An
epoxide and a furan with attractant activity were also
isolated and identified. They are β-caryophyllene oxide[127],
(0.4%) and bisabolene oxide (0.2%)[56].

Besides these major attractant compounds, several more
may be present. The multiplicity is similar to that involving
the boll weevil feeding stimulant complex in the cotton
plant[53]. Subsequent screening of 37 commercially available
terpenoids and related plant constituents gave 15 which
possessed at least 50% potency at 1-10 ppm[39].

As described in the feeding stimulant investigations,
efforts were made to formulate mixtures of greater potency.
A mixture of 5 constituents from the cotton plant, (+)-α-
pinene, (+)-limonene, (-)-β-caryophyllene oxide, and (+)-
β-bisabolol slightly exceeded the attractancy of the standard
(Minyard et al., 1969). Two mixtures from the commercially
available compounds, (I = borneol, rose oxide, (+)-α-pinene,
linalool oxide) (II = neryl acetate, rose oxide, nerol,
farnesol, isovaleraldehyde, dihydroterpinyl acetate) were as
potent as the standard, but slightly less potent than the
formulation of cotton plant components[39].

These mixtures were also evaluated, in field trials,
as additives to a cottonseed oil bait[118]. The cotton
plant component mixture increased the response two-fold
before mid-September, but did not elicit a significant
increase during the late season phase of the experiment[117].
Failure to obtain increased feeding in the second phase
suggests decreased discrimination by the insect as the food
supply decreased.

Stoner (1968) compiled a list of malvaceous plants which are considered host plants of the boll weevil, and divided them according to their use only for food, or for both food and oviposition. The infestation of certain plants has often been attributed to the odors transpired[36]. However, no one concept has been found which can explain the variations in the degree of preference of the boll weevil for its hosts. When the essential oils, of several host and nonhost species from the Malvaceae and also a variety of angiosperms and gymnosperms were tested for attractancy, all were as attractive as cotton oil. Moreover, the insect actually favored essential oils of 2 alternate host plants, *Hisbiscus militaris* Cav. and *H. lasiocarpus* Cav. in comparative preference tests. Thus, in laboratory tests, the essential oils did not control insect preference exclusively. Morphological and environmental conditions may influence the concentrations of volatile components released by the plants[193]. Five of these oils have been investigated by gas liquid chromatography, to determine the extent of qualitative and quantitative component differences from those of cotton oil.

The essential oils of *Gossypium hirsutum* L. and *Cienfuegosia heterophylla* (Vent.) Garcke contained 67.0% and 58.2% hydrocarbons, respectively. Those of *Hibiscus lasiocarpus* (Cav.), and *Callirhoe involucrata* (Nutt.)   Gray, each contained 14%, and that of *Cienfuegosia sulphurea* (H.B.K.) Hassel   only 2.9%. The major oxygenated component in *G. hirsutum* (13.7%), *H. militaris* (55%) and *H. lasiocarpus* (32.6%), was β-bisabolol. The major constituents in the essential oil of *C. involucrata* were myrtenal (32%) and trans-2-cis-6-nonadienal (12%). The major oxygenated component (17.4%) in *C. heterophylla* was an unknown sesquiterpene alcohol similar to β-bisabolol. The concentration of oil was 150.0 ppm in *G. hirsutum*, 50.0 ppm in *H. militaris*, 46.3 ppm in *C. heterophylla*, 41.0 ppm in *H. lasiocarpus*, 25.4 in *C. sulphurea*, and 10.0 ppm in *C. involucrata*[195].

These results must be considered in perspective, however, because the intact cotton plant is the prime food source of the boll weevil. While plant odors no doubt contribute to plant infestations, the preference for cotton over other malvaceous food hosts apparently cannot be explained by laboratory attractancy bioassays. It is recognized that the essential oils extracted by steam distillation may differ, both qualitatively and quantitatively, from the volatile

profile emitted during transpiration. Attempts were made
to assess the extent of any differences by a GLC study of
the condensate from the atmosphere around cotton plants
in the field[196]. The total essential oil from the condensate
was found to consist 50-60% of β-bisabolol and γ-bisabolone
and 39-40% of geraniol, myrtenal, nerolidol, and β-
caryophyllene oxide. As the plant matured, trans-2-hexenol
was produced in concentrations of 7 to 27%. Before fruiting,
β-bisabolol comprised as much as 60% of the oil, and as the
concentration of β-bisabolol increased, that of γ-bisabolene
decreased. The condensate oil thus does differ from the
steam distillate, particularly in the absence of monoterpene
hydrocarbons, the paucity of the C_5 and C_6 alcohols and
sesquiterpene hydrocarbons, and the high content of β-
bisabolol and caryophyllene oxide. The identification of
the latter two as attractants[127] and major components of the
cotton field aroma gives added credence to the expectation
that they may play some role in the movement of boll weevils
to cotton.

PLANT CONSTITUENTS

In the course of investigations by us and others, on
constituents of the plant (boll, lint and seed), which possess
economic value or biological activity, a multitude of
compounds have been characterized. The following series of
tables (Tables 1-9) include 34 hydrocarbons, 62 alcohols,
polyols, phenols, sterols, 26 amino acids, 6 classes of
neutral and polar lipids, 34 pigments including flavonoids,
carotenoids and triterpenoids, 40 carboxylic acids, 38
carbonyl compounds, oxides and esters, and 12 miscellaneous
nitrogenous compounds; a total of 252. A table is also
included on proximate analysis of cotton plants and lint.

INSECT CONSTITUENTS

Insect volatiles. While reports on insect constituents
date back to 1900, most structural work, chiefly on
attractants and repellents, has been done in the past 10
years. Recent reviews[6,77] suggest that perhaps 500-1000
compounds have been identified. However, exhaustive
investigations of the entire volatile profile of insects,

Table 1. Hydrocarbons Isolated from the Cotton Plant

Hydrocarbon	Source	Composition	Chrom.	IR	NMR	MS	Ref.
Ethylene*	Plant	C_2H_4	X	–	–	–	41
α-Pinene	Bud oil	$C_{10}H_{16}$	X	X	X	–	122
Camphene	Bud oil	$C_{10}H_{16}$	X	–	X	–	122
β-Pinene	Bud oil	$C_{10}H_{16}$	X	X	X	–	122
Myrcene	Bud oil	$C_{10}H_{16}$	X	X	X	–	122
α-Phellandrene	Bud oil	$C_{10}H_{16}$	X	–	–	–	122
α-Terpinene	Bud oil	$C_{10}H_{16}$	X	–	–	–	122
Limonene*	Bud oil	$C_{10}H_{16}$	X	X	X	–	122
β-Phellandrene	Bud oil	$C_{10}H_{16}$	X	–	–	–	122
trans-β-Ocimene	Bud oil	$C_{10}H_{16}$	X	X	X	–	122
γ-Terpinenene	Bud oil	$C_{10}H_{16}$	X	–	–	–	122
Terpinolene	Bud oil	$C_{10}H_{16}$	X	X	X	–	122
α-Fenchene	Bud oil	$C_{10}H_{16}$	X	–	–	X	63
Undecane	Seedlings	$C_{11}H_{24}$	X	–	–	X	62
Dodecane	Seedlings	$C_{12}H_{26}$	X	–	–	X	62
Tetradecane	Seedlings	$C_{14}H_{30}$	X	–	–	X	62
Copaene	Bud oil	$C_{15}H_{24}$	X	X	X	X	123
1-trans-α-Bergamotene	Bud oil	$C_{15}H_{24}$	X	X	X	–	123
1-Caryophyllene*	Bud oil	$C_{15}H_{24}$	X	–	X	–	123
Farnesene	Bud oil	$C_{15}H_{24}$	X	–	–	–	123
α-Humulene	Bud oil	$C_{15}H_{24}$	X	X	X	–	123
cis-γ-Bisabolene	Bud oil	$C_{15}H_{24}$	X	X	X	–	123
1-δ-Guaiene	Bud oil	$C_{15}H_{24}$	X	X	X	–	123
1-δ-Cadinene	Bud oil	$C_{15}H_{24}$	X	X	X	–	123
γ-Cadinene	Seedlings	$C_{15}H_{24}$	X	–	–	X	43
γ-Muurolene	Bud oil	$C_{15}H_{24}$	X	–	–	X	63
n-Pentadecene	Bud oil	$C_{15}H_{30}$	X	–	–	X	62
Tricosane	Buds	$C_{23}H_{48}$	X	–	–	X	186
Pentacosane	Buds	$C_{25}H_{52}$	X	–	–	X	186
Heptocosane	Buds	$C_{27}H_{56}$	X	–	–	X	186
Nonacosane	Buds	$C_{29}H_{60}$	X	–	–	X	186
Triacontane	Leaves	$C_{30}H_{62}$	–	–	–	–	175
Untricontane	Buds	$C_{31}H_{64}$	X	–	–	X	186
C_{24}-C_{32} (trace)		$C_{24}H_{50}$-$C_{32}H_{66}$	X	–	–	–	186 175
Hexatricontane	Leaves	$C_{36}H_{74}$	–	–	–	–	175

*Attractant

Table 2.   Alcohols, Polyols, Phenols and Sterols Isolated from the Cotton Plant

Compound	Source	Composition	Chrom.	IR	NMR	MS	Ref.
1-Penten-3-ol	Bud	$C_5H_{10}O$	X	–	–	X	55
Cyclopentanol	Bud	$C_5H_{10}O$	X	–	–	X	55
Isoamyl alcohol	Bud	$C_5H_{12}O$	X	–	–	–	55
2-Methyl butanol	Bud	$C_5H_{12}O$	X	–	–	–	55
3-Methyl butanol	Bud	$C_5H_{12}O$	X	–	–	X	55
1-Pentanol	Bud	$C_5H_{12}O$	X	–	–	X	55
α-Furfuryl alcohol	Bud	$C_5H_6O_2$	X	–	–	X	63
1-Catechol	Plant	$C_6H_6O_2$	X	–	–	–	171
trans-2-Hexen-1-ol	Bud	$C_6H_{12}O$	X	–	–	X	55
cis-3-Hexen-1-ol	Bud	$C_6H_{12}O$	X	–	–	X	55
4-Hexen-1-ol	Bud	$C_6H_{12}O$	X	–	–	–	55
Cyclohexanol	Bud	$C_6H_{12}O$	X	–	–	X	55
1-Hexanol	Bud	$C_6H_{14}O$	X	–	–	X	55
2-Ethyl-1-butanol	Bud	$C_6H_{14}O$	X	–	–	X	62
Glucose	Leaves	$C_6H_{12}O_6$	X	–	–	–	172
Fructose	Leaves	$C_6H_{12}O_6$	X	–	–	–	172
Inositol	Plant	$C_6H_{12}O_6$	–	–	–	–	177
Benzyl alcohol	Bud	C_7H_8O	X	–	–	X	55
2-Phenyl ethanol	Bud	$C_8H_{10}O$	X	–	–	X	55
6-Octen-4-ol	Bud	$C_8H_{16}O$	X	–	–	X	55
1-Octanol	Bud	$C_8H_{18}O$	X	–	–	X	54
1-Nonanol	Bud	$C_8H_{18}O$	X	–	–	X	54
2-Phenoxyethanol	Bud	$C_8H_{10}O_2$	X	–	–	X	54
3-Nonanol	Bud	$C_9H_{20}O$	X	–	–	X	54
Caffeic acid	Plant	$C_9H_8O_4$	–	–	–	–	150
Cuminyl alcohol	Bud	$C_{10}H_{14}O$	X	–	–	X	62
Carveol	Bud	$C_{10}H_{16}O$	X	–	X	X	55
Myrtenol	Bud	$C_{10}H_{18}O$	X	–	–	X	62
Nerol	Bud	$C_{10}H_{18}O$	X	–	–	X	55
Geraniol	Bud	$C_{10}H_{18}O$	X	–	–	–	55
α-Terpineol	Bud	$C_{10}H_{18}O$	X	–	–	X	55
Isoborneol	Bud	$C_{10}H_{18}O$	X	–	–	X	55
Linalool	Bud	$C_{10}H_{18}O$	X	–	–	X	55
Citronellol	Bud	$C_{10}H_{20}O$	X	–	–	X	55
2-Methyl nonanol	Bud	$C_{10}H_{22}O$	X	–	–	X	62

Table 2 - continued

Compound	Source	Composition	Chrom.	IR	NMR	MS	Ref.
3-Methyl nonanol	Bud	$C_{10}H_{22}O$	X	–	–	X	62
6-Undecanol	Bud	$C_{11}H_{24}O$	X	–	–	X	63
α-Copaene alchol	Bud	$C_{15}H_{26}O$	X	–	–	X	63
Cis, trans- Farnesol	Bud	$C_{15}H_{26}O$	X	–	–	X	63
trans, trans- Farnesol	Bud	$C_{15}H_{26}O$	X	–	–	X	63
Caryophyllene alcohol	Bud	$C_{15}H_{26}O$	X	–	–	X	43
Nerolidol	Bud	$C_{15}H_{26}O$	X	–	–	X	55
β-Bisabolol*	Bud	$C_{15}H_{26}O$	X	X	X	X	126 127
α-Bisabolol*	Bud	$C_{15}H_{26}O$	X	X	X	X	126 54
l-epicatechol	Plant	$C_{15}H_{14}O_6$	X	–	–	–	171
l-epigallocatec- hol	Plant	$C_{15}H_{14}O_7$	X	–	–	–	171
dl-gallocatechol	Plant	$C_{15}H_{14}O_7$	X	–	–	–	171
Chlorogenic acid	Plant	$C_{16}H_{18}O_9$	–	–	–	–	150
l-epicatechol gallate	Plant	$C_{22}H_{18}O_{10}$	X	–	–	–	171
l-gallocatechol gallate	Plant	$C_{22}H_{18}O_{11}$	X	–	–	–	171
Cholesterol	Buds	$C_{27}H_{46}O$	X	–	–	–	194
Campesterol	Buds	$C_{28}H_{48}O$	X	–	–	–	194
Octacosanol	Leaves	$C_{28}H_{58}O$	–	–	–	–	175
Montanyl alcohol	Lint	$C_{28}H_{58}O$	–	–	–	–	3,31
β-Sistostanol	Plant	$C_{29}H_{52}O$	–	–	–	–	177
β-Sitosterol	Plant	$C_{29}H_{50}O$	–	–	–	X	177
β-Amyrin	Plant	$C_{30}H_{50}O$	–	–	–	–	177
Gossypol alcohol	Lint	$C_{30}H_{62}O$	–	–	–	–	3,31
Dotriacontanol	Leaves	$C_{32}H_{66}O$	–	–	–	–	175
Campesterol galactoside	Plant	$C_{34}H_{58}O_6$	X	–	–	–	194
Stigmasterol galactoside	Plant	$C_{35}H_{58}O_6$	X	–	–	–	194

*Attractant

Table 3. Amino Acids Isolated from the Cotton Plant

Amino Acid	Source	Reference
Aspartic acid	Cotton, thuberia	16,157,158
Asparagine	" "	" " "
Threonine	" "	" " "
Serine	" "	" " "
Glutamine	" "	" " "
Glutamic acid	" "	" " "
Proline	" "	" " "
Glycine	" "	" " "
Alanine	" "	" " "
Valine	" "	" " "
Methionine	" "	" " "
Isoleucine	" "	" " "
Leucine	" "	" " "
Tyrosine	" "	" " "
Phenylalanine	" "	" " "
Lysine	" "	" " "
Histidine	" "	" " "
Arginine	" "	" " "
Cysteic acid	Cotton, Malv.	157
γ-amino-n-butyric acid	" "	"
β-alanine	" "	"
Ornithine	" "	157,158
Taurine	Cotton	158
2-methyl-β-alanine	"	"
Hydroxylysine	"	"
Citrulline	"	"

such as is common with plant essential oils, have been very limited.

The first known report of boll weevil volatiles, other than the sex attractant, was that of Hedin et al. (1972b). Forty-two GLC maxima were observed with tentative identification of 34 components from a Carbowax 4000 column. Twenty-three of these maxima were observed in sufficient quantity to warrant mass spectral analysis. Of the 34 compounds for which a structure was suggested, 13 have been found in the

Table 4.   Pigments Isolated from the Cotton Plant

Pigment	Source	Composition	Chrom	UV	NMR	Ref.
Cyanidin	fls	$C_{15}H_{10}O_6 \cdot HCl$	X	X	–	38
Populnetin	fls	$C_{15}H_{10}O_6$	X	X	X	53 37 187
Herbacetin	fls	$C_{15}H_{10}O_7$	X	X	–	37 146
Quercetin*	pl	$C_{15}H_{10}O_7$	X	X	X	53,37 106,163
Gossypetin	gl, fls	$C_{15}H_{10}O_8$	X	X	–	3,37
Leucodelphinidin	sd hls	$C_{15}H_{14}O_8$	X	X	–	21,37
Vitamin A_1[#]	pl, sd	$C_{20}H_{30}O$	X	X	–	190
Trifolin	pl	$C_{21}H_{20}O_{11}$	X	X	–	155
Cyanidin-3-glucoside*	fls	$C_{21}H_{20}O_{11}$	HCl X	X	–	53,52
Quercetin-3-glucoside*	fls	$C_{21}H_{20}O_{12}$	X	X	–	53,153
Herbacitrin	fls	$C_{21}H_{20}O_{12}$	X	X	–	37,177 146
Isoquercitrin*	pl	$C_{21}H_{20}O_{12}$	X	X	–	53,155 37,162
Quercimeritrin*	fls	$C_{21}H_{20}O_{12}$	X	X	–	37,53 2,155
Gossypitrin	gl	$C_{21}H_{20}O_{13}$	X	X	–	37,155 177
Gossypin	pl	$C_{21}H_{20}O_{13}$	X	X	–	155,177
Kaempferol-3-rutinoside	pl	$C_{27}H_{30}O_{15}$	X	X	–	155
Rutin	pl	$C_{27}H_{30}O_{16}$	X	X	X	155,159 37,106
Gossypol+	sd, gl	$C_{30}H_{30}O_8$		X	–	1,3
Gossycaerulin	gl, fls	$C_{30}H_{30}O_8$	–	X	–	3
Gossyfulvin	gl	$C_{34}H_{34}N_2O_8$	–	X	–	3,121
Lycopene	lvs	$C_{40}H_{56}$	X	X	–	17
α-Carotene#	pl, sd	$C_{40}H_{56}$	X	X	–	190 17
β-Carotene	pl, sd	$C_{40}H_{56}$	X	X	–	190
Phytoene	pl, sd	$C_{40}H_{64}$	X	X	–	190
Phytofluene	pl, sd	$C_{40}H_{64}$	X	X	–	190
Lutein	pl, sd	$C_{40}H_{56}O_2$	X	X	–	190,17

Table 4 – continued

Pigment	Source	Composition	Chrom	UV	NMR	Ref.
Isolutein	pl, sd	$C_{40}H_{56}O_2$	X	X	–	190
Flavoxanthin	pl, sd	$C_{40}H_{56}O_3$	X	X	–	190
						17
Violaxanthin	pl, sd	$C_{40}H_{56}O_4$	X	X	–	190
						17
Auroxanthin	pl, sd	$C_{40}H_{56}O_4$	X	X	–	190
Neoxanthin	pl, sd	$C_{40}H_{56}O_5$	X	X	–	190
Neochrome	pl, sd	$C_{40}H_{56}O_5$	X	X	–	190
Gossyverdurin	pl	(CHO)	–	–	–	105
Gossypurpurin	gl	(CHON)	–	–	–	3

* Insect feeding stimulant   fls – flowers   sd – seed
+ Toxic   pl – plant   lvs – leaves
# Vitamin   gl – gland   hls – hulls

Table 5. Lipids Isolated from the Cotton Plant

Compound	Source	Composition	Chrom.	UV	IR	NMR	Ref.
Phosphatidyl choline	Buds	Mixed	X	–	–	–	191
Digalactosyl diglyceride	Buds	Mixed	X	–	–	–	191
Sterol glycoside	Buds	Mixed	X	–	–	–	191
Monogalactosyl diglyceride	Buds	Mixed	X	–	–	–	191
Phosphatidyl ethanolamine	Buds	Mixed	X	–	–	–	191
Phosphatidyl inositol	Buds	Mixed	X	–	–	–	191

Table 6. Acids Isolated from the Cotton Plant

Compound	Source	Composition	Chrom	Ref.
Formic acid[*]	lvs	CH_2O_2	–	169,170
Acetic acid	lvs	$C_2H_4O_2$	–	169
Oxalic acid	lvs	$C_2H_2O_4$	–	170
Pyruvic acid[*]	lvs	$C_3H_4O_3$	–	160
Lactic acid[*]	lvs	$C_3H_6O_3$	–	170
Fumaric acid	lvs	$C_4H_4O_4$	–	169
Succinic acid[*]	lvs, bls	$C_4H_6O_4$	–	169,30
Maleic acid	lvs	$C_4H_4O_4$	–	90
Malic acid[*]	lvs	$C_4H_6O_5$	–	174
Tartaric acid	lvs	$C_4H_6O_6$	–	170
Valeric acid	lvs	$C_5H_{10}O_2$	–	169
Glutaric acid	pl	$C_5H_8O_4$	–	177
α-Ketoglutaric acid	lvs	$C_5H_6O_5$	–	174
Caproic acid	bud	$C_6H_{12}O_2$	X	189
Ascorbic acid[#]	lvs	$C_6H_8O_6$	–	169,13,145
Citric acid	lvs	$C_6H_8O_7$	–	169,90
Salicylic acid	lvs	$C_7H_6O_3$	–	169
Caprylic acid	bud	$C_8H_{16}O_2$	X	189
Capric acid	bud	$C_{10}H_{20}O_2$	X	189
Lauric acid	bud	$C_{12}H_{24}O_2$	X	189
Myristic acid	pl,buds	$C_{14}H_{28}O_2$	X	189,172
Abscisin II[o]	pl	$C_{14}H_{20}O_4$	X	149
Pentadecenoic acid	bud	$C_{15}H_{28}O_2$	X	189
Pentadecanoic acid	bud	$C_{15}H_{30}O_2$	X	189
Palmitoleic acid	bud	$C_{16}H_{30}O_2$	X	189
Palmitic acid	bud	$C_{16}H_{32}O_2$	X	189
Heptadecanoic acid	bud	$C_{17}H_{34}O_2$	X	189
Octadecatetraenoic acid	bud	$C_{18}H_{28}O_2$	X	189
Linolenic acid[#]	bud	$C_{18}H_{30}O_2$	X	189
Linoleic acid[#]	bud	$C_{18}H_{32}O_2$	X	189
Malvalic acid[+]	sd	$C_{18}H_{32}O_2$	–	179
Oleic acid	bud	$C_{18}H_{34}O_2$	X	189
Stearic acid	bud	$C_{18}H_{36}O_2$	X	189
Sterculic acid[+]	sd	$C_{19}H_{34}O_2$	–	179
Gibberellic acid[o]	pl	$C_{19}H_{22}O_6$	–	18
Eicosatrienonic acid	pl	$C_{20}H_{34}O_2$	–	177

Table 6 - continued

Compound	Source	Composition	Chrom	Ref.
Eicosadienoic acid	pl	$C_{20}H_{36}O_2$	–	177
Arachidic acid	bud	$C_{20}H_{40}O_2$	X	189,172
Dicosadienoic acid	pl	$C_{22}H_{40}O_2$	–	177
Behenic acid	pl	$C_{22}H_{44}O_2$	–	177

* Insect feeding stimulant   lvs - leaves
+ Toxic   bls - bolls
# Vitamin   pl - plant
o Hormone   sd - seed

Table 7. Nitrogenous Compounds Isolated from the Cotton Plant

Compound	Source	Composition	Chrom	UV	NMR	MS	Ref.
Ammonia	pl	NH_3	–	–	–	–	162
Triethylamine	pl	$C_6H_{15}N$	–	–	–	–	162
Nicotinic acid#	pl	$C_6H_5NO_2$	–	–	–	–	177
Indole	Bud	C_8H_7N	X	–	–	X	63
Scopoline	pl	$C_8H_{13}NO_2$	–	–	–	–	150
3-Ethyl-5-methoxy-indole	Bud	C_8H_7N	X	–	–	X	62
Kinetin o	pl	$C_{10}H_9N_5O$	–	–	–	–	18
3-Indoleacetic acid o	pl	$C_{10}H_9NO_2$	–	–	–	–	18
Adenine glucoside	lvs	$C_{11}H_{17}N_5O_6$	–	–	–	–	160
Riboflavin#	pl	$C_{17}H_{20}N_4O_6$	–	–	–	–	177
Pheophytin a*	pl	$C_{55}H_{72}N_4O_5$	X	X	X	X	188
Pheophytin b*	pl	$C_{55}H_{70}N_4O_6$	X	X	X	X	188

Vitamin   lvs - leaves
o Hormone   pl - plant
* Insect feeding stimulant

Table 8. Carbonyls and Other Oxygenated Compounds Isolated from the Cotton Plant

Compound	Source	Composition	Chrom	UV*	IR	NMR	MS	Ref.
Acetone	Buds	C_3H_6O	X	X	-	-	-	125
Butyraldehyde	Buds	C_4H_8O	X	X	-	-	-	125
Isobutyraldehyde	Buds	C_4H_8O	X	X	-	-	-	125
Ethyl acetate	Buds	$C_4H_8O_2$	X	-	-	-	X	63
Methyl furan	Buds	C_5H_6O	X	-	-	-	X	62
Isovaleraldehyde	Buds	$C_5H_{10}O$	X	X	-	-	-	125
trans-2-Hexenal	Buds	$C_6H_{10}O$	X	X	X	-	-	125
Hexanal	Buds	$C_6H_{12}O$	X	X	-	-	-	125
2-Ethyl butyraldehyde	Buds	$C_6H_{12}O$	X	-	-	-	X	62
Benzaldehyde	Buds	C_7H_6O	X	X	-	-	-	125
Heptanal	Buds	$C_7H_{14}O$	X	X	-	-	-	125
Acetophenone	Buds	C_8H_8O	X	-	-	-	X	62,63
Methyl-tolyl ketone	Buds	$C_8H_{10}O$	X	-	-	-	X	63
ρ-tolualdehyde	Buds	C_8H_9O	X	X	-	-	-	125
2-Octenal	Buds	$C_8H_{14}O$	X	X	-	-	-	125
Ar-ethyl benzaldehyde	Buds	$C_9H_{10}O$	X	-	-	-	X	62
2,4-Dimethyl-2,4-heptadienal	Sdls	$C_9H_{14}O$	X	-	-	-	X	43
trans-2-cis-6-nonadienal	Buds	$C_9H_{14}O$	X	X	-	-	-	125
2-Nonenal	Buds	$C_9H_{16}O$	X	X	-	-	-	125
Nonanal	Buds	$C_9H_{18}O$	X	X	-	-	-	125
Cumic aldehyde	Buds	$C_{10}H_{12}O$	X	-	-	-	X	62
Isopulegone	Buds	$C_{10}H_{16}O$	X	-	-	-	X	62
1-ρ-Menthen-1-al	Buds	$C_{10}H_{16}O$	X	-	-	-	X	63

Table 8 - continued

Compound	Source	Composition	Chrom	UV*	IR	NMR	MS	Ref.
α-Pinene oxide	sdls	$C_{10}H_{16}O$	X	-	-	-	X	43
Myrtenal	Buds	$C_{10}H_{16}O$	X	X	-	-	-	125
n-Decanal	Buds	$C_{10}H_{20}O$	X	-	-	-	X	62
2-Decanone	Buds	$C_{10}H_{20}O$	X	-	-	-	X	63
Hexyl crotonate	sdls	$C_{10}H_{18}O_2$	X	-	-	-	X	43
Methyl nonanate	sdls	$C_{10}H_{20}O_2$	X	-	-	-	X	43
β-Ionone	Buds	$C_{13}H_{20}O$	X	-	-	-	X	55
Methyl laurate	sdls	$C_{13}H_{26}O_2$	X	-	-	-	X	43
β-Caryophyllene oxide	Buds	$C_{15}H_{24}O$	X	X	X	X	X	127
Bisabolene oxide	Buds	$C_{15}H_{24}O$	X	X	X	X	X	56
Methyl myristate	sdls	$C_{15}H_{30}O_2$	X	-	-	-	X	43
Methyl palmitate	sdls	$C_{17}H_{34}O_2$	X	-	-	-	X	43
Methyl linoleate	sdls	$C_{19}H_{36}O_2$	X	-	-	-	X	43
Octacosone	lvs	$C_{28}H_{56}O$	-	-	-	-	X	175
β-Amyrin montanate	pl	$C_{56}H_{104}O$	-	-	-	-	-	177

* of 2,4-dinitrophenylhydrazone

sdls - seedlings
lvs - leaves
pl   - plant

Table 9. Proximate Analysis of the Cotton Plant

Compound	Source	Reference
Reducing sugars	Boll, plant	164,32
Water	Boll, plant	164,180
Oligosaccharides	Boll	164
Protein	Boll	164
Oil	Boll	164
Cellulose	Boll, plant	164,180
Tannins	Boll, plant	164,173
Cellulose	Lint	206
Pentosans	Lint, plant	180,206
Nitrogen	Lint, plant	32,180,206
Wax	Lint	206
Ash	Lint, plant	180,206
Minerals	Leaves, stems, roots	26,208
Galactan	Plant	180
Calcium pectate	Plant	180
Oligosaccharides	Leaves	170
Dextrin	Leaves	170
Mucilages	Leaves	170
Starch	Leaves	170
Hemicellulose	Leaves	170
Pectins	Lint	32
Xylan	Lint	32

cotton bud. These, not unexpectedly, include several C_5 and C_6 alcohols, and mono- and sesquiterpene hydrocarbons. The major source of these components is the gut, however, storage in fatty tissue, and even biosynthesis is not precluded, since several species of Coleoptera biosynthesize terpenoids which act as pheromones[181]. The volatile constituents from adult, mixed sex boll weevils are given (Table 10) along with pertinent chemical information and references. Only 35% of the oil could be accounted for, in terms of chemical identification by GC-MS analysis.

In a later more extensive study[61] of the volatile constituents of male boll weevils, females, and their frass, the compounds found included 26 carbonyls, 23 hydrocarbons,

Table 10.   Volatile Constituents Identified From Mixed Sex
Adult Boll Weevils

Compound[a]	% Total oil	MX	Chrom.[b]	Ref.
[ ]-Pentanol	2.3		X	57
β-Pinene	0.5	X	X	57
α-Phellandrene	1.0		X	123,57
Limonene	1.4		X	57
1-Hexanol	1.6	X	X	123,57
cis-Hexen-1-ol	1.9		X	57
[ ]-Hexenol	1.7		X	57
Cyclohexanol	2.6		X	57
α-Muurolene	1.9	X	X	57
[ ]-Octenol	0.2	X	X	57
Acylic C_{10} alcohol	0.4	X	X	57
Caryophyllene	0.6	X	X	57
$β_2$-Bisabolene	0.2	X	X	57
α-Humulene	0.7	X	X	57
α-Bisabolene	0.3	X	X	57
(Z)- and (E)-3,3-dimethyl-$Δ^1$,α-cyclohexaneacetaldehyde	0.1	X	X	202
(+)-(Z)-isopropenyl-1-methyl-cyclobutaneethanol	0.2	X	X	202
(Z)-3,3-dimethyl-$Δ^1$,β-cyclohexane-ethanol	0.2	X	X	202
o-Toluidine	0.5	X	X	57
m or ρ-Toluidine	0.1	X	X	57
2,6-dimethylaniline	0.2	X	X	57
2,4-dimethylaniline	0.6	X	X	57
1-Dodecanol	0.8	X	X	57
5-Dodecen-1-ol	0.6	X	X	57
10-Methyleicosane	2.6	X	X	57
Monocyclic C_{15} alcohol	1.3	X	X	57
1-Tetradecanol	3.5	X	X	57
11-Methyl-7-tridecen-1-ol	2.1	X	X	57
1-Pentadecanol	3.0	X	X	57
Dibutyl phthalate	2.0	X	X	57

[a] Compounds are listed in the elution order from a Carbowax
    4000 GC column.
[b] Includes GC, TLC and column chromatography.

12 alcohols, 6 phenols, 4 esters, 3 furans, 1 ether, and 1
lactone. Also found were 2 compounds containing nitrogen,
1 halogen, and 1 sulphur. Of the 33 terpenoid compounds,
5 were hydrocarbons, 17 carbonyls, and 11 alcohols. Of the
24 aromatic compounds, 13 were hydrocarbons, 5 carbonyls,
and 6 phenols (Table 11).

The major components of the male insect volatile oil,
in addition to the pheromones, were 2 isomers of methyl
sorbate, dodecanol, butylmethoxybenzylpropyl ether, and a
halogen. The methyl sorbate isomers probably were synthesized
in the insect from potassium sorbate, a dietary preservative.
Two other dietary additives, methyl p-hydroxybenzoate and
busulfan were not found, presumably because they had been
metabolized. The ether is a solvent contaminant and was
observed in two other fractions. The halogen possessed a
fragmentation pattern that is consistent with the structure
of 2-Bromo-2-chloroaziridine (C_2H_3N br Cl; M^+ 155). Two
alternative acyclic structures also consistent with the
fragmentation pattern were (Z)-2-bromo-2-chlorovinylamine
and 2-bromo-2-chloroethylidenimine. The relative fragment
intensities at m/e 155, 157, and 159 are indicative of a
compound with one chlorine and one bromine. The lower mass
fragments are consistent with the aziridine ring structure.
The proposed structure is, of course, very unusual and
requires rigorous confirmation. The three other oils also
contained this halogen compound.

The male frass oil contained, in addition to the 4
pheromones, relatively large amounts of 1-(4-methyl-3-furyl)-
2-propanone, the halogen, β-caryophyllene, and 3,7-dimethyl-
1-octanol, a saturated, acyclic monoterpene alcohol. The
furan may be produced from carbohydrates, during the auto-
claving of the diet, or by the insect. The β-caroyophyllene
may represent incompletely metabolized plant material. The
monoterpene alcohol is apparently biosynthesized by the insect
since it was not found in the alcohol fraction of the cotton
bud essential oil[55].

The major components in the female insect volatile oil
were ρ-ethyl toluene, 4-(2-furyl)-2-butanone, methylnaphtha-
lene, carvone, dodecanal, 2,6-di-*tert*-butyl-ρ-cresol,
dodecahydro-2-methylpyrido 2,1,6-*de* quinolizine, the
halogenated compound, and $C_{10}H_{14}O$. ρ-Ethyltoluene and methyl-
naphthalene were only two of several aromatic hydrocarbons

Table 11. The Volatile Compounds in Male and Female Boll Weevils and Their Frass

Compound	M+	I_k OV-17*	Reference or Fragmentation#	% Male	% Male frass	% Female	% Female frass#
Toluene	92	1035	AMSD		0.1	0.4	
Phenol	94	1230	AMSD	0.2	1.0	1.9	0.6
1-Methylcyclohexene	96	1035	AMSD			0.4	
Ethylbenzene	106	1075	AMSD				0.4
( )-Dimethylbenzene	106	1080	AMSD	0.6	1.4	0.9	0.1
Benzaldehyde	106	1172	AMSD		0.3		
Phenylacetonitrile	117	1385				0.1	
( )-Amino-( )-methoxy-cyclohexane	117	1528	57,41,87,117, 91,44,39			0.5	
Isopropylbenzene	120	1110	AMSD		0.3		
o-Ethyltoluene	120	1120	AMSD		0.3	1.5	0.1
m-Ethyltoluene	120	1142	AMSD		0.3		
ρ-Ethyltoluene	120	1172	AMSD	0.1	1.2	3.9	0.3
Phenylacetaldehyde	120	1248	AMSD			0.3	
Acetophenone	120	1270	AMSD	0.1		0.4	
Methyl-(Z,Z)-sorbate	126	1167	AMSD	2.3	0.6		1.1
Methyl-(E,E)-sorbate	126	1189	AMSD	13.0			0.4
Naphthalene	128	1386	AMSD	0.4	0.6		
( )-Dimethyl-( )-ethyl benzene	134	1152	AMSD				1.0

Table 11 – continued

Compound	M+	I_k OV-17*	Reference or Fragmentation#	% Male	% Male frass	% Female	% Female frass#
o-Propyltoluene	134	1175	AMSD		0.2		
4'-Methylacetophenone	134	1385	AMSD			0.2	
Benzothiazole	135	1480	AMSD	0.4	1.0		
$C_{10}H_{16}$	136	1110	81,67,95,54, 109,121				
$C_{10}H_{16}$	136	1152	41,39,55,136, 94,121,107				0.9
Limonene	136	1168	AMSD		1.6		4.8
( )-Methoxybenzaldehyde	136	1507	AMSD	0.1			
ρ-Mentha-1,8-diene	136	1172	AMSD				0.8
2-( )-Butyl-4-methyl furan 1-(4-Methyl-3-furyl)-2-propanone	138	1195	67,68,95,123, 109,138	0.4			
4-(2-furyl)-2-butanone	138	1290	95,58,138,67, 123,53,71	0.7		6.2	
( )-Trimethylcyclohexenone	138	1308	AMSD			0.8	
n-Decane	142	1020	AMSD		0.1		
n-Nonanal	142	1225	AMSD	1.4	1.8	0.7	1.0
( )-Methylnaphthalene	142	1504	AMSD	0.6	1.0	4.5	
Carvone	150	1436	AMSD			5.4	6.8
ρ-Mentha-1,8-dien-3-one	150	1478	AMSD				0.6

Table 11 – continued

Compound	M^+	I_k OV-17*	Reference or Fragmentation#	% Male	% Male frass	% Female	% Female frass#
$C_{10}H_{14}O$ cyclic ketone	150	1480	135,108,107, 69,136,150	2.5			
Tolyl acetate	150	1481	AMSD		1.3		
ρ-Mentha-5,8-dien-2-one	150	1492	135,108,43, 41,39,69,82				1.2
$C_{10}H_{14}O$	150	1512	43,107,67,58, 39,82,95				0.6
$C_{10}H_{14}O$	150	1520	58,135,54,150			5.4	
$C_{10}H_{14}O$ Alcohol	150	1532	91,105,53,51, 135,54,50				0.1
$C_{10}H_{16}O$ Alcohol	152	1258	43,41,57,109, 81,134,137				1.1
ρ-Menth-1-en-9-al	152	1280	AMSD				1.7
ρ-Mentha-1,4-dien-7-ol	152	1295	AMSD				0.9
$C_{10}H_{16}O$	152	1336	69,41,84,39, 67,97,123				1.2
$C_{10}H_{16}O$	152	1362	41,39,60,109, 67,134,119				0.9
$C_{10}H_{16}O$	52	1367	84,41,83,39, 69,67,95,108				3.4
$C_{10}H_{16}O$	152	1385	41,94,84,39, 79,83,67,69				7.3

Table 11 - continued

Compound	M^+	I_k OV-17*	Reference or Fragmentation#	% Male	% Male frass	% Female	% Female frass≠
$C_{10}H_{16}O$	152	1390	109,84,41,39, 83,69,53				4.3
$C_{10}H_{16}O$	152	1404	84,41,39,134, 109,69,83				7.8
(Z)-3,3-Dimethyl-$\Delta^{1,\alpha}$-cyclohexaneacetaldehyde (Compound II)	152	1428	Tumlinson et al. (1969)	0.1	1.6		
(E)-3,3-Dimethyl-$\Delta^{1,\alpha}$-cyclo-hexaneacetaldehyde (Compound IV)	152	1439	Tumlinson et al. (1969)	0.1	1.4		
$C_{10}H_{16}O$	152	1473	67,41,53,68, 79,119,106				0.2
$C_{10}H_{16}O$	152	1485	67,68,81,79. 95,123				0.6
Linalool	154	1316	AMSD				0.4
α-Terpineol	154	1328	AMSD				0.5
(+)-cis-2-Isopropenyl-1-Methylcyclobutaneethanol (Compound I)	154	1363	Tumlinson et al. (1969)	0.9	14.6	0.2	
(Z)-3,3-dimethyl-$\Delta^{1,\beta}$-cyclohexaneethanol-(Compound II)	154	1383	Tumlinson et al. (1969)	2.8	21.3	0.3	

Table 11 – continued

Compound	M^+	I_k OV-17*	Reference or Fragmentation#	% Male	% Male frass	% Female	% Female frass#
C$_2$H$_3$N Br Cl (2-Bromo-2-chloroaziridine?)	155	1282	157,155,75,39,49,38	5.1	8.6	8.0	0.3
C$_{10}$H$_{20}$O	156	1302	41,57,44,70,82,95,112	0.5			
Trans-ρ-menthan-7-ol	156	1305	AMSD				
n-Decanal	156	1309	AMSD				0.4
Citronellol	156	1326	AMSD			1.2	0.2
( )-Dimethylnaphthalene	156	1572	AMSD				2.0
γ-Nonalactone	156	1576	AMSD	0.2	1.9	0.3	
( )-Dimethylnaphthalene	156	1597	AMSD	0.2		0.2	
3,7-Dimethyl-1-octanol	158	1565	AMSD		15.6	0.5	
tert-Butyl-o-cresol	164	1542	AMSD	0.4		0.5	
C$_{12}$H$_{18}$O	178	1540	57,45,39,75,85,101,100			1.9	
tert-Butyl-2-methoxyphenol	180	1590	AMSD	10.5			
n-Dodecanal	184	1505	AMSD			0.9	
n-Dodecanol	186	1565	AMSD			2.3	
Dodecahydro-2-methylpyrido 2,1,6-de quinolizine	193	1518	192,150,151,193,41,164,136	1.4	1.4	3.6	3.4
C$_{12}$H$_{18}$O$_2$	194	1529	91,71,69,149,105,194	0.6			1.3
	194	1533	57,41,43,87,91,117,147	0.2			

Table 11 – continued

Compound	M^+	I_k OV-17*	Reference or Fragmentation#	% Male	% Male frass	% Female	% Female frass#
( )-Decyl acetate	200	1570	AMSD				0.4
β-Caryophyllene	204	1520	AMSD		3.2		0.8
2,4-Di-tert-butylphenol	206	1655	AMSD	0.2	0.1		
( )-Pentadecene	210	1522	AMSD	2.9			
$C_{14}H_{26}O$	210	1528	43,41,57,29, 55,69,71	0.1			
$C_{15}H_{22}O$	218	1581	161,41,141, 175,218,203	1.3			
ρ-Nonylphenol	220	1588	AMSD	1.6			
2,6-Di-tert-butyl-ρ-cresol	220	1625	AMSD	1.3		8.7	4.4
( )-Hexadecene	224	1637	AMSD	0.9	0.5		0.8
n-Hexadecane	226	1600	AMSD		0.9		5.4
( )-Butyl-( )-methoxy-benzyl propyl ether	236	1570	165,177,180, 137,91,221,205	8.1		8.9	
Totals#				61.8	93.7	71.5	70.4

* Kovats (1961) indices: e.g. n-pentadecane = 1500, n-hexadecane = 1600.

AMSD = Atlas of mass spectral data, Stenhagen et al. (1969), Tumlinson et al. (1969), or most abundant fragment ion values in descending order.

≠ Percentage of total oil; the totals are less than 100 because not all the GLC maxima were successfully analyzed.

which were found in the three other oils. They are generally
found in volatile plant and animal systems and are probably
formed from amino acids and/or plant phenols. The furan and
two others in compounds in these oils are probably formed
from carbohydrates. Carvone is also found in the female
frass, but not in the male oils; thus it may be sex specific.
The quinolizine alkaloid was first observed as a defensive
compound in the so-called European lady beetle (*Propylea
quatuordecimpunctata* (L.)[205]. It was first isolated, in this
laboratory, from the lady bird beetle (*Coleomegilla maculata*
(Deg.) by Henson (private communication) as a major component,
and its structure was subsequently determined by MS, NMR, and
I.R. to be the same as that of the European lady beetle.
Dodecanal is probably derived from the insect fatty acids.

The major components in the female frass are limonene,
carvone, citronellol, dodecanol, 2,6-di-tert-butyl-p-cresol,
the M^+ 236 ether, and a series of $C_{10}H_{14}O$ and $C_{10}H_{16}O$ compound
Limonene and citronellol are probably unmetabolized dietary
components, dodecanol is probably derived from the insect
fatty acids, and the cresol and ether may be solvent contaminants. The most interesting constituents are a series of
monoterpene aldehydes or ketones with the elemental formula
$C_{10}H_{16}O$, M^+ 152 that gave mass spectra very similar to the
isomers of 3,7-dimethyl-2,6-octadienal. They are unique
to the female frass, and it is tempting to speculate that
some of them may be involved in communication. Alternatively,
they may be common precursors of the male pheromones which
accumulate in the absence of the appropriate enzyme systems
constitutive in the male. Larger quantities of frass are
being collected and processed, to obtain more definitive
information on these monoterpene compounds.

Both differences and similarities were found, when the
data from the present study were compared with those from
our previously reported work on the volatiles of mixed
sexes of boll weevils[57]. Several monoterpene and sesquiterpene hydrocarbons, alkyl alcohols, and the pheromones were
found in both investigations. However, in the earlier work,
several C_5 and C_6 alcohols and $C_{15/20}$ alcohols and hydrocarbons were identified. They were undoubtedly present in
the insect and frass volatile oils, but the conditions for
gas chromatographic introduction to the mass spectrometer
were selected for best resolution of the monoterpenes.
Therefore, the C_5 and C_6 compounds were eluted with the

solvent, and the analysis was terminated prior to elution of the higher molecular weight compounds. Also absent in the present analysis were two toluidines and N-methylanilines. The major additions were several aromatic compounds and a large number of oxygen containing monoterpenes.

Insect lipids. The large-scale lipid deposition during the transition of the adult boll weevil into diapause is a subject of great interest. An understanding of the biochemistry of this physiological event is fundamental to efforts aimed at the modification of the life cycle thereby affecting the ability of a colony to survive the winter. Thus a thorough knowledge of the components of the insect's lipids becomes paramount.

The fatty acid fraction of the adult boll weevil is a complex mixture of 23 fatty acids, ranging from 6 to 20 carbon atoms. Eight major acids account for 98% of the total. These are: myristic (14:0) (1.4%), palmitic (16:0) (30.7%), palmitoleic (16:1) (5.9%), heptadecanoic (17:0) (1.0%), stearic (18:0) (5.0%), oleic (18:1) (29.8%), linoleic (18:2) (12.3%), and linolenic acids (18:3) (13.1%)[92]. The relative percent composition is similar in body fat and in isolated triglycerides. A diet of cotton bolls, which is high in glycogen contributes to the buildup of lipids prior to diapause. Nettles and Betz (1965) found that the glycogen content (on a dry weight basis), of the weevil is maximum in the egg (11%). It decreases to 1.5% through most of the larval period, rises to 6% late in the last larval instar, and declines rapidly during the first half of the pupal period, but more slowly in the second half. The highest glycogen content in adults is at 6 to 15 days of age, when fed on cotton bolls or buds, but the concentration is several times higher in boll-fed weevils. Tingle et al. (1971) have shown that the addition of carbohydrate to a low protein diet will cause a large increase in body fat of adult boll weevils entering diapause.

The dietary requirements for fatty acids during sexual maturation and egg production have been described[207]. Low concentrations of lipids and fat-soluble vitamins in the diet reduced oviposition. In diets lacking inositol, choline, or cholesterol, oviposition was less than half that obtained with the complete diet. When the premature stages of the insect were fed diets low in lipids, the resulting female adults had a long pre-oviposition period and low mortality. Even though the weevil is capable of synthesizing fatty acids from carbohydrates, it cannot dehydrogenate saturated acids to the di and tri-unsaturated fatty acids[96,97]. The weevil is capable of synthesizing fatty acids from ^{14}C-acetate and can produce palmitic and stearic acids from the corresponding mono-unsaturated fatty acids, palmitoleic (16:1), and oleic acids (18:1). However, in fatty acid synthesis studies of larvae, pupae and newly molted, unfed adults, 60% of the ^{14}C-acetate activity was incorporated into oleic acid (18:1). Thus, the boll weevil requires polyunsaturated fatty acids in the diet for normal oviposition and egg hatch. However, for diapausing purposes, these polyunsaturated acids are apparently not required[95,199].

Bumgarner and Lambremont (1966) found that the lipid-class spectrum of the boll weevil egg, adult male, and adult female resembles that of the diet. Females feeding on low fat diets deposit in the egg yolk, lipids which contain primarily those fatty acids synthesized by the adult from non-lipid components.

Carbon-14 studies show that the boll weevil does not synthesize the sterol nucleus[98]; this is in agreement with studies on other insects and invertebrates. However, interconversion of the functional groups and side chains does occur.

Upon injection of hexadecane-1-^{14}C into the boll weevil, the principal products are ^{14}CO$_2$ and labeled palmitic acid[82]. This suggests direct conversion of hexadecane to the fatty acid by terminal oxidation. Palmitoleic acid was the main unsaturated acid.

Henson et al. (1971) found phosphatidyl choline and phosphatidyl ethanolamine to be the major phospholipids of newly-emerged adult weevils. Table 12 shows the percentages of each of the phospholipids which were isolated. After enzymatic hydrolysis of phosphatidyl choline and phosphatidyl ethanolamine, the 2-position of both compounds contained large percentages of unsaturated fatty acids, the major one (62%) being oleic acid. The 1-position also had a preponderance of unsaturated fatty acids (51 to 75%).

Table 12. Phospholipid Content of the Boll Weevil

Phospholipid	Percentage of total phospholipid
Cardiolipin	8.0
Lysophosphatidyl choline	2.5
Lysophosphatidyl ethanolamine	1.8
Phosphatidyl choline	35.5
Phosphatidyl ethanolamine	32.7
Phosphatidyl inositol	5.2
Phosphatidyl serine	5.1
Sphingomyelin	7.5
Unknown	1.2

In our studies of weevil lipids during the larval-pupal ecdysis[62], the concentration of neutral lipids was found to increase until the pupal stage; after which it decreased slightly. The opposite was true for phospholipids. The neutral lipids were high in palmitic (16:0), palmitoleic (16:1) and oleic acids (18:1). Whereas the phospholipids had large amounts of stearic (18:0) and linoleic (18:2) acids. The composition of phospholipids was similar to all metamorphic stages, though the larval stage had an abundance of phosphatidyl ethanolamine.

Neutral lipids predominated in the nuclear and mito-chondrial fractions of pharate pupae and phospholipids were highest in the microsomes[198]. The nuclear and mitochondrial lipids contained more linoleic acid than the microsomes. A decrease in content of 18:2 in the microsomes, coincided with increased 18:3. The average number of unsaturations per mole-cule was highest in the phospholipids. Phosphatidyl ethanolamine and phosphatidyl choline comprised 50-70% of the phospholipids in all subcellular components of pharate pupae. Sterols and sterol derivatives were evenly distributed in the nuclei, mitochondria and microsomes of pharate pupae, pupae and adult stages.

In subcellular particles of pupae it was found that phospholipids predominate in the lipids of microsomes, nuclei and mitochondria[197]. Phosphatidyl choline and phosphatidyl ethanolamine were equally distributed in the mitochondria, and phosphatidyl choline was most abundant in the remaining subcellular fractions. The neutral lipids contained approximately 50% triglycerides in nuclei, microsomes and cytoplasm; whereas monoglycerides and diglycerides were most abundant in the mitochondria.

The glycerol ethers have also been isolated and identified from the phospholipids[99]. They were isolated by LiAlH reduction of the lipid extract, followed by TLC of the reduced mixture[214]. The adult weevil was found to contain a 10 times greater concentration of alkyl ethers than pupae. The concentration of the alk-1-enyl ethers is about three times higher in adults than in pupae. The presence of these compounds has been observed in the egg and larvae but could not be quantitated because of interfering compounds.

A limited number of boll weevil enzymes have been investigated. Lambremont (1962) studied the dehydrogenases of the brain and related neural and secretory structures. Endogenous dehydrogenase activity, as measured on fresh tissue, was found to produce the pattern of formazan deposition similar to that of tissue from other organisms. The presence of several Krebs cycle enzymes was also shown. Of seven glycolytic intermediates and end-products, α-glycerophosphate was dehydrogenated most actively. Glucose-1-phosphate and lactate were active; but acetate, ethanol, pyruvate and glycerol were unreactive. A study of coenzyme dependence showed that lactate production was increased by NAD and isocitrate by NADP. Malate and glutamate were increased equally with both coenzymes, whereas, glycerol, ethanol, beta-hydroxybutyrate, pyruvate and choline were not influenced by either.

Four alkaline inorganic phosphatases capable of producing orthophosphate from pyrophosphate have been reported[94]. The greatest activity was found in a band which migrated toward the anode. The enzyme showed zero order kinetics for up to one hour, under optimal conditions of 2.5×10^{-3} M pyrophosphate, 10^{-3} M $MgCl_2$, at pH 8.0. The activation energy was 15.6 Kcal/mole, and the maximum activity temperature was 50°C. Magnesium was the only activator.

Insect nitrogenous compounds. The increasing use of
chemicals, which affect the genetic components of the cell,
particularly those which induce sterility in insects, led
to an investigation of the normal complement of nucleic acids
during different developmental stages of the boll weevil[209].
The amounts of nucleic acids present tended to parallel the
growth-curve. The egg and first-instar larvae contained
little nucleic acid. The increase of RNA in the second-
instar through the fourth-instar larvae was paralleled by
increased protein synthesis. However, when feeding and
rapid growth ceased, the amount of RNA continued to increase,
since the epidermal cells were continually stimulated by
the moulting hormone, throughout the pupal molt[212]. After
emergence into adults, the amount of DNA was greater in
males, and the RNA greater in females. This difference was
attributed to the quality and quantity of gametes produced
by the two sexes[209].

Mitlin and Vickers (1964) observed that in addition to
uric acid, which is the principle nitrogenous excretory
product, guanine was also present in boll weevil feces (3%
of the total nitrogen). The appearance of guanine in the
excreta might be attributed to a deficiency of the enzyme
guanase. In another study, guanine was found to be present
in highest concentration in feces of insects fed cotton buds,
and lowest in the feces of insects fed cotton bolls or an
artificial diet of acetone extracted cotton buds[132]. Since
the insects, used in these studies, were not reared asep-
tically the results may be biased by effects produced from
the microflora of the gut.

Fecal uric acid, from boll weevils injected with
thymidine 2-^{14}C and uridine 2-^{14}C, was radioactive while the
guanine present was not[141]. Injected uric acid-2-^{14}C was
metabolized to RNA, CO_2, and amino acids. [14]C incorporation
into DNA was slight. Cytidylic acid of RNA showed the
greatest incorporation[142].

Mitlin et al. (1966) studied the free and bound amino
acids (Table 13), during metamorphosis, and observed 23
free amino acids, plus ammonia, with 3 additional amino
acids produced upon acid hydrolysis. In both instances, glutamic acid, tyrosine and histidine were found in abundance.
In the bound state, aspartic, glycine and alanine also
occurred in high concentrations. There was little

Table 13.   Amino Acids of Boll Weevil Prepupae and Feces
(μ mole/mg N)[a]

Amino Acid	Prepupae		Feces	
	Bound	Unbound	Bound	Unbound
Cystic acid	T[b]	T	0.08	0
Taurine	T	T	T	0
Methionine sulfoxide	0.03	0.02	0	0
Aspartic acid	0.09	0.02	0.72	0.07
Threonine	0.04	0.10	0.33	0.09
Serine	0.06	0.06	0.17	0.22
Glutamic acid	0.26	0.08	1.73	0.22
Proline	0.11	0.05	0.53	0.27
Glycine	0.09	0.03	0.47	0.09
Alanine	0.08	0.04	0.51	0.13
Cystine	0.01	T	0.06	0
Valine	0.06	0.03	0.32	0.14
Methionine	0.01	0.01	0.03	0.09
Isoleucine	0.03	0.02	0.31	0.11
Leucine	0.03	0.02	0.29	0.21
Tyrosine	0.10	0.13	0.08	0.04
Phenylalanine	0.02	0.01	0.13	0.09
Ammonia	0.26	0.11	3.16	0.64
Lysine	0.04	0.02	0.24	0.08
Histidine	0.06	0.06	0.19	0.08
Arginine	0.02	0.02	0.13	0.05
α-amino-n-butyric acid	0		0.06	0.01
Tryptophan	0	0.01	0	0
β-amino-n-butyric acid	0	0	0.09	0
β-Alanine	T	T		
Hydroxylysine	0	0	0.07	0.02
α-Amino butyric acid	0	0	0	0.01
Ethanolamine	0.27	0	0	0
Hydroxyproline	T	0	0	0
Ornithine	0	0	0.02	0.02

a/ Figures derived from tables in references, Mitlin et al.
   1964b, 1966
b/ T = Trace

fluctuation, of either individual amino or protein-bound acids, during metamorphosis.

A similar study of amino acids from feces shows a wide variation when compared with that of the insect (Table 13)[133]. The amino acids in boll weevil feces appear to be more similar to those of the male reproductive tissue of the cotton plant[158]. The amides of aspartic and glutamic acids appear in large quantities in both the male reproductive tissue and feces, indicating an excess of these acids in the diet. Table 13 expresses the amino acids of prepupae and feces as μ-mole/mg nitrogen. Also investigated were the free amino acid levels in the adult male as affected by feeding buds, bolls, and laboratory diet[213].

In an electrophoretic study of the proteins in the haemolymph throughout the life cycle, 14 proteins plus several glycoproteins and lipoproteins were found, including one unique to the larval stage and one to the pupal stage[137]. Some proteins were common to all stages of the developing insect, however, no definite correlation could be made between proteins in the egg and in the haemolymph. During the larval stages, the 14 protein bands were divided, almost evenly, in their electrophoretic movement toward the cathode and anode. However, some of the cathodic movement may be due to electrosmosis characteristic for cellulose acetate film.

Mitlin et al. (1968a) isolated 30 and identified 25 ninhydrin-positive compounds from the free and hydrolyzed amino acid fractions of the haemolymph. These amino acids were quantitatively similar to those of the whole insect, with the exception of 2 basic components eluting between tryptophan and arginine. It is interesting to note that the nitrogen titer of haemolymph increases in the female and decreases in the male in the first day after emergence (Table 14). The reverse occurs during the second day. After the third day, variations in nitrogen titer are similar for both sexes (Table 14).

The implication of lipoprotein as an intermediate in protein synthesis is indicated by the work of Mitlin et al. (1968b). The incorporation of labeled ^{14}C-lysine into protein was followed for 24 hrs after injection into the haemocoele. As the concentration of free ^{14}C-lysine diminished, protein ^{14}C-lysine increased. However, the

Table 14.   Nitrogen Titer in Haemolymph of Boll Weevil

Age (day)	μmoles/μl	
	Males	Females
Newly emerged	0.73	0.62
1	0.51	0.68
2	0.65	0.41
3	0.57	0.58
4	0.47	0.51
5	0.51	0.56

concentration in the pool tended to reach equilibrium at a relatively high level after 4 hours; whereas, that in the protein fraction leveled off at about 8 hours. The lipoprotein fraction was labeled within the first hour but lost its activity within the next hour. This indicates that much of the labeled lysine was incorporated into lipoprotein prior to incorporation into the protein fraction.

In another study of the utilization of amino acids in the synthesis of more complex nitrogenous compounds, the incorporation of radiolabeled tryptophan into the lipoproteins, nucleic acids, and proteins was examined[140]. Injection of $3-^{14}C$-tryptophan resulted in an initial rapid drop of this acid in the free amino acid pool and then a gradual decline until 12 hrs post-injection. Incorporation into the lipoprotein and protein fractions peaked at 12 hours, but incorporation into the nucleic acids, increased continually Table 15 shows the relative rates of incorporation into RNA and DNA. All RNA bases were labeled in 4 hours, but complete labeling of the DNA bases did not occur until 18 hours after injection.

Twenty-three amino acids were separated and isolated, but only 10 contained ^{14}C label. These included: aspartic, threonine, serine, glutamic, proline, alanine, valine, isoleucine, leucine and phenylalanine. The lipid fraction was labeled, but activity was too low to determine the extent of ^{14}C incorporation into the individual fatty acids.

Table 15. Relative Rates of Incorporation of Labeled Carbon into the Constituent Bases of the Nucleic Acids

	RNA			DNA		
Hours	Guanine Adenine	Cytosine Uracil	Purines Pyrimidines	Guanine Adenine	Cytosine Thymine	Purines Pyrimidines
4	1.28[a/]	1.81	1.36	–	–	–
6	1.35	1.70	1.14	–	–	–
18	0.91	1.27	0.56	0.83	0.61	0.63
24	1.21	1.16	0.87	0.66	0.82	0.87
48	0.68	1.60	0.63	1.00	0.74	0.95

Uric acid is stored, in the tissues of boll weevils, throughout its life cycle, but guanine is not[136]. Since uric acid-2-^{14}C injected into the metamorphosing weevils causes labeling of at least 6 of the free amino acids, uric acid appears to be in a dynamic state and is evidently involved in nitrogen metabolism.

Sex pheromone. Pheromones are chemical messengers which allow individuals of a species to communicate with one another. Since, by means of pheromone communication, insects find each other and propagation can ensue, the term sex attractant has become widely used to describe these chemicals. However, a pheromone may have more than one function, as in the case of the queen substance (9-Oxo-2-decenoic acid) of the honey bee *Apis mellifera* L., which, in addition to its function as a sex attractant and aphrodisiac, functions also as an inhibitor of queen cell construction[8].

In the earliest record of sexual attraction in the boll weevil, Hunter and Hinds (1905) concluded that females were not attractive to males and that "...instead of seeking widely for the females, the males are content to wait for them to come their way." It was not until 1962 that it was shown conclusively that the male boll weevil produces a wind-borne sex attractant (pheromone) attractive to females[22,85]. It was later shown that the male pheromone is not only a sex pheromone for females but also acts as an aggregating pheromone for both sexes, primarily in the spring

and fall, and to a lesser degree in mid-season[10,23,47].
Hardee et al. (1970) confirmed the aggregating
characteristic of the pheromone in the field and the
influence of diet on pheromone production was reported
later[48.49].

The first isolation of the sex attractant was accomplished in 1964[85]. Of all isolation methods tested, steam
distillation of dichloromethane extracts was the most
suitable[201]. Extractions and purifications from fecal
material and insects were done and bioassays were made of
each fraction[45]. Insects (67,000 males and 4,500,000 weevils
of mixed sexes) or fecal material (54.7 kg) were extracted
with dichloromethane; the extract was concentrated under
vacuum and steam-distilled. The distillate was reextracted
with dichloromethane and the solvent again removed under
vacuum.

The extract of the steam distillate was fractionated by
column chromatography on Carbowax 20M coated silica gel.
None of the individual fractions from this column were
attractive to females, but the combination of two of the
fractions was as active as the original distillate. Each
of these two fractions was then separately fractionated on
a column containing Adsorbosil-CABN (25% $AgNO_3$ on silica
gel).

Various recombinations of all the fractions, from both
$AgNO_3$-silica gel columns, yielded two fractions, one from
each column, which were attractive together but unattractive
alone. Each of these latter two active fractions was then
fractionated by glpc on Carbowax 4000 and SE-30. Three
components, which were attractive when all three were combine
but unattractive alone or in pairs, was collected. Rechromatography on Carbowax 4000, SE-30, and a 50-ft supported
coated open tubular (SCOT) column showed two of these
components to be pure (I, II) and the third to consist of
two compounds (III, IV). Concentrations of compounds I, II,
III, and IV in fecal material, determined by glpc, were
0.76, 0.57, 0.06, and 0.06 ppm, respectively. Concentrations
in weevils were about tenfold less. Compound I was identifie
as (+)-cis-2-isopropenyl-1-methylcyclobutaneethanol. The
cis configuration was assigned by comparison with the NMR
spectrum of the synthetic cis isomer.

The structure of Compound II ($\underline{Z}$-3,3-dimethyl-$\Delta^{1,\beta}$-cyclohexaneethanl) was elucidated on the basis of its mass, NMR, and IR spectra, and with further evidence obtained from its palladium catalyst reduction and ozonolysis. The structures of Compounds III ($\underline{Z}$-3,3-dimethyl-$\Delta^{1,\alpha}$-cyclohexaneacetaldehyde) and IV ($\underline{E}$-3,3-dimethyl-$\Delta^{1,\alpha}$-cyclohexaneacetaldehyde) were determined as described by Tumlinson et al. (1969, 1971). The synthesis of each compound has been done by several laboratories[7,40,202,203] (Fig. 1).

Tumlinson et al. (1969) calculated that Compounds I, II, III, and IV were present, in mixed sex feces, at concentrations of 0.76, 0.57, 0.06, and 0.06 ppm, respectively. The total content of the components in males is 200 ng or less, but the average content in the frass produced during 1 day is 1268 ng, and the lifetime production may be 40,000 ng[60]. The ratio of the four components, in frass, was 6:6:2:1/ I:II:III:IV. None of the four components was normally found in females; however in three analyses traces were found. The synthesis rate is initially negligible, but increases to a maximum by day 8 and is maintained for the remainder of the insect's life.

Before the pheromones had been identified, it was established that males needed to feed on cotton to become attractive. The peak sexual activity of both males and females does not occur until the weevils are 4 to 6 days old; comparisons between laboratory-reared and wild weevils indicated food to be of greater importance than culture, in determining female response[4,45,46].

Additional laboratory and field diet studies showed that: (1) males fed cotton squares, bolls, and blooms were considerably more attractive than males fed terminals, cotyledons and leaves; (2) after food was removed pheromone production by males was reduced by about 50 percent within one hour and over 90 percent after twenty-four hours; (3) males survived and produced pheromone on a variety of foods (50-70 percent as much as on cotton squares) such as apples, bananas , okra, peaches, and string beans, but the most favorable diet/was cotton squares; and (4) overwintered male boll weevils survived longer without food than laboratory-reared males, but both needed some food before pheromone production began[49]. In field tests, however, Cross et al. (unpublished data) were not able to show response to male boll weevils fed on

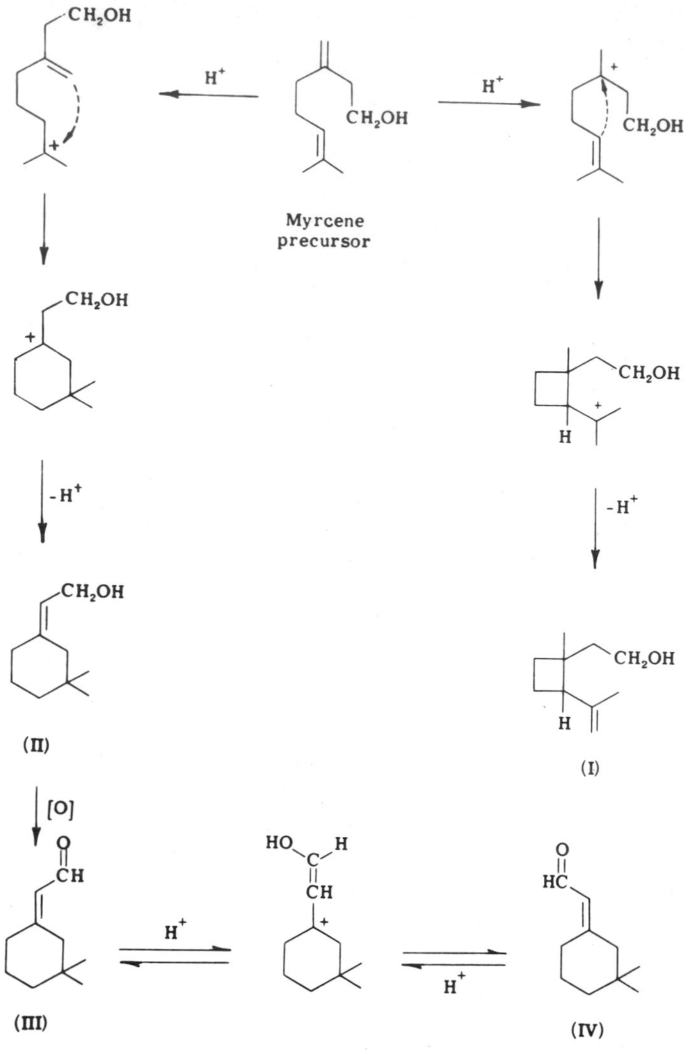

Figure 1.   Hypothetical scheme for biosynthesis of the four pheromone components.

any diet except cotton. These results indicated that an
adequate supply of food, preferably cotton squares or small
cotton bolls, is essential to continued production of a
high level of pheromone by males.

Even though males must feed to attract females, it was
shown that the biosynthesis of the pheromone components was
de novo[143]. The steam distilled feces of adult males, which
had been injected with acetate-1[14]C, acetate-2[14]C, mevalonic
acid-2[14]C or glucose [14]C (U), was shown to contain approximately 0.02% of the administered radioactivity. This is a
typical incorporation of radioactivity into monoterpenes
by plants. The incorporation was also quite selective,
since the 4 components comprised 57 to 80% of the radioactivity of the volatiles, but only 39% of the total content
of the volatiles. Although the boll weevil is essentially
an obligate insect of cotton, it does not appear to require
any specific plant component for synthesis of the pheromone.

Nevertheless, some constituent in cotton may be efficiently converted to the pheromones. A hypothetical scheme,
in which a myrcene precursor such as geraniol could be (Fig. 1)
converted into all four active components has been suggested[204]. Myrcene and β-ocimene are major constituents
of the cotton bud essential oil[122]. Ten other monoterpene
hydrocarbons, 7 monoterpene alcohols[55], and myrtenal[125] have
also been found in buds. To resolve this question, the
administration of ^{14}C-monoterpenoids should be studied.

In recent work, it was shown that cotton buds promoted
a higher level of pheromone biosynthesis than did the
laboratory diet, mostly because weevils fed on the laboratory
diet produced lesser amounts of the aldehydes (III and IV)[65].
Unpublished work with P. P. Sikorowski showed, for a period
in the late fall of 1974, that germ free insects fed the
laboratory diet produced as much pheromone as insects fed
buds, whereas contaminated insects (5,000 colonies per
insect) fed on the laboratory diet produced much more.
However, during January-April, 1975, very little pheromone
was biosynthesized by insects on any diet. In May and June,
pheromone biosynthesis returned to the levels of previous
summers. Although the insect strain used here does not
attain firm diapause as do wild insects, laboratory bioassays
have always proved erratic in winter, and therefore it is
possible that the apparent seasonal fluctuations are real.

Another complication is the rate of biosynthesis of pheromone by males in mixed colonies, versus those which have been separated. Preliminary data indicates that isolated males may produce more pheromone.

It has also been of interest to determine whether insects which have been sterilized, biosynthesize as much pheromone as do normal males. Bartlett et al. (1968) showed that males sterilized with apholate or gamma radiation, were as attractive as untreated males, when both were fed cotton buds. Klassen and Earle (1970) showed that the treatment of male boll weevils with the chemosterilant, busulfan, does not reduce pheromone production. However, extensive field tests in which males, which had been chemosterilized in the mass rearing facility, were released for the purpose of mating with untreated females, were generally not competitive with untreated males and experienced a rapid mortality[28]. This mortality and loss of competitiveness has been attribute to injury of mid-gut cells from both irradiation[166] and chemosterilants[165]. The effects which a heavy bacterial mid-gut contamination contribute to the sterilization damage is currently being studied (P. P. Sikorowski, private communication). The expectation is that germ free weevils should tolerate sterilization better than contaminated weevils.

This matter of bacterial gut contamination is also of interest because R. McLaughlin (BWRL, private communication) has shown that larvae residing in fallen cotton squares (buds are sterile. Adults which emerge from and feed on squares generally have little contamination, but adults fed squares, after having been reared on a contaminated laboratory diet, remain contaminated. Extracts prepared from cotton buds will suppress the growth of *Bacillus thuringensis* Berliner in petri plate tests, however when added to larvae and adult die they are only marginal in their ability to suppress the gut bacterial flora (P. A. Hedin, P. P. Sikorowski, O. H. Lindig, unpublished data).

There are, therefore, many factors which may modify pheromone production. Diet appears to be the main factor, but bacterial contamination of the gut, age, presence of females, season of the year, and treatment with drugs are some of the other contributing factors.

THE PILOT BOLL WEEVIL ERADICATION EXPERIMENT

The aspects of plant-insect interactions, discussed in previous sections of this review, have dealt primarily with laboratory or field experiments of that aspect. The thrust of the boll weevil research program of the past 15 years was to develop a multidisciplinary technology to eliminate the insect. From 1970 to 1974 a considerable number of field tests were done, including the South Mississippi Pilot Boll Weevil Eradication Experiment (1971-1973). Elements of plant-interaction have been included in most of the test regimens, so it is on this basis that the results will be discussed. The literature on earlier field tests (1962-1970) was reviewed by Hedin et al. (1973).

The 2-year Pilot Boll Weevil Eradication Experiment begun in July 1971, was an execution of plans made in 1958 by Federal, State, and industry researchers, to develop technology capable of eliminating the boll weevil. In 1969, a site selection subcommittee determined that southern Mississippi was the portion of the country where it would be most difficult to eliminate the boll weevil. The experiment was therefore planned for this area and the adjoining areas in Louisiana and Alabama. The purpose of the experiment was to determine whether it was technically and operationally feasible to eliminate a boll weevil population from an isolated area, and to further develop the different suppression measures for use on an operational size program.

The supression measures employed were:
I.   In-season control of the boll weevil.
II.   Reproduction-diapause control of the boll weevil in late summer and fall.
III. Pheromone trapping in the spring with:
      a. Grandlure baited traps.
      b. Grandlure baited trap crops.
IV.   Insecticide treatment at pinhead square stage in the spring.
V.   Release of sterile male boll weevils.

During the first year of the Pilot Experiment, two major problems were encountered which resulted in larger than anticipated populations of boll weevils in the spring of 1972. These were (1) an ineffective volunteer, in-season control program by growers in 1971, and (2) physical obstructions which prevented thorough coverage of insecticide applied by aircraft. In 1972, these problems were corrected by in-season insecticide treatments and by supplemental application of insecticide with ground equipment. By mid-October 1972, weevils were not detected by field surveys or woods trash examinations.

In the spring of 1973, the capture of weevils, in baited traps, indicated that a substantial number of boll weevils immigrated into the northern 1/3 of the core area presumed to be isolated. However, the movement of the insects into this area, from infested cotton 10 miles away, clearly showed that complete isolation did not exist. During the last month of the Pilot Experiment, 33 of a total of 236 cotton fields, all located in the northern 1/3 of the eradication area, received supplemental insecticide treatments to eliminate low level infestations detected when cotton plants began fruiting. It can be established from several criteria, that the majority of these infestations probably developed from eggs laid by previously mated female weevils which had migrated up to 25 miles from moderately infested fields. However, no infestations were detected in any of the 170 cotton fields located in the lower 2/3 of the eradication area and which were more than 25 miles from infested cotton.

Upon completion of the Pilot Experiment, the Technical Guidance Committee concluded that it is technically and operationally feasible to eliminate the boll weevil as an economic pest (E. P. Lloyd, BWRL, unpublished data).

Several aspects of this test relate to plant-insect interactions. The fall reproduction-diapause control is predictable since the lipid content of weevils increases in September prerequisite to entering diapause, and thus they are sensitive to phosphate insecticides at this time[11,12]. The early spring pheromone trapping is effective since males leave wood trash hibernation sites in search of early fruiting cotton on which they must feed to biosynthesize pheromone. In the spring and fall, both sexes respond readily to the male pheromone, but the efficiency of the pheromone traps varies inversely with the number of overwintered weevils[24,50,101].

As early as 1901, Malley (1901) suggested the use of an early planted trap crop and Isely (1950) reported that trap crops, 3-30 rows wide, concentrated overwintered boll weevils which were subsequently killed with insecticides. Lloyd et al. (1972) used natural sources of the boll weevil pheromone to lure low density overwintered boll weevil populations into aldicarb treated strips on margins of cotton fields. With the increasing availability of grandlure, in-field traps (traps placed in the cotton rows so as to not disturb cultivation) were designed so that several times more insects were attracted to traps or nearby cotton than to cotton only a short distance away. Moreover, significant numbers of females were captured in mid-season, whereas previous trap studied failed to obtain this mid-season response[131].

The early season insecticide treatment at pinhead square stage is based on the requirement for males to feed and for females to feed and oviposit. This treatment thus provides the final opportunity for suppression of the overwintered boll weevil population prior to plant fruiting.

The production of sterile males which are competitive with wild males has been perhaps the most difficult element of the entire program. The laboratory and field studies through 1972 have been reviewed[58]. The objective is to completely sterilize both sexes of the insect without decreasing its vigor, pheromone production, mating capability, or life expectancy. Until recently, a series of chemosterilants were either fed or administered by dipping or fumigation, such as apholate, busulfan, and hempa. Efforts to utilize these agents with large numbers of insects have encountered difficulties. Nevertheless, controlled releases, under proper conditions, achieved marked to apparently total population control. Recently, Haynes and Mitlin (BWRL, unpublished data) have developed a very promising procedure in which mixed sexes of early pupae are administered 6250 rads in 25 doses of 250 rads gamma irradiation at 4 hour intervals. Initially, 97% of the adult males and 98% of the females are sterile. Return of fecundity occurs to only a limited extent, and the mortality rate is decreased in comparison with a single dose of 6250 rads. The limited mortality has been attributed to the decreased mid-gut cell damage, and increased opportunity for cell regeneration with the modified irradiation regimen. This new sterilization technique could permit releases of mixed sexes to be made

soon after emergence, and eliminate the previous requirements that insects first be sexed and then fed a chemosterilant for 6 days before release. Another recent development
involves the application of TH-6040, a non-mutagenic agent
that inhibits cuticle formation of larvae, to cotton plants
(Lloyd, unpublished data).

Recently, Congress has passed enabling legislation to
proceed with a Belt-wide boll weevil elimination program,
contingent upon feasible technology and favorable propects
for elimination. At this time, final plans have not been
made, nor have funds been appropriated. However, plans have
been made to conduct a trial eradication program. This program could be extended to other areas of the Cotton Belt.

REFERENCES

1. Adams, R., R. C. Morris, T. A. Geissman, D. J.
        Butterbaugh, E. C. Kirkpatrick. 1938. Structure
        of Gossypol XV An interpretation of its reactions.
        J. Am. Chem. Soc. 60:2193-2204.
2. Attree, G. F., A. G. Perkin. 1927. Position of the
        sugar nucleus in the quercetin glucosides. *J. Am.
        Chem. Soc.* 234-40.
3. Bailey, A. E. 1948. Cottonseed and Cottonseed Products
        Interscience Pub. Inc., New York, p. 213.
4. Bartlett, A. C., P. A. Hooker, D. D. Hardee. 1968.
        Behavior of irradiated boll weevils. I. Feeding,
        attraction, mating, and mortality.
        61:1677-80.
5. Beck, S. D. 1965. Resistance of plants to insects.
        Rev. Entomol. 10:207-32.
6. Beroza, M. 1970. Chemicals Controlling Insect Behavior
        Academic Press, New York. 170 **pp**.
7. Billups, W., J. H. Cross, C. V. Smith. 1973. A Synthes:
        of (+) - Grandisol. *J. Am. Chem. Soc. 95*:3438-9.
8. Blum, M. S. 1970. The Chemical Basis of Insect Sociali
        In Chemicals Controlling Insect Behavior. (Ed.)
        M. Beroza, Academic Press, N. Y. pp. 61-94.
9. Bottger, G. T., E. T. Sheehan, M. J. Lukefahr. 1964.
        Relation of gossypol content of cotton plants to
        insect resistance. *J. Econ. Entomol. 57*:238-5.

10. Bradley, J. R., D. F. Clower, J. B. Graves. 1968. Field studies of sex attraction in the boll weevil. *J. Econ. Entomol. 61*:1457-8.

11. Brazzel, J. 1959. The effect of late-season applications of insecticides on diapausing boll weevils. *J. Econ. Entomol. 52*:1042-5.

12. Brazzel, J. R., L. D. Newsom. 1959. Diapause in *Anthonomus grandis* Boh. *J. Econ. Entomol. 52*:603-11.

13. Bregetova, L. G. 1951. Ascorbic acid content of cotton plant leaves. *Botan. Zhur. 36*:34-8.

14. Buford, W. T., J. N. Jenkins, F. G. Maxwell. 1968. A boll weevil oviposition suppression factor in cotton. *Crop Sci. 8*:647-8.

15. Bumgarner, J. E., E. N. Lambremont. 1966. The lipid class spectrum and fatty acid content of the boll weevil egg. *Comp. Biochem. Physiol. 18*:975-81.

16. Burks, M. L., N. W. Earle. 1965. Amino acid composition of upland cotton squares and Arizona wild cotton bolls. *J. Agr. Food Chem. 13*:40-3.

17. Buzitskova, E. P., A. S. Sadykov, A. I. Ismailov, D. Kh. Rasuleva. 1966. Cotton leaf carotenoids. D. Kh. Rasuleva Narek Tr. Tashkent Gos. Univ. No. 286: 38-41.

18. Carns, H. R. 1966. Abscission and its control. *Ann. Rev. Plant Physiol. 17*:295-314.

19. Coakley, J. M., F. G. Maxwell, J. N. Jenkins. 1969. Influence of feeding, oviposition, and egg and larval development of the boll weevil on abscission of cotton squares. *J. Econ. Entomol. 62*:244-5.

20. Coker, R. R. 1958. The impact of the boll weevil on cotton production costs. *Cotton Gin and Oil Mill Press 59*:22-24.

21. Chandler, K., T. R. Seshadri. 1957. Leucodelphinden in cottonseed. *J. Sci. Ind. Res. 16A*:319-20.

22. Cross, W. H., H. C. Mitchell. 1966. Mating behavior of the female boll weevil. *J. Econ. Entomol. 59*: 1503-7.

23. Cross, W. H., D. D. Hardee. 1968. Traps for survey of overwintered boll weevil populations. *Miss. Coop. Econ. Ins. Rpt. 18*(20):430.

24. Cross, W. H., D. D. Hardee, F. Nichols, H. C. Mitchell, E. B. Mitchell, P. M. Huddleston, J. H. Tumlinson. 1969. Attraction of female boll weevils to traps baited with males or extracts of males. *J. Econ. Entomol. 62*:154-61.

25. Cross, W. H., M. J. Lukefahr, P. A. Fryxell, H. R. Burke 1975. Host plants of the boll weevil. *Environ. Entomol.* 4:19-26.

26. Dastur, R. H., A. H. Ahad. 1941. Periodic partial failures of Punjab-American cottons in the Punjab. III Uptake and distribution of minerals in the cotton plant. *Ind. J. Agr. Sci.* 11:297-300.

27. Daum, R. J., G. H. McKibben, T. B. Davich, R. M. McLaughlin. 1969. Development of the bait principle for boll weevil control: Calco Oil Red N-1700 dye for measuring ingestion. *J. Econ. Entomol.* 62:370-5.

28. Davich, T. B., J. C. Keller, E. B. Mitchell, P. Huddlestc R. Hill, D. A. Lindquist, G. McKibben, W. H. Cross. 1965. Preliminary field experiments with sterile males for eradication of the boll weevil. *J. Econ. Entomol.* 58:127-31.

29. Dethier, V. G., L. B. Browne, C. N. Smith. 1960. The designation of chemicals in terms of responses they elicit from insects. *J. Econ. Entomol.* 53:134-6.

30. Ergle, D. R., F. M. Eaton. 1951. Succinic acid content of the cotton plant. *Plant Physiol.* 26:186-8.

31. Fargher, R. G., M. E. Probert. 1924. Alcohols present in the wax of American cotton. *J. Textile Inst.* 15:337-46.

32. Fieser, I. E. 1961. Determination of the non-cellulosic constituents of vegetable fibers. Determination of fats and waxes in cotton. *Texil-Rundschau* 16:78.

33. Folsom, J. W. 1931. A chemotropometer. *J. Econ. Entomol.* 24:827-33.

34. Frankel, G. S. 1959. The raison d'etre of secondary plant substances. *Science* 129:1466-70.

35. Fryxell, P. A. 1967. *Hampea* and the Boll Weevil: A correction. *Science* 156:1770.

36. Fryxell, P. A., M. J. Lukefahr. 1967. *Hampea* Schlecht: Possible Primary Host of the Boll Weevil. *Science* 155:1568-9.

37. Geissman, T. A. 1962. The Chemistry of Flavonoid Compounds. McMillan Co., New York, p. 1.

38. Ghosh, D., H. E. Johan. 1964. Leaf anthocyanin content of *Gossypium hirsutum* as influenced by magnesium and nitrogen deficiencies. *Plant Physio.* 39, suppl. XXI.

39. Gueldner, R. C., A. C. Thompson, D. D. Hardee, P. A. Hedin. 1970. Constituents of the cotton bud. XIX. Attractancy to the boll weevil of terpenoids and related plant constituents. *J. Econ. Entomol.* 63:1819-21.

40. Gueldner, R. C., A. C. Thompson, P. A. Hedin. 1972. Steroselective synthesis of Racemic Grandisol. *J. Org. Chem.* 37:1854-6.

41. Hall, W. C., G. B. Truchelut, C. L. Leinweber, F. A. Herrero. 1957. Ethylene production by the cotton plant and its effects under experimental and field conditions. *Plant Physiol.* 10:306-17.

42. Hamomura, Y., K. Hayashiya, K. Naito, K. Matsuura, J. Nishida. 1962. Food selection by the silkworm larvae. *Nature* 194:754-5.

43. Hanny, B. W., A. C. Thompson, R. C. gueldner, P. A. Hedin. 1973. Constituents of cotton seedlings: An investigation of the preference of male boll weevils for the epicotyl tips. *J. Agr. Food Chem.* 21:1004-6.

44. Hardee, D. D., E. B. Mitchell, P. M. Huddleston, T. B. Davich. 1966. A laboratory technique for bioassay of plant attractants for the boll weevil. *J. Econ. Entomol.* 59:240-1.

45. Hardee, D. D., E. B. Mitchell, P. M. Huddleston. 1967a. Procedure for bioassaying the sex attractant of the boll weevil. *J. Econ. Entomol.* 60:196-71.

46. Hardee, D. D., E. B. Mitchell, P. M. Huddleston. 1967b. Laboratory studies of sex attraction in the boll weevil. *J. Econ. Entomol.* 60:1221-4.

47. Hardee, D. D., W. H. Cross, E. B. Mitchell. 1969. Male boll weevils are more attractive than cotton plants to boll weevils. *J. Econ. Entomol.* 62:165-9.

48. Hardee, D. D., T. C. Cleveland, J. W. Davis, W. H. Cross. 1970a. Attraction of boll weevils to cotton plants and to males fed on 3 diets. *J. Econ. Entomol.* 63:990-1.

49. Hardee, D. D. 1970b. Pheromone production by male boll weevils as affected by food and host factors. Contrib. Boyce Thompson Inst. 24(13):315-22.

50. Hardee, D. D., O. H. Lindig, T. B. Davich. 1971. Suppression of populations of boll weevils over a large area in West Texas with pheromone traps in 1969. *J. Econ. Entomol.* 64:928-33.

51. Hedin, P. A., A. C. Thompson, J. P. Minyard. 1966. Constituents of the cotton bud III. Factors that stimulate feeding by the boll weevil. *J. Econ. Entomol.* 59:181-5.

52. Hedin, P. A., J. P. Minyard, A. C. Thompson, R. F. Struc
    J. Frye. 1967. Constituents of the cotton bud. VI
    Identification of the anthocyanin as chrysanthimin.
    Phytochemistry 6:1165-7.
53. Hedin, P. A., L. R. Miles, A. C. Thompson, J. P. Minyard
    1968. Constituents of the cotton bud. X. Formula-
    tion of a boll weevil feeding stimulant mixture.
    J. Agr. Food Chem. 16:505-13.
54. Hedin, P. A., A. C. Thompson, R. C. Gueldner, J. P.
    Minyard. 1971a. Constituents of the cotton bud.
    XX. Isolation of alpha-bisabolol. *Phytochemistry*
    10:1693-4.
55. Hedin, P. A., A. C. Thompson, R. C. Gueldner, J. P.
    Minyard. 1971b. Constituents of the cotton bud.
    The alcohols. *Phytochem.* 10:3316-8.
56. Hedin, P. A., A. C. Thompson, R. C. Gueldner, J. M. Ruth
    1972a. Isolation of bisabolene oxide from the cotto
    bud. *Phytochem.* 11:2118-9.
57. Hedin, P. A., A. C. Thompson, R. C. Gueldner, J. P.
    Minyard. 1972b. Volatile constituents of the boll
    weevil. *J. Insect Physiol.* 18:79-86.
58. Hedin, P. A., A. C. Thompson, R. C. Gueldner. 1973.
    The boll weevil-cotton plant complex. *Toxicological
    and Environ. Chem. Rev.* 1:291-351.
59. Hedin, P. A., F. G. Maxwell, J. N. Jenkins. 1974a.
    Insect Plant Attractants, Feeding Stimulants, Repel-
    lents, Deterrents, and Other Related Factors Affecti
    Insect Behavior, *In* Proc. of the Inst. on Biol.
    Control of Plant Insects and Diseases, Univ. Press
    Inc., pp. 494-527.
60. Hedin, P. A., D. D. Hardee, A. C. Thompson, R. C.
    Gueldner. 1974b. An assessment of the lifetime
    biosynthesis potential of the male boll weevil.
    J. Insect Physiol. 20:1707-12.
61. Hedin, P. A., R. C. Gueldner, R. D. Henson, A. C.
    Thompson. 1974c. Volatile constituents of male
    and female boll weevils and their frass. *J. Insect
    Physiol.* 20:2135-42.
62. Hedin, P. A., A. C. Thompson, R. C. Gueldner. 1975a.
    A survey of the air space volatiles of the cotton
    plant. *Phytochem.* In Press. 14:2088-90.
63. Hedin, P. A., A. C. Thompson, R. C. Gueldner. 1975b.
    Constituents of the cotton bud, an updated list.
    Phytochem. In Press. 14:2087-8.

64. Hedin, P. A., C. S. Rollins, A. C. Thompson, R. C. Gueldner. 1975c. Pheromone production of male boll weevils treated with chemosterilants. *J. Econ. Entomol.* In Press. *68*:587-91.

65. Henson, R. D., A. C. Thompson, R. C. Gueldner, P. A. Hedin. 1971. Phospholipid composition of the boll weevil. *Lipids* 6:352-5.

66. Henson, R. D., A. C. Thompson, R. C. Gueldner, P. A. Hedin. 1972. Variations in lipid content of the boll weevil during metamorphosis. *J. Insect Physiol.* *18*:161-7.

67. Holder, D. G., P. A. Hedin, W. L. Parrott, F. G. Maxwell, J. N. Jenkins 1973-74. Sugars in the leaves of Frego bract and Deltapine 16 varieties of cotton. *Miss. Acad. of Sci.*, In Press. *19*:178-80.

68. Howard, L. O. 1896. The Mexican Cotton Boll Weevil, USDA Bull. Ent. Circ. 18 Second Series, Revision of 14, 8 pp.

69. Howard, L. O. 1918. Report of the Entomologist. USDA Bur. Ent. Rpt., 24 pp.

70. Hunter, W. D. 1904. The use of Paris green in controlling the boll weevil. USDA Farmer's Bull. 211, 23 pp.

71. Hunter, W. D. and W. E. Hinds. 1905. The Mexican cotton boll weevil. USDA Agr. Bur. Ent. Bull. No. 51, 181 pp.

72. Hunter, W. D. and W. D. Pierce. 1912. The Mexican cotton boll weevil. USDA Agr. Bur. Ent. Bull. No. 114, 188 pp.

73. Hunter, W. D. and B. R. Coad. 1923. The boll weevil problem. USDA Farmer's Bull. No. 1329, 30 pp.

74. Isely, D. 1932. Abundance of the boll weevil in relation to summer weather and to food. Ark. Agr. Exp. Sta. Bull. No. 271, 34 pp.

75. Isely, D. 1934. Relationship between early varieties of cotton and boll weevil injury. *J. Econ. Entomol.* 27:762-6.

76. Isely, D. 1950. Trapping weevils in spots with early cotton. Ark. Agr. Exp. Sta. Bull. No. 496. 42 pp.

77. Jacobson, M. 1965. Insect Sex Attractants. Interscience, New York. 154 pp.

78. Jenkins, J. N., F. G. Maxwell, J. C. Keller, W. L. Parrott. 1963. Investigations of the water extracts of *Gossypium*, *Abelmoschus*, *Cucumiis*, and *Phaeseolus* for an arrestant and feeding stimulant for *Anthonomus grandis* Boh. *Crop Science* 3:215-9.

79. Jenkins, J. N., F. G. Maxwell, H. N. Lafever. 1966. The
     comparative preference of insects for glanded and
     glandless cotton. *J. Econ. Entomol. 59:*352-6.
80. Jenkins, J. N., F. G. Maxwell, W. L. Parrott. 1967.
     Field evaluation of glanded and glandless cotton
     lines for the boll weevil. *Crop Sci. 4:*437-40.
81. Jenkins, J. N., W. L. Parrott, J. C. McCarty, Jr. 1973.
     The role of a boll weevil resistant cotton in pest
     management research. *J. Environ. Quality 2:*337-40.
82. Joiner, R. L., E. N. Lambremont. 1969. Hydrocarbon
     metabolism in insects: Oxidation of hexadecane-1-
     ^{14}C in the boll weevil and the house fly. *Ann.
     Entomol. Soc. Am. 62:*891-4.
83. Keller, J. C., F. G. Maxwell, J. N. Jenkins. 1962.
     Cotton extracts as arrestants and feeding stimulants
     for the boll weevil. *J. Econ. Entomol. 55:*800-1.
84. Keller, J. C., F. G. Maxwell, J. N. Jenkins, T. B. Davich.
     1963. A boll weevil attractant from cotton. *J.
     Econ. Entomol. 56:*110-1.
85. Keller, J. C., E. B. Mitchell, G. McKibben, T. B. Davich.
     1964. A sex attractant for female boll weevils from
     males. *J. Econ. Entomol. 57:*609-10.
86. King, E. E., H. C. Lane. 1969. Abscission of cotton
     flower buds and petioles caused by protein from boll
     weevil larvae. *Plant Physiol. 44:*903-6.
87. King, E. E. 1973. Endo-polymethylgalacturonase of boll
     weevil larvae: An initiator of cotton flower bud
     abscission. *J. Insect Physiol. 19:*2433-7.
88. Klassen, W., N. W. Earle. 1970. Permanent sterility
     induced in boll weevils with busulfan without reduc-
     ing production of pheromone. *J. Econ. Entomol.
     63:*1195-8.
89. Knipling, E. F. 1964. The potential role of the steri-
     lity method for insect population control with
     special reference to combining this method with
     conventional methods. USDA-ARS Rpt. No. 33-98,
     54 pp.
90. Kriventsov, V. I., E. G. Belkina, Y. R. Malikova. 1959.
     Citric and malic acid content of the leaves of local
     cotton varities and of waste products of cotton
     cleaning plants. Trudy Inst. Khim Akad Nauk Turkmen
     SSR. 18-31.
91. Lambremont, E. N. 1962. Enzymes in boll weevil. I.
     Dehydrogenases of the brain and related structures.
     *J. Insect Physiol. 8:*181-90.

92. Lambremont, E. N., M. S. Blum. 1963. Fatty acids of
     the boll weevil. *Ann. Entomol. Soc. Amer.* *56*:612-6.
93. Lambremont, E. N., R. M. Shrader. 1964a. Electrophore-
     is of an insect in organic pyrophosphatase. *Nature*
     204:833-4.
94. Lambremont, E. N., R. M. Shrader. 1964b. Enzymes of the
     boll weevil. II. Inorganic pyrophosphatases. *J.*
     Insect Physiol. *10*:37-52.
95. Lambremont, E. N., M. S. Blum, R. M. Shrader. 1964.
     Storage and fatty acid composition of triglycerides
     during adult diapause of the boll weevil. *Ann.*
     Entomol. Soc. Amer. *57*:526-32.
96. Lambremont, E. N., C. I. Stein, A. F. Bennett. 1965a.
     Synthesis and metabolic conversion of fatty acids
     by the larval boll weevil. *Comp. Biochem. Physiol.*
     16:289-302.
97. Lambremont, E. N. 1965b. Lipid biosynthesis in the
     boll weevil: Distribution of radioactivity in the
     principal lipid classes synthesized from ^{14}C-1-
     acetate. *Comp. Biochem. Physiol.* *14*:417-29.
98. Lambremont, E. N., J. E. Bumgarner, A. F. Bennett. 1966.
     The lipid-class spectrum and fatty acid content of
     the boll weevil egg. *Comp. Biochem. Physiol.* *18*:
     975-81.
99. Lambremont, E. N., R. Wood. 1968. Glyceryl ethers in
     insects: Identification of alkyl and alk-1-enyl
     glyceryl ether phospholipids. *Lipids* *3*:503.
100. Lloyd, E. P. 1971. Digging a grave for the boll weevil.
     Cotton International (Meister Pub. Co.) Memphis,
     Tenn., Ed. 38th Ann. p. 70-1.
101. Lloyd, E. P., W. P. Scott, K. K. Shaunak, F. C. Tingle,
     T. B. Davich. 1972. A modified trapping system for
     suppressing low-density populations of over-wintered
     boll weevils. *J. Econ. Entomol.* *64*:1144-7.
102. Lukefahr, M. J., D. F. Martin. 1966. Cotton plant
     pigments as a source of resistance to the bollworm
     and tobacco budworm. *J. Econ. Entomol.* *59*:176-9.
103. Lukefahr, M. J., L. W. Noble, J. E. Houghtaling. 1966.
     Growth and infestations of bollworm and other insects
     on glanded and glandless strains of cotton. *J. Econ.*
     Entomol. *59*:817-20.
104. Lukefahr, M. J., F. G. Maxwell. 1969. The differential
     resistance mechanism in female *Hampea* sp. trees to
     the boll weevil. *Ann. Entomol. Soc. Am.* *62*:542-4.

105. Lyman, C. M., A. S. El-Nockrasky, J. W. Ballehite. 1963. Gossyverdurin. A newly isolated pigment from cotton-seed pigment glands. *J. Am. Oil. Chem. Soc. 40*:571.

106. Mabry, T. J., J. Kagan, H. Rosler. 1964. Nuclear magnetic resonance analysis of flavonoids. Univ. of Texas Pub. #6418, 38 pp.

107. Malley, F. W. 1901. The Mexican cotton-boll weevil. USDA Farmer's Bull. No. 1930. p. 11-12.

108. Maxwell, F. G., J. N. Jenkins, J. C. Keller. 1963a. A boll weevil repellent from the volatile substance of cotton. *J. Econ. Entomol. 56*:894-5.

109. Maxwell, F. G., J. N. Jenkins, J. C. Keller, W. L. Parrott. 1963b. An arrestant and feeding stimulant for the boll weevil in water extracts of cotton plant parts. *J. Econ. Entomol. 56*:449-54.

110. Maxwell, F. G., H. N. Lefever, J. N. Jenkins. 1965a. Blister beetles on glandless cotton. *J. Econ. Entomol. 58*:792-3.

111. Maxwell, F. G., W. L. Parrott, J. N. Jenkins, H. N. Lafever. 1965b. A boll weevil feeding deterrent from the calyxes of an alternate host, *Hibiscus syriacus* L. *Ibid. 58*:985-8.

112. Maxwell, F. G., H. N. Lafever, J. N. Jenkins. 1966. Influence of the glandless genes in cotton on feeding oviposition, and development of the boll weevil in the laboratory. *Ibid. 59*:585-8.

113. Maxwell, F. G., D. D. Hardee, W. L. Parrott, J. N. Jenkins, J. M. Lukefahr. 1969a. *Hampea* sp., host of the boll weevil. I. Laboratory preference studies. *Ann. Entomol. Soc. Am. 62*:315-8.

114. Maxwell, F. G., J. N. Jenkins, W. L. Parrott, W. T. Buford. 1969b. Factors contributing to resistance and susceptibility of cotton and other hosts to the boll weevil. *Ent. exp. and appl. 12*:801-10.

115. McIndoo, N. E. 1926. Senses of the cotton boll weevil. An attempt to explain how plants attract insects by smell. *Agr. Res. J. 33*:1095-1140.

116. McKibben, G. H., P. A. Hedin, T. B. Davich, R. J. Daum, M. W. Laseter. 1971a. Development of the bait principle for boll weevil control: Addition of food acidulants to increase the attractiveness of cottonseed oil baits. *J. Econ. Entomol. 64*:583-5.

117. McKibben, G. H., P. A. Hedin, R. E. McLaughlin, T. B. Davich. 1971b. Development of the bait principle for boll weevil control. Addition of terpenoids and related constituents. *J. Econ. Entomol. 64:* 1493–5.

118. McLaughlin, R. E. 1967. Development of the bait principle for boll weevil control. II. Field cage tests with a feeding stimulant and the protazoan *Mattesia grandis*. *J. Invert. Pathol. 9:70–77.*

119. McMichael, S. C. 1960. Combined effects of glandless genes gl_2 and gl_3 on pigment glands in the cotton plant. *Agron. J. 52:385–7.*

120. Michel-Wolwertz, M. R., C. Sironval. 1965. Chlorophylls separated by paper chromatography from *Chlorella* extracts. *Biochim. Biophys. Acta 94:330–43.*

121. Miller, R. F., R. Adams. 1937. Structure of Gossypol IV. Anhydrogossypol and its derivatives. *Am. Chem. Soc. 59:1736–8.*

122. Minyard, J. P., J. H. Tumlinson, P. A. Hedin, A. C. Thompson. 1965. Constituents of the cotton bud. Terpene hydrocarbons. *J. Agr. Food Chem. 13:599–602.*

123. Minyard, J. P., J. H. Tumlinson, P. A. Hedin, A. C. Thompson. 1966. Constituents of the cottonbud. Sesquiterpene hydrocarbons. *J. Agr. Food Chem. 14:* 332–6.

124. Minyard, J. P. 1967. Constituents of the cotton plant. The volatile fractions. Ph.D. Thesis, Miss. State Univ., State College MS, 172 pp.

125. Minyard, J. P., J. H. Tumlinson, A. C. Thompson, P. A. Hedin. 1967. Constituents of the cotton bud. The carbonyl compounds. *J. Agr. Food Chem. 15:517–24.*

126. Minyard, J. P., A. C. Thompson, P. A. Hedin. 1968. Constituents of the cotton bud. VIII. beta-Bisabolol, a new sesquiterpene alcohol. *J. Org. Chem. 33:909–11.*

127. Minyard, J. P., D. D. Hardee, R. C. Gueldner, A. C. Thompson, G. Wiygul, P. A. Hedin. 1969. Constituents of the cotton bud. Compounds attractive to the boll weevil. *J. Agr. Food Chem. 17:1093–7.*

128. Mistric, W. J., E. R. Mitchell. 1966. Effects of low dosages of insecticidal seed-treatments on cotton and cotton insects. *J. Econ. Entomol. 59:57–60.*

129. Mitchell, E. R., H. M. Taft. 1966. Host plant selection by migrating boll weevils. *J. Econ. Entomol. 59:390–* 2.

130. Mitchell, E. G. 1971. Manipulation and reduction of
      boll weevil field populations with plant and sex
      attractants. Ph.D. Thesis, Miss. State Univ.,
      State College Miss., 68 pp.
131. Mitchell, E. B., D. D. Hardee. 1974. In-field traps:
      A new concept in survey and suppression of low
      populations of boll weevils. *J. Econ. Entomol. 67:*
      506-8.
132. Mitlin, N., D. H. Vickers, R. T. Gast. 1964a. Estima-
      tion of nitrogenous compounds in the feces of boll
      weevils fed different diets. *Ann. Entomol. Soc.*
      *Amer. 57:*757-9.
133. Mitlin, N. D., H. Vickers, P. A. Hedin. 1964b. End
      products of metabolism in the boll weevil: Non-
      protein amino acids in the feces. *J. Insect Physiol.*
      *10:*393-7.
134. Mitlin, N., D. H. Vickers. 1964. Guanine in the excret
      of the boll weevil. *Nature 203:*1403-4.
135. Mitlin, N., J. K. Mauldin, P. A. Hedin. 1966. Free
      and protein-bound amino acids in the tissues of the
      boll weevil during metamorphosis. *Comp. Biochem.*
      *Physiol. 19:*35-43.
136. Mitlin, N., J. K. Mauldin. 1966. Uric acid in nitrogen
      metabolism of the boll weevil. *Ann. Entomol. Soc.*
      *Amer. 59:*651-3.
137. Mitlin, N., G. J. Lusk, G. Wiygul. 1967. An electro-
      phoretic study of the changes in proteins in the
      haemolymph during the life cycle of the boll weevil.
      *Ann. Entomol. Soc. Amer. 60:*1155-8.
138. Mitlin, N., G. Wiygul, J. K. Mauldin. 1968a. Free
      amino acids in the haemolymph of the maturing boll
      weevil. *Comp. Biochem. Physiol. 25:*139-48.
139. Mitlin, N., G. Wiygul, G. J. Lusk. 1968b. Incorporatio
      of lysine 6-^{14}C into the protein of the adult boll
      weevil. *J. Insect Physiol. 14:*1277-1284.
140. Mitlin, N., G. Wiygul. 1969. Incorporation and meta-
      bolism of ^{14}C-labeled tryptophan-3 in the boll weevil
      *Comp. Biochem. Physiol. 30:*375-82.
141. Mitlin, N., G. Wiygul. 1972. Contribution of pyrimidin
      to the biosynthesis of fecal uric acid in normal and
      busulfan treated weevils. *Ann. Entomol. Soc. Amer.*
      *65:*612-3.
142. Mitlin, N., G. Wiygul. 1973. Uric acid in nucleic and
      amino acid synthesis in the boll weevil. *J. Insect*
      *Physiol. 19:*1569-74.

143. Mitlin, N., P. A. Hedin. 1974. Biosynthesis of grand-lure, the pheromone of the boll weevil from acetate, mevalonate and glucose. *J. Insect Physiol*. *20*:1825-31.

144. Murray, J. C., L. M. Verhalem, D. E. Bryan. 1965. Observations on the feeding preference of the striped blister beetle to glanded and glandless cotton. *Crop Sci*. *5*:189.

145. Naaber, L. Rh. 1956. The content of ascorbic acid in the leaves of the cotton plant. *Izvest. Akad Nauk Uzbek SSR No*. *5*:11-16.

146. Neelakantan, K., T. R. Seshadri. 1936. Pigments of cotton flowers. IV. Constitution of herbacitrin and herbacetin. *Proc. Ind. Acad. Sci*. *4*:357.

147. Neff, D. L., E. S. Vanderzant. 1963. Methods of evaluating the chemotropic response of boll weevils to extracts of the cotton plant and various other substances. *J. Econ. Entomol*. *56*:761-6.

148. Nettles, W. C., N. L. Betz. 1965. Glycogen in the boll weevil with respect to diapause, age, and diet. *Ann. Entomol. Soc. Amer*. *58*:721-6.

149. Ohkuma, K., F. T. Addicott, O. E. Smith, W. E. Theisson. 1965. The structure of abscission II. *Tetrahedron Lett*. *29*:2529-35.

150. Ozeretskovskaya, O. L., N. N. Guseva. 1966. Effect of phenols on cotton resistance to vercillium wilt. Tr. Vses. Nauk-Issled. Inst. Zashch. Ruts #26, 132-5.

151. Painter, R. H. 1951. Insect Resistance in Crop Plants. (McMillan, New York) 520 pp.

152. Painter, R. H. 1958. Resistance of plants to insects. *Ann. Rev. Entomol*. *3*:267-90.

153. Pakudina, Z. P., A. S. Sadykov, P. K. Denliev. 1965. Flavonols of *Gossypium hirsutum* flowers (cotton growth). Khim. Prirodn. Soedin Akad. Narek Uzbek, SSR 67-70.

154. Parencia, C. R., J. W. Davis, C. B. Cowan. 1964. Studies on the ability of overwintered boll weevils to find fruiting cotton plants. *J. Econ. Entomol*. *57*:162.

155. Parks, C. R. 1965. Floral Pigmentation studies in the genus *Gossypium*. II. Chemotaxonomic analysis of diploid *Gossypium* species. *Am. J. Botany 52*: 849-56.

156. Parrott, W. L., F. G. Maxwell, J. N. Jenkins, J. K. Mauldin. 1969a. Amino acids in hosts and nonhosts of the boll weevil. *Ann. Entomol. Soc. Amer. 62*: 255-61.

157. Parrott, W. L., F. G. Maxwell, J. N. Jenkins, D. D.
      Hardee. 1969b. Preference studies with hosts and
      nonhosts of the boll weevil. *Ann. Entomol. Soc.
      Amer. 62*:261-4.

158. Patterson, W. J., R. A. Scott, H. R. Carns, P. A. Hedin.
      1966. The male sterile response in cotton. The
      effect of sodium 2,3-dichloro-2-methyl-propionate
      9FW-450 on the incorporation of amino acids in floral
      tissue. *Phyton 23*:43-8.

159. Perkin, A. G. 1896. XXII Luteolin, Part I. *J. Chem.
      Soc. 69*:206-12.

160. Plaisted, P. H., R. B. Reggio. 1962. Adenine-glucose
      compound from cotton. *Nature 193*:685-6.

161. Power, F. B., H. Browning. 1914. Chemical examination
      of cotton root bark. *Pharm. J. 93* Series 4, 39
      420-3.

162. Power, F. B., V. K. Chesnut. 1925. The odorous consti-
      tuents of the cotton plant. Emanation of ammonia
      and trimethylamine from the living plant. *J. Am.
      Chem. Soc. 47*:1751-74.

163. Pratt, C., S. H. Wender. 1959. Identification of rutin
      and isoquercitrin in cottonseed. *J. Am. Oil Chem.
      36*:392-4.

164. Rainey, R. C. 1939. Observations on the development
      of the cotton boll, with particular reference to
      changes in susceptibility to pests and diseases.
      Emprie Cotton Growing Corp., Proc. Rpt. 1939-40
      (CA 36-6586).

165. Reinecke, L. H., W. Klassen and J. F. Norland. 1969.
      Damage to testes and recovery of fertility in boll
      weevils fed chemosterilants. *Ann. Entomol. Soc. Am.
      62*:511-25.

166. Riemann, J. G. and H. M. Flint. 1967. Irradiation of
      effects on mid-guts and testes of the adult boll
      weevil determined by histological and shielding
      experiments. *Ann. Entomol. Soc. Am. 60*:298-308.

167. Robbins, W. E., M. J. Thompson, R. T. Yamamoto, T.
      Shortine. 1965. Feeding stimulants for the female
      house fly. *Science 147*:628-30.

168. Roussel, J. S., and D. F. Clower. 1957. Resistance to
      the chlorinated hydrocarbon insecticides in the boll
      weevil. *J. Econ. Entomol. 50*:463-8.

169. Sadykov, A. S., Z. P. Pakudina, E. P. Buzitskova, A.
      Sh. Guli-Kevkhyan, A. Karimdzhanov, Kh. Isaev. 1958.
      Uzbek Khim Zhur Akad Nauk Uzbek SSR. 41 - CA 53-
      22292.

170. Sadykov, A. S., Z. P. Pakudina. 1959. Carbohydrates of
      cotton leaves. Khim Khlop. Akad. Nauk Uzbek. SSR.
      58-63; CA 55-15631.

171. Sadykov, A. S., A. K. Karimdzhanov. 1960. Characteris-
      tics of cotton tanning matter determined by paper
      chromatography. Uzbek Khim. ZH. #1, 52-6 CA
      54-25083.

172. Sadykov, A. S., Z. P. Pakudina. 1960. Sterols of
      cotton flowers. Dokl. Akad. Nauk. Uzbek SSR. No.
      4, 25-8.

173. Sadykov, A. S., A. K. Karimdzhanov, A. I. Ishmailov.
      1962. Quantitative and qualitative composition of
      the tannings of the thin fibered cotton plant.
      Uzbek Khim. Zh. 6, 60-3, CA 57-12905.

174. Sadykov, A. S., N. I. Salit. 1962. Hydroxy acids from
      the cotton plant.   Uzbek Khim. Zh. 6, 68-74.

175. Sadykov, A. S., K. I. Isaeu, H. I. Ishmailov. 1963.
      Isolation by extraction and separation of some
      substances from the cotton plant. Uzbek Khim. Ah.
      #7, 53-6, CA 59-10473.

176. Sadykov, A. S., Z. P. Pakudina. 1963. Substances from
      the flowers of the cotton plant. Nauchn. Tr.
      Tashkentsk Gos. Univ., No. 263:88-93.

177. Sadykov, A. S. 1965. Recent advances in the chemistry
      of the cotton plant. *J. Sci. Ind. Res.* 24:77-81.

178. Sanderson, E. D. 1904. The cotton boll weevil in
      Texas. *Soc. Prom. Agr. Sci. Proc.* 25:157-70.

179. Shenstone, F. S., J. R. Vickery. 1961. Occurrence of
      cyclopropene acids in some plants of the order
      Malvales. *Nature 190*:168-9.

180. Shimbo, K. 1935. Cellulose resources. IV. Chemical
      composition of cotton stalks. *J. Agr. Chem. Soc.
      Japan 11*:1075-6.

181. Silverstein, R. M. 1970. Chemicals Controlling Insect
      Behavior, ed. M. Beroza, Acad. Press, New York, N.Y.
      p. 212-40.

182. Smith, G. L., T. G. Cleveland, J. C. Clark. 1965.
      Boll weevil movement from hibernation sites to
      fruiting cotton. *J. Econ. Entomol.* 58:257-8.

183. Stoner, A. 1968. *Sphaeralcea* spp. as hosts of the boll
      weevil in Arizona. *J. Econ. Entomol.* 61:1100-2.

184. Struck, R. F., J. Frye, Y. F. Shealy, P. A. Hedin, A. C. Thompson, J. P. Minyard. 1968a. Constituents of the cotton bud. IX. Further studies on a polar feeding stimulant complex. *J. Econ. Entomol.* *61*: 270-4.

185. Struck, R. F., J. Frye, Y. F. Shealy, P. A. Hedin, A. C. Thompson, J. P. Minyard. 1968b. Constituents of the cotton bud. XI. Studies on a feeding stimulant complex from flower petals for the boll weevil. *J. Econ. Entomol.* *61*:664-7.

186. Struck, R. F., J. Frye, Y. F. Shealy. 1968c. Constituents of the cotton bud. Mass spectrometric identification of the major high-molecular weight hydrocarbons in buds and flowers. *J. Agr. Food Chem.* *16*:1028-30.

187. Suryprakasa, P. R., T. R. Seshadri. 1943. Pigments of cotton flowers. IV. Occurrence of populetin in Indian cotton flowers. *Proc. Ind. Acad. Sci.* *18A*: 204-5.

188. Temple, C., E. C. Roberts, J. Frye, R. F. Struck, Y. F. Shealy, A. C. Thompson, J. P. Minyard, P. A. Hedin. 1968. Constituents of the cotton bud. XIII. Further studies on a nonpolar feeding stimulant for the boll weevil. *J. Econ. Entomol.* *61*:1388-93.

189. Thompson, A. C., P. A. Hedin. 1965. Volatile and non-volatile constituents of the cotton bud, the fatty acid composition. *Crop Sci.* *5*:133-5.

190. Thompson, A. C., R. D. Henson, P. A. Hedin, J. P. Minyar 1968b. Constituents of the cotton bud. XII. The carotenoids in buds, seeds, and other tissue. *Lipids* *3*:495-7.

191. Thompson, A. C., R. D. Henson, J. P. Minyard, P. A. Hedi 1968a. Fatty acid composition of polar lipids of cotton buds. *Lipids* *3*:373-4.

192. Thompson, A. C., J. R. Pratt, J. P. Minyard, P. A. Hedin 1970a. Constituents of the cotton bud. XVII. Lipid and fatty acid studies of glanded cotton as a function of boll weevil non-preference. *J. Econ. Entomol.* *63*:753-6.

193. Thompson, A. C., B. J. Wright, D. D. Hardee, R. C. Gueldner, P. A. Hedin. 1970b. Constituents of the cotton bud. XVI. A comparison of the attractancy of host plants and non-host plants to the boll weevil *J. Econ. Entomol.* *63*:751-3.

194. Thompson, A. C., R. D. Henson, R. C. Gueldner, P. A.
        Hedin. 1970c. Constituents of the cotton bud.
        XVIII. Sterols and sterol derivatives. *Lipids*
        5:283-4.
195. Thompson, A. C., B. W. Hanny, R. C. Gueldner, P. A.
        Hedin. 1971a. Pytochemical studies in the
        family Malvaceae. I. Comparison of essential oils
        of five species of gas-liquid chromatography. *Am.*
        J. Botany 58:803-7.
196. Thompson, A. C., R. D. Baker, R. C. Gueldner, P. A.
        Hedin. 1971b. Identification and quantitative
        analysis of the volatiles transpired by maturing
        cotton in field. *Plant Physiol. 48*:50-2.
197. Thompson, A. C., R. D. Henson, R. C. Gueldner, P. A.
        Hedin. 1972. Constituents of the boll weevil. III.
        Lipids and fatty acids of pupae subcellular particles.
        Comp. Biochem. Physiol. 43B:883-90.
198. Thompson, A. C., R. D. Henson, R. C. Gueldner, P. A.
        Hedin. 1973. Constituents of the boll weevil.
        VIII. Lipids and fatty acids of prepupae subcellular
        particles. *Comp. Biochem. Physiol. 45B*:233-9.
199. Tingle, F. C., H. C. Lane, E. E. King, E. P. Lloyd.
        1971. Influence of nutrients in the adult diet on
        diapause in the boll weevil. *J. Econ. Entomol. 64*:
        812-4.
200. Townsend, C. H. T. 1895. Report on the Mexican cotton
        boll weevil in Texas. *Insect Life 7*:295-309.
201. Tumlinson, J. H., D. D. Hardee, J. P. Minyard, A. C.
        Thompson, R. T. Gast, P. A. Hedin. 1968. Boll
        weevil sex attractant: Isolation studies. *J. Econ.*
        Entomol. 61:470-4.
202. Tumlinson, J. H., D. D. Hardee, R. C. Gueldner, A. C.
        Thompson, P. A. Hedin, J. P. Minyard. 1969. Sex
        pheromones produced by the male boll weevil: Isola-
        tion, identification and synthesis. *Science 166*:
        1010-12.
203. Tumlinson, J. H., R. C. Gueldner, D. D. Hardee, A. C.
        Thompson, P. A. Hedin, J. P. Minyard. 1971. Identi-
        fication and synthesis of the four compounds compri-
        sing the boll weevil sex attractant. *J. Org. Chem.*
        36:2616-21.
204. Tumlinson, J. H., R. C. Gueldner, D. D. Hardee, A. C.
        Thompson, P. A. Hedin, J. P. Minyard. 1970. "The
        Boll Weevil Sex Attractant" *In* Chemicals Controlling
        Insect Behavior, ed. M. Beroza, Acad. Press, NY pp. 41-
        59.

205. Tursch, B., D. Daloze, C. Hootele. 1972. The alkaloid of *Propylaea quatuordecipunctata* L. *Chimia 26:* 74-75.

206. Vadimovich, I. I. 1938. Investigation of the chemical composition of the Ukrainian cotton crops of 1934 and 1935. Mem. Inst. Chem Tech. Akad. Sci. Ukrain, SSR No. *7*, 83-94.

207. Vanderzant, E. S., C. D. Richardson. 1964. Nutrition of the adult boll weevil: Lipid requirements. *J. Insect Physiol. 10:*267-72.

208. van Schoar, G. H. J. 1964. Mineral content of the cott plant in relation to the concentration and ratios of nutritive elements. *Ann. Physiol. Veg. Univ. Bruxell 9:*81-216.

209. Vickers, D. H., N. Mitlin. 1966. Changes in nucleic acid content of the boll weevil during its development. *Physiol. Zool. 39:*70-76.

210. Viehoever, A., L. H. Chernoff, C. O. Johns. 1918. Chemistry of the cotton plant with special reference to upland cotton. *J. Agr. Res. 13:*345-52.

211. Walker, K. L., L. C. Fife, F. F. Bondy. 1949. Comparative effectiveness of chlorinated hydrocarbons agains the boll weevil. *J. Econ. Entomol. 42:*685-6.

212. Wigglesworth, V. G. 1963. The action of moulting hormone and juvenile hormone at the cellular level in *Rhodnius prolixus. J. Exp. Biol. 40:*231-45.

213. Wiygul, G., N. Mitlin, A. C. Thompson, O. H. Lindig. 1974. Free amino acid levels in boll weevil: The effect of five different diets. *Comp. Biochem. Physiol. 49B:*663-7.

214. Wood, R., f. Snyder, 1968. Quantitative determination of alk-1-enyl and akyl-glyceryl ethers in neutral lipids. *Lipids 3:*129-135.

Chapter Seven

CHEMICAL MESSENGERS IN INSECTS AND PLANTS

L. B. HENDRY, J. G. KOSTELC, D. M. HINDENLANG,
J. K. WICHMANN, C. J. FIX, AND S. H. KORZENIOWSKI

*Department of Chemistry, The Pennsylvania State
University, University Park, Pennsylvania 16802*

INTRODUCTION

Studies of the chemistry and behavior of living organisms are often limited by the narrow focus of the researcher. Research efforts are frequently concentrated on one particular aspect of several diverse organisms rather than on a comprehensive view of one organism and its environment. The significance of such efforts is further limited by our lack of appreciation of the interrelationships within groups of organisms and more importantly by our insistence on studying static (artificial) rather than dynamic systems. Although it is unlikely that experiments can be designed which completely eliminate those pitfalls, it is clear that "human-centric" reasoning must be minimized. For the scientist, the view that "man is the master of the environment" must be reversed, i.e., "the environment is the master of man." It is from this frame of reference that this report on chemical messenger systems in insects and plants is written.

In describing the behavior of insects and plants, it is important to recognize that they have evolved together with

very complex and essential ties which are constantly being
modified. Certain insects and plants are incredibly
adaptable, consequently their interactions are likely to
involve a complex series of parameters. We will attempt
to describe some of these parameters and then theorize as
to what evolutionary mechanisms insure the survival of insects
and plants.

It is clear that insects and plants have evolved chemical
messenger systems which are essential to their survival. In
this report three basic types of messengers will be discussed
allomones, kairomones, and pheromones[5,40]. Although much is
known about messenger systems in regard to the identification
and biological activity of chemical cues in insects, little
is known about the origin of the signals. This is not surprising since it has been generally accepted that most cues
are synthesized de novo by insects. Recently, our laboratory
has found that, in many cases, these agents may be sequestered
by insects from food plants[22,23], and it is from these recent
results that our discussion will proceed.

ALLOMONES IN PLANTS

Sequestering of certain defensive substances (allomones)
by insects, from plants, has been established by several
researchers[4]. Two cases that have been studied in detail
are described below.

Pine Sawfly (Neodiprion sertifer). The larvae of the
pine sawfly (Fig. 1), a destructive pest that inhabits the
Northeastern United States and Canada, have a unique means
of defending themselves[8]. When attacked by a potential
predator, they rear up on their larval pods and emit a mouth
droplet (Fig. 2). The droplet has repellent qualities and
normally wards off predators. In some cases, the larvae
will dab the droplet upon a persistent attacker and this
frequently frustrates the assailant. For example, after being
christened in this manner, an attacking ant spends considerabl
time removing the viscous material before attempting to continue the battle.

The chemistry of the regurgitate of the sawfly larvae
is rather simple (Fig. 3), i.e., the terpenes, α and β-
pinene, and a series of resin acids[3-9]. Interestingly,

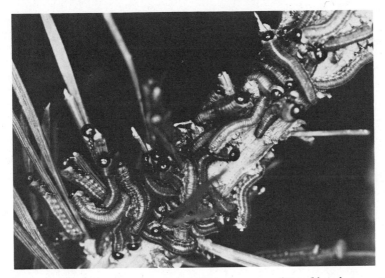

Figure 1.   Cluster of pine sawfly larvae (*Neodiprion sertifer*) feeding on *Pinus sylvestris*.

Figure 2.   Defensive posture of a pine sawfly larva.

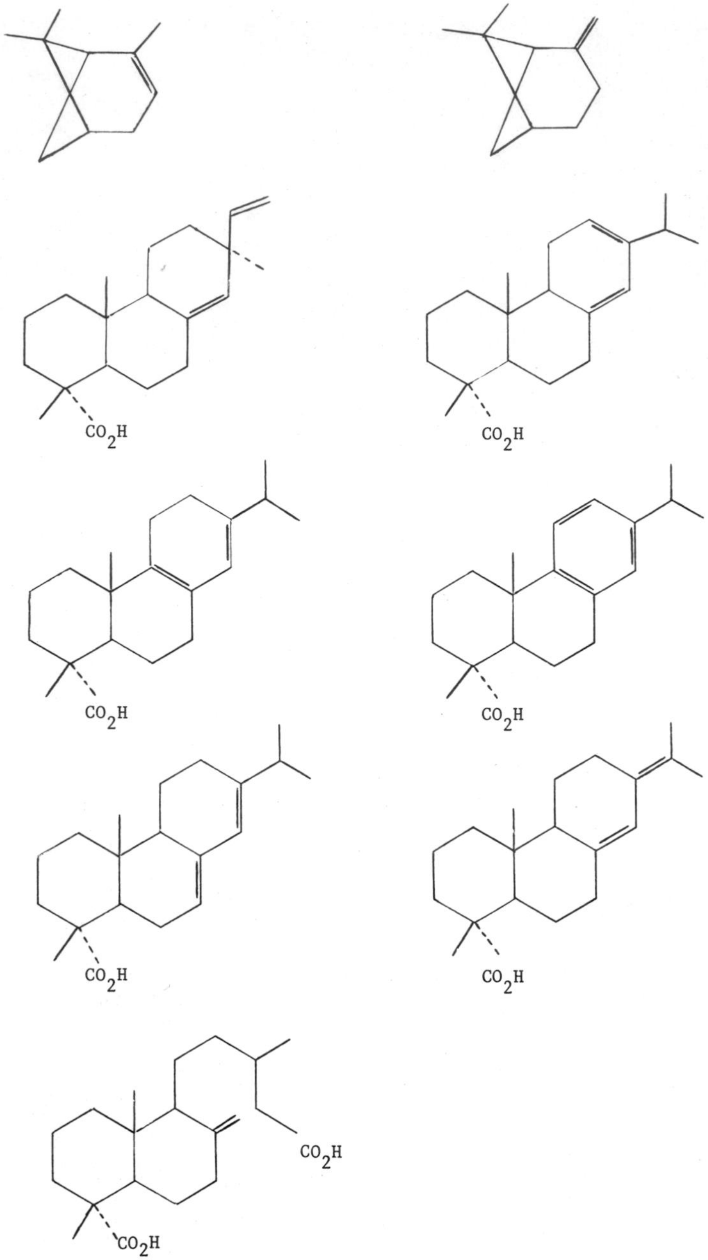

Figure 3.   Chemicals identified in the defensive effluent
of pine sawfly larvae.

the chemistry of the regurgitate matches that of pine needles and resin on which the larvae feed. A close examination of the morphology of pine sawfly larvae reveals two diverticular pouches attached to the foregut; these pouches are responsible for sequestering the defensive chemicals. None of the terpenes were found in the midgut or fecal material of the larvae indicating the diverticulae are remarkably efficient in their ability to sequester defensive chemicals from the ingested plant material.

Pine trees are known to produce terpenes as a means of defense against predation. Sawfly larvae are unusual in their ability to penetrate the pine tree's natural chemical protection. However, the most surprising consequence of this relationship is that sawfly larvae use the tree's chemical weaponry for their own defense.

Grasshopper (*Romalea microptera*). Another example of the sequestering of chemicals from food sources for defensive purposes is that of the prolific grasshopper, *Romalea microptera*[7]. The grasshopper expels a froth, from its anterior thoracic respiratory spiracles (Fig. 4), which is an effective repellent against potential predators. Although the compounds isolated from the froth (Fig. 5) appear to be mostly of plant origin[11-18], 2,5-dichlorophenol (10) probably is not, as few terrestrial organisms are known to produce chlorinated compounds. Due to the structural similarity of (10) to commercial herbicides, it is believed that (10) is a breakdown product of 2,4,5-T (2,4,5-trichlorophenoxy acetic acid) or 2,4-D (2,4-dichlorophenoxyacetic acid).

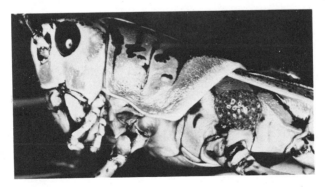

Figure 4. Grasshopper *Romalea microptera* emitting froth from anterior thoracic spiracles.

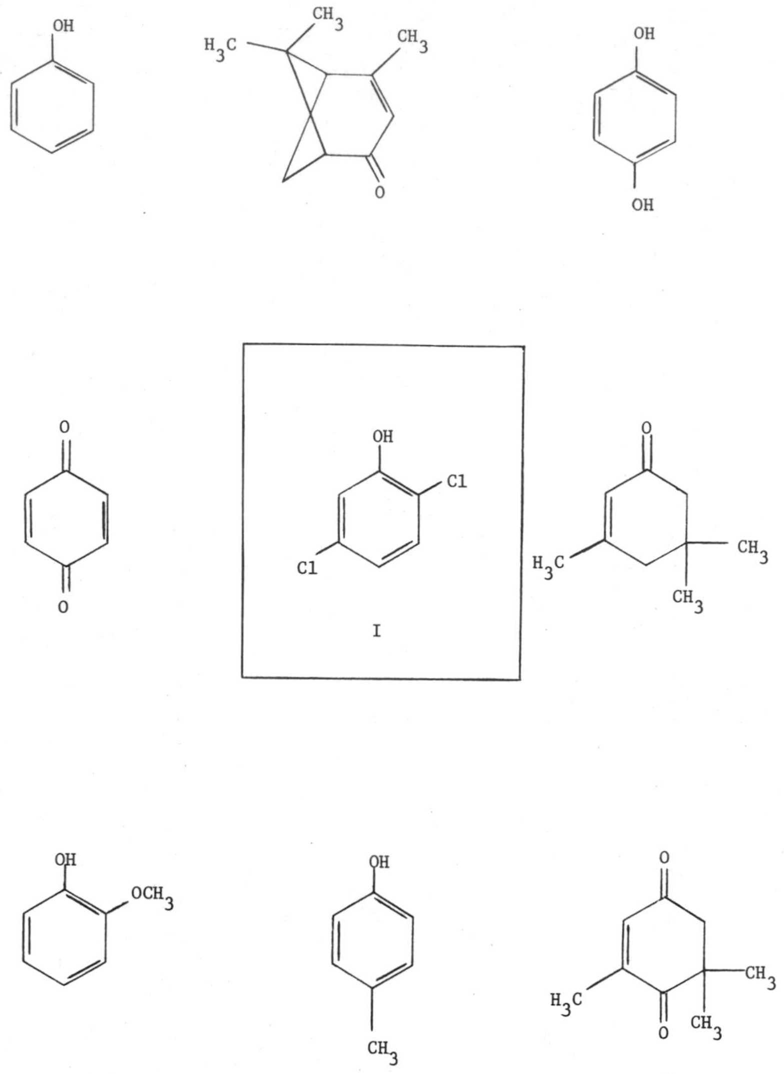

Figure 5.   Some components of the defensive secretion of
the grasshopper *Romalea microptera*.

Biological testing of (10) indicates that it is in effective
deterrent to ants and may have been sequestered by the grasshopper for that purpose.

The use of defensive compounds, unaltered from diet,
is an efficient evolutionary mechanism for protection of
particular species. Clearly, the ability of an insect species
to selectively concentrate, from its diet, a broad spectrum
of compounds which have toxic or repellent properties, would
help insure the survival of the species even if it is forced
to make abrupt changes in food plants.

The raison d'être of plants producing secondary substances
which insects utilize for their own defense is not obvious.
However, it appears that secondary plant chemicals, such as
phenols and terpenes, have excellent defensive utility for
the plants themselves. During coevolution of insects and
plants only a few insects appear to have been able to break
through the chemical defense systems of plants. However,
the overall statistical benefit of these chemicals in protecting plants is quite good.

KAIROMONES IN PLANTS

Kairomones are defined as "transpecific chemical
messengers, the adaptive benefit of which falls on the
recipient rather than on the emitter"[5,40]. Insect parasites
are known to locate host insects by kairomones emitted from
the host. Recently, several kairomones have been identified
in host-parasite communication systems[11,27,28], however, the
origin of these messengers remains unknown. To determine if
plant components may have an effect on these chemical cues,
two unrelated host-parasite communication systems were
chosen for study (Table I).

Table I

Host Insect	Parasite	Kairomone
Corn Earworm Moth (*Heliothis zea*)	*Trichogramma evanescens*	Tricosane ($C_{23}H_{48}$)
Potato Tuberworm Moth (*Phthorimaea operculella* Zeller)	*Orgilus lepidus*	Heptanoic Acid (19)

Corn Earworm (HOST) - *Trichogramma evanescens (PARASITE)*. *Trichogramma evanescens* is an egg parasite which attacks the corn earworm moth (*Heliothis zea*). Tricosane ($C_{23}H_{48}$), the major kairomone associated with corn earworm eggs, elicits intense host searching behavior from *T. evanescens*[27]. A computerized analysis of corn, the primary food source of the corn earworm, gas chromotography-mass spectrometry (G (GC-MS) (Finnigan 3200 GC-MS with 6000 computer) indicated that tricosane (Fig. 6) was present in a significant concentration. Chemical analyses of related plants showed that they contained less tricosane than corn[23].

Potato Tuberworm Moth (HOST) - *Orgilus lipidus (PARASITE)* In a similar fashion, potatoes, the major food source of the potato tuberworm moth (*Phthorimaea opercullella* Zeller), were examined for the presence of kairomones. Heptanoic acid, (19) a kairomone which elicits host searching responses from the parasitic wasp *Orgilus lepidus,* was previously found in substantial quantities in the frass (fecal material) of the potato tuberworm[11]. Figure 7 depicts a typical response of a female *O. lepidus* to 10 ng of heptanoic acid spotted on filter paper; identical behavior was elicited when *O. lepidus* contacted the potato tuberworm frass. Heptanoic acid was found in substantial quantities in the potato by GC-MS analyses; less heptanoic acid was found in other plants[23].

The evidence that kairomones are present in plants does not conclusively show that the host insect concentrates kairomones from its diet. In order to elucidate this point, heptanoic acid was labelled with deuterium in the following manner.

$$CH_3(CH_2)_4CH_2\overset{\overset{\text{O}}{\|}}{C}\text{-OH} \xrightarrow{\begin{array}{l}(1)\ CH_3OH/HCl \\ (2)\ LTMP*\ (1\ eq.) \\ \hline (3)\ D_2O \\ (4)\ aq.\ HCl\end{array}} CH_3(CH_2)_4CHD\overset{\overset{\text{O}}{\|}}{C}\text{-OH}$$

(19)   (20)

*LTMP

(21)

*LTMP = Lithium tetramethyl piperidine

Deuterated heptanoic acid (20) was incorporated into potato slices in approximately the same concentration as naturally occurring heptanoic acid (10 ppb). Potato tuberworm larvae were then reared on the potato slices, and frass was collected and analyzed by computerized GC-MS aided by mass fragmento-graphy [16,17,22,23]. Ions m/e 60 and 61 were monitored (Fig. 8); these ions arise from the respective McClafferty rearrangements of labelled and non-labelled heptanoic acid as shown below.

Incorporation of deuteroheptanoic acid in the frass exceeded 70% (Table II); this result is rather surprising since it indicates very little acid was absorbed by the insect for biosynthetic activity.

In searching for a host some insect parasites have adapted to sensing plant chemicals which are concentrated in

---

[*]Mass fragmentography is a technique which focuses the mass spectrometer on a few key ions of interest. Since it is no longer necessary for the mass spectrometer to scan over an entire mass range, the sensitivity increases dramatically, e.g., from normal operation of 1-10 ng to 25-150 pg.

TRICOSANE
CORN HC PREP
SPEC.#   95

35	•04	•00	•00	•04	1•92	1•60	28•08
42	7•65	62•16	4•42	•00	•14	•00	•00
49	•00	•00	•00	•04	1•29	2•33	23•39
56	13•17	100•00	4•73	•04	•00	•00	•00
63	•35	•00	•25	•35	3•06	2•85	15•36
70	10•46	73•31	4•84	•56	•14	•00	•46
77	•67	•00	•25	•35	2•23	2•85	12•03
84	6•71	47•67	3•27	•35	•14	•04	•04
91	•35	•00	•25	•04	•77	1•50	8•80
98	4•42	15•67	•98	•14	•14	•00	•00
105	•35	•00	•14	•00	•35	•98	3•90
112	3•17	9•00	•46	•46	•14	•14	•00
119	•00	•00	•00	•14	•25	•35	1•81
126	2•44	6•09	•67	•00	•00	•00	•04
133	•00	•04	•00	•00	•25	•46	•87
140	1•71	3•90	•46	•00	•00	•00	•00
147	•00	•04	•14	•04	•00	•14	•14
154	1•55	2•70	•29	•03	•00	•00	•00
161	•00	•00	•00	•03	•00	•05	•09
168	1•27	2•14	•20	•00	•00	•00	•00
175	•00	•00	•00	•00	•03	•03	•11
182	•68	1•42	•24	•14	•00	•00	•00
189	•03	•00	•11	•00	•09	•00	•05
196	•56	1•15	•09	•00	•00	•00	•00
203	•00	•11	•05	•05	•03	•00	•00
210	•41	•89	•09	•00	•00	•03	•00
217	•05	•00	•02	•00	•00	•00	•02
224	•21	•59	•06	•00	•00	•02	•00
231	•00	•00	•00	•00	•00	•00	•00
238	•26	•48	•09	•00	•00	•00	•02
245	•00	•09	•00	•00	•00	•00	•00
252	•18	•33	•09	•00	•0C	•00	•00
259	•00	•05	•00	•00	•02	•00	•00
266	•09	•35	•12	•00	•00	•00	•03
273	•00	•02	•03	•00	•00	•00	•00
280	•00	•29	•00	•00	•00	•05	•00
287	•00	•00	•00	•09	•00	•00	•00
294	•00	•05	•00	•00	•00	•00	•00
301	•02	•00	•05	•00	•00	•00	•00
308	•00	•00	•00	•00	•05	•00	•00
315	•00	•00	•00	•02	•00	•00	•00
322	•05	•05	1•45	•41	•00	•00	•00
329	•02	•00	•00	•00	•00	•00	•00

Figure 6.   Mass spectrum of a component of corn hydrocarbons
identified as tricosane.

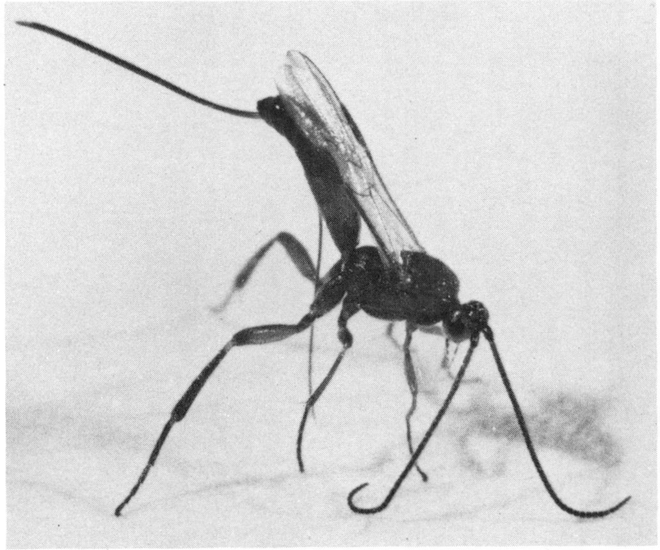

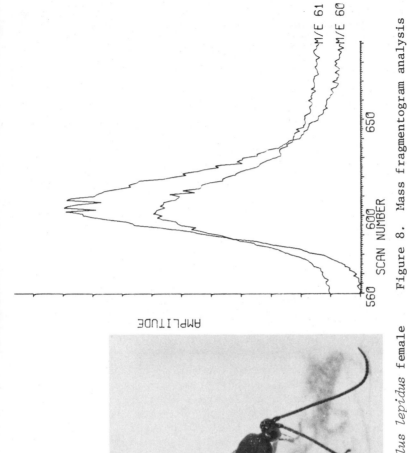

AMPLITUDE

SCAN NUMBER

560   600   650

M/E 61

M/E 60

Figure 8. Mass fragmentogram analysis using ions m/e 60 and 61 of potato tuberworm frass incorporated with α-deutero heptanoic acid.

Figure 7. Response of an *Orgilus lepidus* female to filter paper spotted with 10 ng of heptanoic acid.

Table II. Heptanoic Acid (HA) Content in Potatoes and Frass of the Potato Tuberworm

	Potato (ng/g)	Frass (ng/g)	% of Total in Frass
Naturally occurring: HA	2.3	150	100
α deutero HA	0	0	0
Feeding study: HA	2.3	84	30
α deutero HA	10	196	70

and released by the host insect. Phytochemicals, which serve as kairomones appear to provide a dual evolutionary role by 1) initially attracting parasites from a distance, to fields having a suitable foliage environment and 2) subsequently eliciting from parasites intense searching behavior for host insects that have concentrated these substances. We believe that some parasites, in their larval stage, may be programmed or "imprinted" to specific plant environments and to host insects in these environments by kairomones to which they are exposed. In some cases, this programming may be detrimental to the parasite. For example, pressures by a parasite population within a particular ecosystem, may force the transferra of the host insect to an alternate food plant. During such a shift, rather than transfer to an alternate chemical environment having a kairomone complex, the parasites are likely to remain in the original environment. This environmental preference may result from larval programming, of the parasite, to specific plant complexes ingested by host insects. Observations tend to support this postulate[1,24,28,33-35,37,39]

The evolutionary benefits to plants, resulting from their production of kairomones, are clear. Plants having evolved the ability to synthesize relatively large amounts of kairomones would insure their survival by encouraging symbiotic relationships with insect parasites.

PHEROMONES IN PLANTS

Insect pheromones have been the subject of extensive research programs for the past twenty years[2,3,26,41]. The main reason for this interest is that pheromones may provide means for surveying, detecting, and controlling insect pest populations. Despite the intense interest, researchers have been baffled by the elusive and complex nature of pheromone systems. Frequently, separate laboratories have isolated and identified different pheromones for the same insect. Further confusion has recently resulted from the general occurrence, within a given species, of temporal and spatial differences in biological responses to pheromones[16,31,36].

The recent isolation of several insect sex pheromones in food plants may resolve this problem[22]. The concentrations and ratios of the insect sex attractants in plants vary with the growth, stage and location of the plant. These plant variations may explain the temporal and spatial differences in apparently identical insect species. However, it has not yet been conclusively proven that insects are storing sex attractants during feeding. Hence, there remain several possible origins of insect pheromones, namely: 1) storage from plants; 2) degradation of plant precursors; 3) "induction" of biosynthesis in the insect via plant pheromones; 4) de novo biosynthesis; 5) others, e.g. microorganisms associated with plants or insects. It is unlikely that all phytophagous insect species will fit into one of the above categories; however, we feel that the presence of insect sex pheromones in plants is not coincidental. De novo biosynthesis of three insect pheromones has been reported[25,29,32]. The remainder of this report will be devoted to describing our intial experiments with insect-plant pheromones.

Oak Leaf Roller. The oak leaf roller moth (OLR), *Archips semiferanus* Walker, has an exceedingly complex sexual communication system[12,14,16,18-20]. The adult female attracts her mate with a series of mono-unsaturated C_{14} acetates (tetradecenyl acetates). Recently, using computerized capillary chemical ionization GC-MS, a considerably greater number of related compounds were preliminarily identified in the female (Fig. 9). The present list of compounds found in active oak leaf roller female extracts is shown in Table III. It is interesting to note that many of these compounds are pheromones in other insect species. The frequent cross

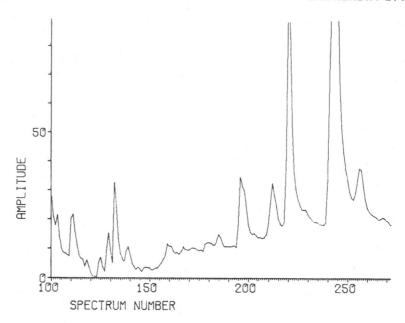

Figure 9. Capillary gas chromatograph-mass spectral trace of the oak leaf roller active female extract; 150,000 theoretical plates.

attractantcy of insect species could be explained if the complexity of sexual message in the OLR is a mineral phenomone

Leaves of various oak tree species, on which OLR feed, were analyzed by CG-MS and found to contain sex attractants[22]. The sex attractant complex varies with the species of oak as well as the growth stage of the leaf. Field trapping studies indicate that different populations of OLR may exist, which correlate with the chemistry of the trees upon which they feed[18]. A close examination for pheromones in eggs, larvae, pupae and adults showed that all growth stages contained sex attractants; however, adult females had greater concentrations of tetradecenyl acetates (TDA) than adult males (M:F = 100:1).

Further chemical analyses using electron impact GC-MS (Fig. 10A,B) revealed that OLR males contain benzaldehyde.

Table III. Some Components of the Oak Leaf Roller Female
Active Fraction

Undecyl Alcohol	Dodecyl Alcohol*
Undecyl Acetate*	Dodecyl Acetates*
Undecenyl Acetates	Dodecenyl Acetates*
Tridecyl Alcohol	Tetradecyl Alcohol*
Tridecyl Acetate	Tetradecyl Acetate*
Tridecenyl Acetates	Tetradecenyl Acetates*
Tridecadienyl Acetates	Tetradecadienyl Acetates*
Misc. Polyunsaturated	Tetradecatrienyl Acetates
C_{13} Compounds	Misc. Polyunsaturated
	C_{14} Compounds
Pentadecyl Alcohol	Hexadecanol*
Pentadecyl Acetate	Hexadecyl Acetate*
Pentadecenyl Acetates	Hexadecenyl Acetates*
Misc. Polyunsaturated	Misc. Polyunsaturated
C_{15} Compounds	C_{16} Compounds

Dihydroambrettalide*
Ambrettalide Isomers
Misc. Macrocyclic Lactones

*Identified as Chemical Messengers in Other Insect Species

Benzaldehyde is a known "aphrodisiac" emitted by over twenty
species of male moths and prepares females for mating[3]. OLR
females contain benzaldehyde, though in lower quantities than
are present in males (F:M = 1:100). A search for benzaldehyde
in oak leaves revealed considerable concentrations of this
male pheromone. Obviously, if OLR are able to store pheromones from their diet, there must be differential sequestering
mechanisms operating in each sex, i. e., females store
primarily tetradecenyl acetates whereas males store benzaldehyde.

A theoretical framework has been constructed which is
consistent with the above findings. It suggests that larvae
of the oak leaf roller store sex pheromones as they feed on
the leaves of a particular tree. The female sequesters

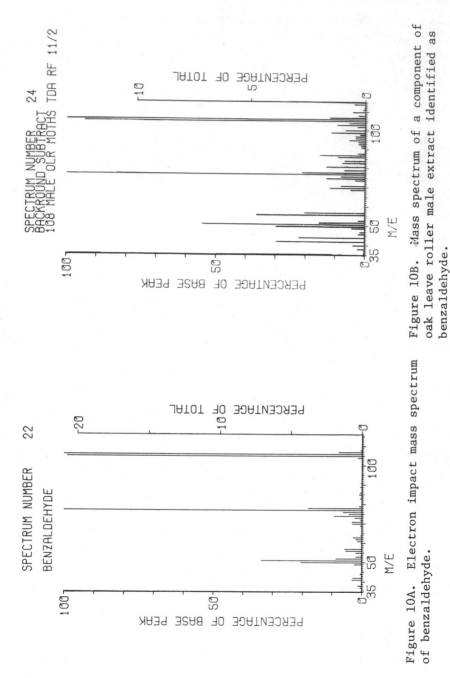

Figure 10A. Electron impact mass spectrum of benzaldehyde.

Figure 10B. Mass spectrum of a component of oak leave roller male extract identified as benzaldehyde.

tetradecenyl acetates in a gland, which becomes her pheromone
gland after metamorphosis. The male on the other hand primarily sequesters benzaldehyde, but relinquishes traces of
tetradecenyl acetates after it has emerged as an adult from
the puparium. During metamorphosis, the male "imprints" or
programs the chemicals left in the meconium. After the odor
of the meconium dissipates, the male flies to a host tree
having the same odor and then locates females that have
concentrated the odor. (Female insects usually release their
pheromone only at a specific time of day). The female
becomes receptive when it detects benzaldehyde (imprinted
during metamorphosis). After copulation, the female lays its
eggs with a complement of pheromone. The next spring when
the eggs hatch, the pheromone complex serves as a feeding cue
to first instar larvae thus aiding in the selection of a host
plant.

The following experiments elucidate this framework. Oak
leaf roller males raised from field collected eggs, on semi-
synthetic diets which do not contain pheromone, did not
respond behaviorally to natural pheromone from field females.
Moreover, OLR male antennae did not respond to field female
pheromone or to any tetradecenyl acetates by the electro-
antennagram (EAG) method[18]. Females reared on the semi-
synthetic diet did not attract field males nor did they
contain detectable quantities of sex attractants when analyzed
by GC-MS. However, when scarlet or black oak leaves were
mixed with the synthetic diets, OLR males responded normally
in EAG and laboratory biological assays; OLR females reared
in this manner were attractive to field males and were found
to contain pheromone.

With regard to pheromones left in the eggs by ovipositing
females, little is known about their possible function as
feeding cues. However, OLR larvae which hatch from eggs laid
on scarlet oak trees prefer scarlet to other species of oak.
If OLR larvae are deprived of scarlet oak upon emergence, their
tree preference for other oak species is not specific. Larvae
reared on scarlet oak leaves will not successfully change to
another oak species in the last instar, in fact most die rather
than eat. On the other hand, the same OLR larvae force fed on
black oak after hatching will not feed on scarlet oak in the
last instar. Several elegant experiments that establish
larval feeding cues as important behavioral messengers in
other species have been conducted by Hanson (1975).

Sciarid mushroom fly pheromone complex. A second insect
which feed on plants containing its sex pheromone complex is
the sciarid mushroom fly, (*Lycoriella mali* Fitch). The
sciarid is a major pest of the cultivated mushroom, *Agaricus
bisporus* Lange. The sex pheromone complex of the female
sciarid fly was identified as a mixture of saturated straight
chain aliphatic hydrocarbons with the general formula, CH_3
$(CH_2)_nCH_3$ where n = 13-30[21]. Heptadecane was statistically
the most active hydrocarbon, when tested on male flies in
a biological assay chamber[9].

To determine if pheromones were present in mushrooms,
six different strains of fresh mushrooms extracted in redistil
led, pesticide grade hexane. The hexane extracts were
chromatographed on silica gel TLC plates using hexane as the
developing solvent. A band having the same R_f value as
authentic samples of n-alkanes was eluted with hexane and
concentrated. These extracts were analyzed by gas chromatography and electron impact GC-MS analyses using mass fragmentography. The hydrocarbons pentadecane to heneitriacontane
were present (Table IV) in all 6 strains. The variation in
concentrations and ratios of each hydrocarbon are shown in
Figure 11.

Since there are discrete behavioral differences in the
sexual responses of several strains of sciarid males reared
on various laboratory diets, we believe that the chemistry of
the mushroom governs in part the sex pheromone complex of
the fly. This postulate is strengthened by the observation
that attempts to rear sciarid flies on hydrocarbon free diets
were unsuccessful. On the other hand, larvae reared on diets
supplemented with excess quantities of hydrocarbons, e.g.,
heptadecane, fed considerably longer than normal. This
supports the hypothesis that sex pheromones derived from
plants serve as larvae feeding cues. In many cases, larvae
fed the excess hydrocarbon diet died before undergoing metamorphosis, indicating that pheromones may have hormonal propertie
as well.

European corn borer. The sex pheromone of the European
corn borer (*Ostrinia nubilalis*) has been reported to be
drastically different. Klun (1973) reported the pheromone
to be a mixture of *cis*- and *trans*-11-tetradecenyl acetates
in a 96:4 ratio, whereas Roelofs (1975) reported that a
ratio of 4:96 was optimal for attraction of male corn borers.

Table IV. Relative Amounts (ppb) of n-alkanes in *Agaricus bisporus* Lange

P.S.U. Strains	15	16	17	18	19	20	21	22	23
310	49.1	164	374	765	964	626	370	298	220
324	97.4*	128	382	611	815	568	452	284	218
340	54.6	108	338	462	636	363	206	175	154
341	64.7*	83.8	344	480	662	464	316	308	315
342	35.0	123	344	696	820	627	275	236	192
D$_{26}$	63.1*	471	458	734	918	612	328	278	204

Table IV. Continued

P.S.U. Strains	24	25	26	27	28	29	30	31	Total
310	142	453	119	114	379	176	104	349	5,626
324	179	452	190	174	426	229	206	284	5,696
340	153	305	125	138	388	129	106	186	4,026
341	319	449	316	338	578	394	401	406	6,238
342	157	218	124	114	328	123	87.3	170	4,669
D$_{26}$	244	414	192	182	464	219	216	380	6,377

*only one replication

The insects used by the two laboratories were collected from widely differing geographical locations, namely Iowa (Klun) and New York (Roelofs).

Our interest was in determing if corn, a major food source for the corn borer, contains tetradecenyl acetates. Initially, sweet corn was extracted with redistilled pesticide methylene chloride. The suspension was filtered, concentrated, and chromatographed on silica gel TLC plates using 50:50 methylene chloride: hexane as a developing solvent. A band having the same Rf as an authentic sample of *cis*-11-tetradecenyl acetate (TDA) was eluted and subjected to GC on a 3% OV-1 column at 130°C. A region, having the same approximate retention time

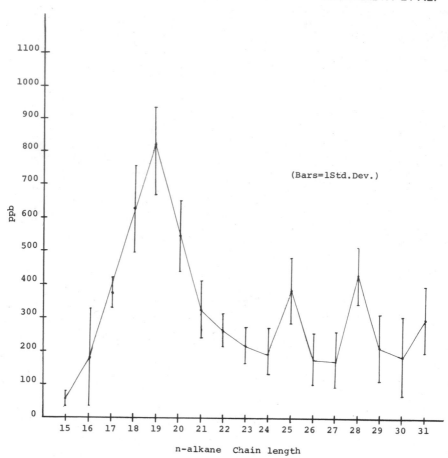

Figure 11. Composite of n-alkanes from Six Strains of
Agaricus bisporus Lange

as TDA, was collected and analyzed by chemical ionization GC-
MS using a 10% diethylene glycol succinate column at 160°C,
and methane as a carrier gas. Peaks at the same retention
times as *cis*- and *trans*-11-tetradecenyl acetates had the
characteristic ions m/e 255 (m+1), 283 (m+29), 295 (m+41)
(Fig. 12). Electron impact GC-MS analyses of the corn
isolate showed peaks having the identical retention times
and mass spectra as *cis*- and *trans*-11-tetradecenyl acetate.
Mass fragmentograms were also identical (Fig. 13A,B).

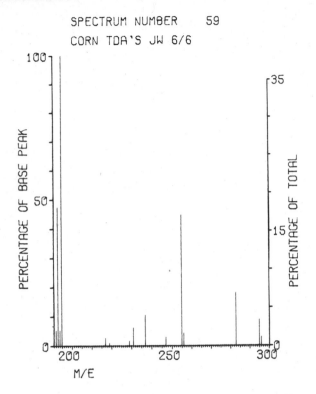

Figure 12.   Chemical ionization mass spectrum of a component
of corn identified as a tetradecenyl acetate.

Different varieties of corn, from several locations,
have been analyzed in this manner, and the ratios of the
tetradecenyl acetate isomers were found to vary considerably.
Differences in *cis/trans* TDA ratios were also noted in corn,
from the same location, which was monitored throughout the
growing season.   It is plausible that these changes are
reflected in the variable TDA chemistry in different strains
of cornborers.   These results may complicate potential survey
and detection programs based upon trapping the cornborer
with a single pheromone complex.

Corn apparently also contains dodecyl acetate (DDA) as
evidenced from chemical ionization spectra of a corn constituent.   DDA is a "synergistic" component of the sexual message
of closely related insect species[36], although it has not as

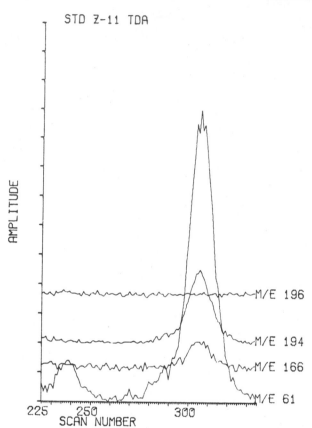

Figure 13A. Mass fragmentogram of standard *cis*-11-tetradecenyl acetate using characteristic ions $\underline{m}/\underline{e}$ 196, 194, 166 and 61.

yet been implicated as part of the cornborer sex pheromone complex. If DDA is eventually found to be active, it would be the first example of a component, initially isolated from a plant, which subsequently was found to have sexual properties in an insect.

   Apple feeding tortricid moths. It has been reported that many Tortricid insects which feed upon apple leaves in the larval stage contain *cis*- and *trans*-11-tetradecenyl acetates as part of their sex attractant complex[3]. Recently, variations in the optimum ratios of the attractants were

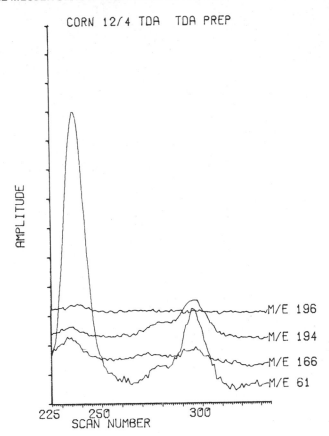

Figure 13B. Mass fragmentogram using identical ions as in
A of corn isolate.

observed within the same species, e.g., red banded leaf roller
pheromone[36]. These variations have not been satisfactorily
explained.

Apple leaves were analyzed for the presence of *cis*- and
trans-11-tetradecenyl acetates in a manner analogous to the
corn analyses. Peaks were observed which had identical
retention times and mass spectra by electron impact and
chemical ionization GC-MS as *cis*- and *trans*-11-TDA (mass
fragmentograms were also identical, Fig 14A,B). These
analyses, when repeated at several stages of apple leaf

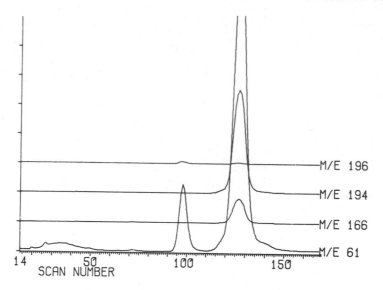

Figure 14A. Mass fragmentogram of *cis*-11-tetradecenyl acetate
using characteristic ions m/e 196, 194, 166 and 61.

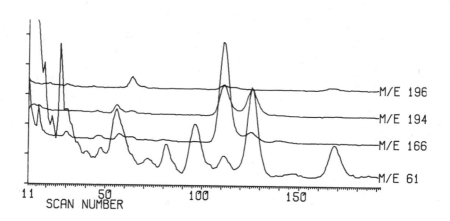

Figure 14B. Mass fragmentography of apple extract under
same conditions as A.

development, showed changes in isomeric *cis:trans* TDA ratios. Dodecyl acetate was also found in apple leaf extracts (Fig. 15). DDA is a part of the pheromone complex of the redbanded leafroller (*Argyrotaenia velutinana*) which feeds upon apple leaves.

Preliminary chemical analyses of apple leaf extracts indicated (Fig. 16) dodecanol (DDOL) was present. DDOL has been reported to be a "synergist" in the pheromone complex of the oriental fruit moth (*Grapholita molesta*) which feeds on apples[6]. Moreover, DDOL elicits close range copulatory behavior from males. The identification of a close range sexual stimulant in a plant could explain the frequent excited behavior that male moths exhibit in the vicinity of host plants.

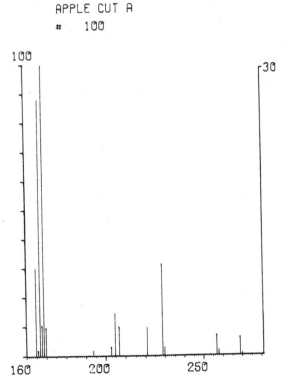

APPLE CUT A

\#   100

Figure 15.   Chemical ionization mass spectrum of a component of apple leaf extract identified as dodecyl acetate.

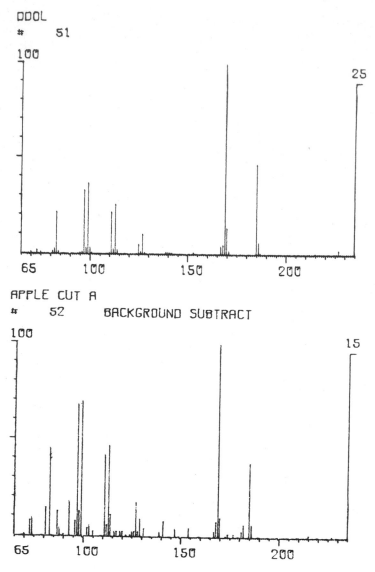

Figure 16.   A.   Chemical ionization mass spectrum of dodecanol.
             B.   Chemical ionization spectra of a component
                  identified as dodecanol in apple leaf extracts.

Table V. summarizes the relationships of plants and insects in regard to sex pheromones studied in our laboratory to date.

CONCLUSIONS

It is clear that numerous insect chemical communicating agents are of plant origin. The reasons why plants produce

Table V. Preliminary Identifications of Sex Attractants Found in Plants

Plant	Insect Predator	Reported Sex Attractants	Plant Pheromone
Cabbage	Cabbage Looper (*Trichoplusia ni*)	Z-7-dodecenyl acetate	Dodecanol dodecenyl acetates
Corn	European Corn Borer (*Ostrinia nubilalis*)	Z- and E-11-tetradecenyl acetate	Z- and E-11 tetradecenyl acetate dodecyl acetate
Mushroom	Sciarid Fly (*Lycoriella mali*)	n-alkanes	n-alkanes
Apple	Apple Feeding Tortricid Moths	Z- and E-11-tetradecenyl acetate dodecyl acetate	Z- and E-11-tetradecenyl acetate dodecanol dodecyl acetate
Cotton	Boll Weevil (*Anthonomus grandis*)	4 terpenoid components	terpenoids having similar mass fragmentograms as sex pheromones

Z - *cis*
E - *trans*

these cues are understandable in the case of allomones (defensive chemicals) and kairomones (host-parasite signals). However, in the case of sex pheromones, plants appear to be at a major evolutionary disadvantage.

From the standpoint of the insect, it is entirely plausible that some phytophagous insects derive their sex attractant complex from plant food sources either by storage or by "induction" of biosynthesis. In this manner, plant pheromones may serve multiple evolutionary roles: 1) to attract adult insects to suitable host plants by evaporating low concentrations of pheromones from leaf surfaces; 2) to aid the adult in locating mates via high concentrations of pheromones selectively sequestered by the opposite partner and 3) to aid larvae in choosing an appropriate host plant via the pheromones left by the female in the eggs. The overall consequence of these multiple functions of plant pheromones is to reproductively sub-structure insect populations into potentially isolated units. For example, if newly hatched oak leaf roller larvae, derived from the same egg mass, feed separately upon apple leaves and oak leaves until pupation, then the adults would mate only with individuals having the same dietary experiences. Larval feeding cues, left by the female in the eggs, would tend to keep the oak and apple insect populations isolated in ecological time: larval cues and adult pheromones in concert with genetic factors would maintain the isolation in evolutionary time. It follows from this example that divergence of species may be intimately related to diet. Hence, "speciation" may be programmed in part by environment and may not be entirely bound by Darwinian evolutionary theory.

The above discussion about plant produced chemical messengers can be encapsulated in a new theoretical framework, the "Complexity, Plasticity, and Continuum Theory" (Hendry in preparation). The theory states that there is a *complexity* of chemical messengers in every species which elicits a discrete behavioral resonse from the organism. The individual organism can be "imprinted" or programmed to a specific range of chemicals it has contacted during feeding. The range, concentrations and ratios of these chemicals are exceedingly important in species recognition and are part of the genetically determined *plasticity* of the organism. In this regard, it is important to recognize that there are constraints to the kinds of chemicals that may be "imprinted" for a given species. For example, it is unlikely that an oak leaf roller moth, which normally uses a mixture of TDA isomers for its sex attractant, can be transformed into a mushroom roller moth by "imprinting" mushroom hydrocarbons. However, since pheromones in apple and oak leaves are very closely related, it is possible that an oak leaf roller could be transformed into an apple leaf roller. Within the genetic plasticity, it is possible to have a *continuum* of organisms that evolve from slightly different diets. The extent of the continuum may be dependent, in part, upon the plasticity of the larval feeding cue and adult pheromones, i.e., the breadth of host range and mate acceptance. The reasons why plants produce secondary substances, which phytophagous insects use for reproduction, must be approached with some caution, since it is possible that microorganisms associated with plants may be producing the pheromones. Nevertheless, insects and plants have evolved via a constant struggle for survival which sometimes produces compromises. Some plants may have developed mechanisms by which they become resistant to predation by producing too little sex attractant, thereby being unattractive to an insect pest. Preliminary studies suggest that this may be the case with certain resistant trees in forests severely defoliated by the oak leaf roller. Alternatively, the production of too much sex pheromone by the plant could also have a deleterious effect on the insect, as evidenced in mushroom flies given excess heptadecane. In these cases, excess plant pheromones may have striking hormonal effects on the insect.

However, the most likely possiblity is that the plant chemistry changes in concert with the insect. A plant that is chemically slightly out of phase with an insect pest may protect itself from future predation. For example, if the pheromone chemistry of an oak tree changes while the oak leaf roller that fed upon the tree is in its pupal stage, then the adult OLR will be programmed to the original pheromone complex upon emergence and will not be attracted to the tree.

One further evolutionary explanation for plant pheromones cannot be totally ruled out, namely, altruism. Plants could have evolved mechanisms by which they control their own species population. For example, repeated forest defoliations by the oak leaf roller cause the eventual production of an abundance of acorns. In this manner, oak trees may insure continuance of the species albeit by sacrificing the adult.

Whatever mechanisms are involved in the co-evolution of insects, plants, and other organisms, it can be assured that they will be complex. In studying any living system, it is well for us to remember that evolution is a continuing process.

ACKNOWLEDGEMENTS

The section entitled "Allomones in Plants" was derived in part from a Ph.D. thesis by L. B. Hendry, Cornell University, 1970. In this regard, we would like to thank Dr. Jerrold Meinwald and Dr. T. Eisner who directed and contributed to this work.

We would also like to acknowledge Dr. Richard Levins, Dr. Barbara Bentley, Dr. Ron Carroll and Dr. Frank Hanson whose insights were very much appreciated.

REFERENCES

1. Arthur, A. P. 1962. Influence of Host Tree Abundance of *Itoplectis conquisitor* (Say) (Hymenoptera:Ichneumonidae), a Polyphagous Parasite of the European Pine Shoot Moth, *Rhyacionia buoliona* (Schiff) (Lepidoptera:Olethreutidae). *Canadian Entomologist 94:*337-347.
2. Beroza, M. ed., *Chemicals Controlling Insect Behavior*, Academic Press, New York and London, 1970.
3. Birch, M. C. ed., *Pheromones*, American Elsevier Publishing Company, Inc., New York, 1974.
4. Brower, L. P. and Glazier, S. C. 1975. "Localizations of Heart Poisons in Monarch Butterfly", *Science 188:* 19-25.
5. Brown, W. L., Jr., Eisner, T. and Whittaker, R. H. 1970. "Allomones and kairomones: transspecific chemical messengers." *BioScience 20:*21-22.
6. Carde, R. T., Baker, T. C. and Roelofs, W. L. 1975. "Behavioural Role of Individual Components of a Multichemical Attractant System in the Oriental Fruit Moth." *Nature 253(5490):*348-349.

7. Eisner, T., Hendry, L., Peakall, D. B. and Meinwald, J. 1971. "2,5-dichlorophenol (from Ingested Herbicide?) in Defensive Secretion of Grasshopper." *Science 172:* 277-278.

8. Eisner, T., Johnessee, J. S., Carrel, J., Hendry, L. B. and Meinwald, J. 1974. "Defensive Uses by an Insect of a Plant Resin." *Science 184:*996-999.

9. Girard, J., Hendry, L. B., and Snetsinger, R. 1973. "Sex Pheromone in a Mushroom Infesting Sciarid, *Lycoriella mali.*" *The Mushroom Journal 13:*29-31.

10. Hanson, F. 1975. "Comparative Studies on the Induction Food Choice Preferences in Lepidopterous Larvae." Symposium on Host Plant in Relation to Insect Behavior and Reproduction. The Hungarian National Academy of Sciences, in press.

11. Hendry, L. B., Greany, P. D. and Gill, R. J. 1973a. "Kairomone Mediated Host-finding Behavior in the Parasitic Wasp *Orgilus lepidus.*" *Entomologia Experimentalis et. Applicata 16:*471-477.

12. Hendry, L. B., Roman, L. and Mumma, R. O. 1973b. "Evidence of a Sex Pheromone in the Oak Leaf Roller, *Archips semiferanus* (Lepidoptera:Tortricidae): Laboratory nad Field Bioassays." *Environmental Entomology 2:*1024-1028.

13. Hendry, L., Capello, L. and Mumma, R. O. 1974. "Sex Attractant Trapping Techniques for the Oak Leaf Roller (*Archips semiferanus* Walker)" *Melsheimer Series*, No. 16:1-9.

14. Hendry, L. B., Gill, R. J., Santora, A. and Mumma, R. O. 1974. "Sex Pheromones in the Oak Leaf Roller (*Archips semiferanus*): Isolation and Field Studies" *Entomologia Experimentalis et. Applicata 17:*459-67.

15. Hendry, L. B., Jugovich, J., Roman, L., Anderson, M. E. and Mumma, R. O. 1974. "*Cis*-10-tetradecenyl Acetate. An Attractant Component in the Sex Pheromone of the Oak Leaf Roller Moth (*Archips semiferanus* Walker)." *Experientia 30:*886-887.

16. Hendry, L. B., Anderson, M. E., Jugovich, J., Mumma, R. O., Robacker, D., and Kosarych, Z. 1975a. "Sex Pheromone of the Oak Leaf Roller - A Complex Chemical Messenger System Identified by Mass Fragmentography." *Science 187:*355-357.

17. Hendry, L. B. and Hindenlang, D. M. 1975b. "Insect Sex Pheromones-Identification by Mass Fragmentograph". *Finnigan Spectra 5*(1):1-3.

18.   Hendry, L. B., Jugovich, J., Mumma, R., Robacker, D.,
      Weaver, K. and Anderson, M. E. 1975c. "The Oak
      Leaf Roller (*Archips semiferanus* Walker) Sex Pheromone
      Complex - Field and Laboratory Evaluation of Requisite
      Behavioral Stimul". *Experientia 31:*629-631.
19.   Hendry, L. B., Korzeniowski, S. H., Hindenlang, D. M.,
      Kosarych, Z., Mumma, R. O. and Jugovich, J. 1975d.
      "An Economical Synthesis of the Major Sex Attractant
      of the Oak Leaf Roller - cis-10-tetradecenyl acetate"
      Chemical Ecology 1(3):317-322.
20.   Hendry, L. B., Korzeniowski, S. H., Jugovich, J. and
      Hindenlang, D. M. 1975e. "A Macrocyclic Lactone
      Isolated from the Sexual Excitant Component of the Oak
      Leaf Roller Pheromone Complex." In preparation.
21.   Hendry, L. B., Kostelc, J. Snetsinger, R. and Girard, J.
      1975f. "A Sex Attractant in the Mushroom Fly,
      Lycoriella mali Fitch: Identification and Possible
      Origin." In preparation.
22.   Hendry, L. B., Wichmann, J. K., Hindenlang, D. M.,
      Mumma, R. O. and Anderson, M. E. 1975g. "Evidence
      of the Origin of Insect Sex Pheromones: Presence in
      Food Plants." *Science 188:*59-63.
23.   Hendry, L. B., Wichmann, J., Hindenlang, D. M., and
      Weaver, K. 1975h. "Plants: the Origin of Chemical
      Signals in Insect Host-Parasite Relationships."
      Submitted for publication.
24.   Herrebout, W. M. and VanderVeer, J. 1969. "Habitat
      selection in *Eucarcelia rutilla* Vill. (Diptera:
      Tachinidae)." *Z. Agnew. Entomol. 64:*55-61.
25.   Inoue, S. and Hamamura, Y. 1972. "The Biosynthesis of
      "Bombykol", Sex Pheromone of *Bombyx mori*." *Pro.
      Japan. Acad. 48*(5):323-26.
26.   Jacobson, M., *Insect Sex Pheromones*, Academic Press,
      New York and London, 1972.
27.   Jones, R. L., Lewis, W. J., Beroza, M., Bierl, B. A.,
      and Sparks, A. N. 1973. "Host-Seeking Stimulants
      (Kairomones) for the Egg Parasite, *Trichogramma
      evanescens*." *Environmental Biology 2:*593-601.
28.   Lewis, W. J., Sparks, A. N., Jones, R. L., Barres, D. J.
      1972. "Efficiency of *Cardiochiles nigriceps* as a
      Parasite of *Heliothis virescens* on Cotton."
      *Environmental Entomology 1:*468-471.

29. Kasang, G., Schneider, D. and Beroza, M. 1974. "Bio-
     synthesis of the sex pheromone disparlure by olefin-
     epoxide conversion" *Naturwissenschaften 61*(3):130-131.
30. Klun, J. A., Chapman, O. L., Mattes, K. C., Wojtkowski,
     D. W., Beroza, M. and Sonnet, P. E. 1973. "Insect
     Sex Pheromones: Minor Amount of Opposite Geometrical
     Isomer Critical to Attraction." *Science 181*:661-663.
31. Kochansky, J., Carde, R. T., Liebherr, J. and Roelofs,
     W. L. 1975. "Sex Pheromone of the European Corn
     Borer, *Ostrinia nubilalis* (Lepidoptera:Pyralidae),
     New York." *Journal of Chemical Ecology 1*(2):225-231.
32. Mitlin, N. and Hedin, P. A. 1974. "Biosynthesis of
     Grandlure, the Pheromone of the Boll Weevil, *Anthonomus
     grandis*, from acetate, mevalonate, and glucose.
     Journal of Insect Physiology 20:1825-1831.
33. Monteith, L. G. 1958. "Influence of Food Plant of Host
     on Attractiveness of the Host to Tachinid Parasite
     with Notes on Preimaginal Conditioning." *Canadian
     Entomologist 90*:478-482.
34. Monteith, L. G. 1967. "Responses by *Diprion hercyniae*
     (Hymenoptera:Diprionidae) to its Food Plant and their
     Influence on its relationship with its Parasite."
     Canadian Entomologist 99:682.
35. Read, D. P., Feeny, P. P., and Root, R. B. 1970.
     "Habitat Selection by Aphid Parasite *Diaeretiella rapae*
     (Hymenoptera:Baraconidae) and Hyperparasite *Charips
     brassicae* (Hymenoptera:Cynipidae)". *Canadian Entomolo-
     gist 102*:1567.
36. Roelofs, W., Hill, A. and Carde, R. 1975. "Sex Pheromone
     Components of the Redbanded Leafroller, *Argyrotoenia
     velutinana* (Lepidoptera:Tortricidae)". *Journal of
     Chemical Ecology 1*(1):83-89.
37. Streams, F. A., Shahjahan, M. and LeMasurier, H. G. 1968.
     "Influence of Plants on Parasitization of Tarnished
     Plant Bug by *Leiophron pallipes*." *Journal of Economic
     Entomology 61*:996.
38. Taylor, T. A. and Stern, V. M. 1971. "Host-preference
     Studies with Egg Parasite *Trichogramma semifumatum*
     (Hymenoptera:Trichogrammatidae)". *Annals of the
     Entomology Society of America 64*:1381.
39. Thorpe, W. H. and Caudle, H. B. 1938. "A Study of the
     Olfactory Responses of Insect Parasites to the Food
     Plant of Their Host." *Parasitology 30*:523-528.
40. Whittaker, R. H. and Feeny, P. P. 1971. "Allelochemics:
     chemical interactions between species." *Science 171*:757-
     770.

41. Wood, D. L., Silverstein, R. M. and Nakajima, M. ed.,
     Control of Insect Behavior by Natural Products;
     Academic Press, New York and London, 1970.

Chapter Eight

SECONDARY PLANT SUBSTANCES AS MATERIALS FOR CHEMICAL HIGH

QUALITY BREEDING IN HIGHER PLANTS

K. MOTHES

Deutsche Akademie der Naturforscher Leopoldiana
August-Bebel-Str. 50a
Halle (Saale)
German Democratic Republic

The term "secondary plant substances" was coined by Czapek [10]. In his general survey of the occurrence of pyridine and quinoline bases in the plant kingdom, he wrote "Perhaps the sporadic occurrence of these bases and the inconstancy of their appearance in closely related plants gives evidence for the concept that the formation of such substances is not a general process belonging to all cell plasma, but that it is of a more secondary character." K. Paech (1950) in his "Biochemie und Physiologie der sekundaren Pflanzenstoffe" mentioned the difficulties encountered in defining this group of substances and pointed out that a sufficient taxonomical separation of these substances is not given by their predominant occurrence in higher plants, since such plant substances sporadically may also appear in animals and in lower plants. This is accounted for by Luckner (1969) in his monograph which he entitled: "Der Sekundarstoffwechsel in Pflanze und Tier". Today we might add: "und den Prokaryoten" because a great many substances have been identified in actinomycetes and bacteria which we call antibiotics and which show clearly that "secondary substances" are a general characteristic of living organisms.

I feel it is conclusive and significant that these secondary substances do not have the character of essentiality, and that they appear to be without physiological importance in the organisms which produce them. Some have the capacity to undergo turnover (to be remetabolized), but even these do not represent typical reserve substances. It is time that

the misconception disappears from the literature that
alkaloids must be nitrogen reserves because they are
occasionally degraded during seed germination or in
the flowering stage. Even if such degradation might
proceed all the way to newly available nitrogen; e.g.,
of the ammonia character (which, however, is not generally
proven), the quantity of such a so-called nitrogen reserve
is without any importance in the total nitrogen balance
of a seedling. Similar relations are known for the
carbohydrate part of the glycosides.

The continuous synthesis and degradation of secondary
substances--even in a diurnal rhythm--has not been
definitely proven; the curves reported are not always
convincing. That metabolism of secondary substances can
lead to interesting products which attain physiological
importance, can certainly not be excluded. Concerning
this problem, numerous, mostly speculative hypotheses
have been advanced which, however, do not demonstrate
real proof. Such metabolism can especially be expected
for water-soluble secondary substances which are deposited
in vacuoles or other membrane bound cellular compartments.
Aging, changing life conditions, or the appearance of
abscisic acid can increase the permeability of these
membranes so that the deposited secondary substances can
come in contact with enzymes of the plasma from which they
are normally protected. But even a problem so easily
studied as the fate of methyl groups, generated by demethy-
lation from secondary substances, has not been investigated
sufficiently[40]. In a considerable number of papers,
however, it is taken for granted that the methyl group as
such is reincorporated into other sites of
metabolism. In particular I should like to point out
that radioactively labeled alkaloids fed to plants can
have a considerably different fate than is found for the
alkaloids formed in the plant itself. Wherever alkaloids
are deposited protected from the protoplasm (e.g. in
latex tubes) they have only a small chance of being
metabolized further. Exogenously applied alkaloids
first reach the cell plasma and are probably exposed to
the action of enzymes. This agrees with the observation
that alkaloids can be tolerated by plants in unusually
high concentrations, if they are formed in them slowly,
apparently deposited in special spaces,and therefore
rendered harmless. However, if such alkaloids are

applied in the same concentrations to an excised leaf through
the petiole or even to the blade via the surface, they are
toxic. We investigated this for the system, tobacco plant--
tobacco leaf cutting--and nicotine[47],[48].

Because of the enormous structural variability
certainly no uniform physiological response can be expected.
This does not even exist in the group of true alkaloids
which appear so uniform[53].

Interactions of secondary metabolism with primary
(fundamental) metabolism are manifold. We must take into
account that during the course of organic evolution,
secondary substances evolved from physiologically unimportant
to biologically important compounds. For example, the
typical growth hormones of the higher plants are likely
to have existed before the higher plants themselves. We
find auxins, gibberellins and cytokinins in lower plants,
in procaryotes and sometimes even in animals. Most
attempts to ascribe a function to these regulators have a
rather contrived appearance. I feel that in the evolution
of life, Nature has prepared a great abundance of substances without any intention and without the probability
of any of them fulfilling a task in further evolution.

The gibberellins, for example, are chemically
derivatives of gibbane. In a strict sense only those
diterpenes with the gibbane skeleton can be called
gibberellins. This function is studied by various biological tests which yield different reactions for different
gibberellins. There are substances which are not gibberellins in the chemical sense that, however, show gibberellin
reactions)e.g., helminthosporal). Many of the 50 or so
documented gibberellins are known to function as hormones.
Others show this capacity only because they are metabolized
to active hormones by the higher plant; and a third group,
although counted chemically among the gibberellins, usually
cannot stimulate physiological responses characteristic of
"the true gibberellins". As our ability to prepare and
detect compounds continues to improve, the number of
gibberellin-like substances will undoubtedly increase.
Such broad variety is characteristic for most of the
secondary substances. So the borderlines between the
essential and the non-essential often are not sharp. In
spite of all, the term "secondary substance" is very
useful[29].

Such cases at the borderline of primary and secondary metabolites also include many of the reserve substances. An instructive example is the fatty acids, e.g., those of Rape *Brassica napus* ssp. *oleifera*, whose seeds appear to be one of the main lipid producers of the northern temperate zone. The oil content of the seeds approaches nearly 50%. More than half of this oil is erucic acid ($C_{22}H_{42}O_2$) and the rest is represented by linoleic ($C_{18}H_{32}O_2$) and linolenic acid ($C_{18}H_{30}O_2$). Because of nutritional reasons erucic acid was not desired. Rape races were bred which were free of this acid, which means that erucic acid behaves in rape like a non-essential (secondary) substance. However, to increase the content of linoleic acid at the expense of linolenic acid turned out to be difficult since linolenic acid appears to take part in the formation of the lipoprotein membrane of the cell, especially of the chloroplasts. Races free of linolenic acid do not become green[54]. Thus certain fatty acids can be dispensible. But since they are readily used as energy sources, they have a great importance as reserve substances, which places them at the borderline between primary and secondary substances.

This is not true for structural compounds of the character of cellulose and the lignins. It is not by chance that in the first phase of the development of modern genetics that scientists studied the hereditary behavior of "simple" alternatively appearing characters. Mendel was occupied with the color and the structure of seed husks or the color of cotyledons of peas: E. Baur studied the flower colors of *Antirrhinum* races. Only because the flower color "red" is not essential for the existence of the species *Antirrhinum majus*, an amber-colored strain was found to exist which Baur could use for crossing with the red one. In the beginning geneticists studied for the most part the inheritance of molecular morphorlogical characters. Later macromorphological characters were examined e.g.,the crossing of *Antirrhinum* varieties with radial and symmetric flowers. Such relatively simple macromolecular characters exist in an almost unlimited abundance. I point to the form of the leaf margin: plain, crenate, serrate, pinnate.

Chemotaxonomy founded almost exclusively on secondary substances and often yields useful information complementing morphological and structural characterization. For example, the Menyanthaceae were once classified among the

Gentianaceae; presently they are treated as independent
families. It seems to me typical that in this separation
anatomical and embryological characters were very important.
However just as noteworthy is the citation in Engler's
Syllabus (12.ed., vol. II, p. 408) under Gentianaceae
"bitter substances" and under Menyanthaceae "bitter substances (but no Gentiopicrine)". Since chemical investigations are very expensive, they can be done only to a
limited extent and therefore they presently cannot be
highly significant as proof for classifying plants.
Certainly a morphologically unambiguous groups of plants
could be ascribed to the Gentianaceae even if it did not
contain gentiopicrine. This is true for the gentianaceaen
genus, *Voyria*, and perhaps also for some species of the
genus *Swertia*.

In lower plants the loss of a secondary chemical
character is much more frequent. Every producer of antibiotics knows how dangerous the appearance of such defective
mutants is and how they can hold their own and outgrow
their competitors in cultures. We cultivated thousands of
strains of the fungus *Claviceps purpurea* (ergot) on rye
and found repeatedly alkaloid-free strains. This character
was genetically fixed and remained constant even after
repeated inocculations. One also finds sclerotia of
Claviceps which are devoid of the violet-black pigment.
This character, too, is hereditarily constant. Such strains
were used to study the occurrence of mixed sclerotia by a
simultaneous inocculation of strains distinguished by violet
and whitish sclerotia[58,59]. These results also made it
probable that a sclerotium must not be an individual in the
strict sense, but that it may be derived from several
spores without sexual hybridization. A true hybridization,
however, is still possible which makes the chemical character of the sclerotia even more difficult to survey. The
fungus *Claviceps* can exist in nature without sclerotia
pigments and without alkaloids[24]. The manifold problems
of the genetics of the secondary substances of procaryotes
and fungi are treated in the volume of Vanek, Hostalek and
Cudlin (1973).

For a parasitic fungus the host plant is the most
important environmental factor. Since the host represents
a specific complex of nutritional conditions, a race of
Claviceps purpurea was investigated which had been bred

chemically constant to determine whether or not it keeps its character when grown on different hosts. Numerous experiments demonstrated that the peptide alkaloids are qualitatively unchanged, but that the water soluble alkaloids are subject to a certain influence by the host. This difference between the "soluble" and the "insoluble" ergoline alkaloids is perhaps a manifestation of the fact that the soluble ones are intermediates for the synthesis of the peptide alkaloids[39]. In later experiments we reached a greater genetic purity of our material and confirmed the earlier results.

The fact that secondary substances to a high degree are hereditarily predetermined and are not subjected to external influences, became the foundation of ergot cultivation. With respect to the peptide alkaloids it was shown, by examining an enormous amount of material from different geographical and ecological origins, that a great number of constant chemical races exist: races which contain numerous alkaloids, others which contain only one alkaloid and always the same one, and races which are free of alkaloids[58].

According to our experience mutations, in the sense of loss and change, of an alkaloidal character appear to be very rare in nature. Thus,if cautiously treated the same strain could be cultivated for years without any change in its alkaloidal character. However, alkaloidal mutations have been induced artificially[38]. In spite of the rarity of such mutations in nature, vast numbers of chemical races of the ergot fungus have originated during the course of evolution and still exist. The high variability is probably primarily a consequence of the fact that these substances are nontoxic to the producing fungus. But to conclude from this in general that ergot is effectively protected by the alkaloids against all of its enemies is surely going a step too far. Certainly one can easily prove the toxicity to many predators, but on the other hand it is stated that a considerable number of insects and other animals nibble at the sclerotia or eat them completely. If in isolated fields, e.g., in a mountain valley, ergot is regularly cultivated on rye, predators will increase from year to year and will probably destroy the sclerotia.

I have repeatedly pointed out that the greatness of nature cannot be understood on the basis of its protecting

mechanisms, the existence of which I do not doubt, but by
the huge number of its creations. Not only do we know more
than 5000 alkaloids but we also know parasites which are
often very host-specific.

    The biosynthesis of ergot alkaloids has not yet been
elucidated in all details[19],[62]. It is,however, clear that
the synthetic chain is very complicated. These metabolic
pathways appear to be as diverse and unusual as those
encountered in the formation of tryptophan-monoterpene
alkaloids in the Apocynaceae, Rubiaceae and Loganiaceae.
We may conceive that repeatedly in the course of evolution,
independently of one another and at different places of the
plant kingdom, enzymes have been formed which catalyze the
same chemical reaction. That, however, complete enzymatic
chains arise convergently is beyond our current concepts.
Yet this must be true, for we know that ergot alkaloids
are not only found in the genus *Claviceps* and in the
related ascomycete genera *Aspergillus* and *Penicillium*, but
also in the phycomycete genus *Rhizopus*, (literature cited
by Groger 1975). But the greatest surprise was the finding of ergoline alkaloids--even of the peptide type--in
several genera of the Convolvulaceae[30]. There are more
examples of such extreme cases of chemical convergence.
Thus the N-containing betalaines (betacyanidine and
betaxanthines) are widespread in some families of the
Centrospermae. Furthermore, these families are characterized by the lack of anthocyanidins.   Other families which
recently were ascribed to the Centrospermae are, however,
lacking the betalaines, e.g., the Caryophyllaceae. There
are also morphological reasons for the separation of these
families from the Centrospermae. Recently we were surprised
by reports that some pigments of *Amanita muscaria* do not
correspond to the structures indicated by Kogl, but are
actually betalaines[13],[14].

    Of course, it would be of great importance, if one
could show that the enzymes catalyzing specific reactions
in such complicated syntheses in extremely convergent cases
are related or identical, at least as far as their active
sites are concerned. Furthermore, the capacity to carry
out single specific reactions can be expected to be much
more common, even if secondary substances are not formed
by a given organism,because the specific substrates for
intermediate reaction steps are not available. For this

observation, indeed, remarkable examples exist. This we
know that the oxidations essential for the synthesis of
cortisone at C_{11} and C_{17} of the sterol system are possible
in higher plants and animals but are rarely encountered;
however, a number of bacteria, actinomycetes and fungi are
capable of such oxidations, if the corresponding substrates
are offered. These discoveries became the basis of a special
branch of modern fermentation industry[8].

Similar conditions can probably be expected to exist
at an even higher degree in tissue or callus cultures of
higher plants; this has already been documented for some
species. Veliky (1972) fed tryptophan to cell suspensions
of *Phaseolus vulgaris* and obtained harmane alkaloids which
have not yet been found in intact *Phaseolus* plants. Tsudo
et. al (1964) discovered that a great number of microlies
are capable of metabolizing thebaine to codeinone, codeine
or morphine, in a manner similar to the poppy plant.

Groger and Schmauder (1969) in our laboratory investigated the genus *Trametes* which had been studied by Tsuda.
They observed in some species a transformation from thebaine
to 14-β-hydroxycodeinone and/or to 14-β-hydroxycodeine,
depending on the concentration of thebaine and the composition of the nutrient solution.

Grutzmann and Schroter (1966) found that not only
callus cultures of *Papaver somniferum* but also those of
Nicotiana alata are capable of such transformations.
Most likely the species of the genus *Nicotiana,* had
neither at the time nor at any time during their phylogeny,
the opportunity to adapt to such an unusual substrate as
thebaine. Thebaine, which is a rare alkaloid, codeine, and
morphine (Figure 1) have been found together so far only
in two *Papaver* species (*P. setigerum* and *P. somniferum*).
With all its strains the latter species probably must be
considered as a vultivated plant which seems to be very
old and whose reduction to the wild form has not yet been
achieved satisfactorily. *Papaver somniferum*, however, is
not uniform either morphologically, in its flower color
or in its alkaloid composition. Remarkable and apparently
characteristic differences, especially in the composition
of opium, have long been well known for different geographic

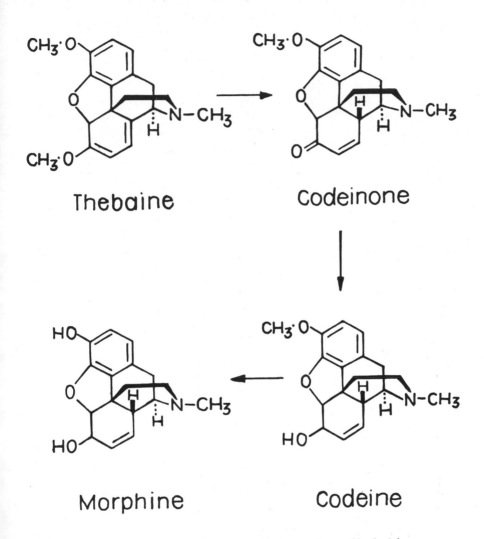

Figure 1. Structures of some of the opium alkaloids.

origins. Possibly these cultivated races correspond to
geographical races of the wild forms from which they were
selected[33]. Thus we are aware from literature of the great
genetic variability of opium poppy. Furthermore, we
analyzed chromatographically huge numbers of this poppy
from widely different origins. These results may be summarized in the following way.

The biosynthetic pathway leads from thebaine via
codeine to morphine. The opium poppy in its ontogenetic
development appears to form thebaine and codeine which in
turn are slowly transformed into morphine as the capsule
ripens. Thus the "morphine poppy" is found to be a
"codeine poppy" in its youth. Besides this ontogenetic
variability there apparently exists an enormous genetically
fixed variability in the spectrum of the secondary alkaloids.
The various poppy races contain 20-30 different benzylisoquinoline derivatives. The secondary alkaloids such as
narcotine can occur in much higher quantities than the
morphinanes:thebaine, codeine and morphine combined. Thus,
such a poppy should really be called a narcotine poppy.
Economically the main compound was morphine; therefore, all
these races were called morphine poppies.

Of course, it is conceivable that the selection of
poppy by man may have been directed very early towards the
sleep-making principle so that the morphine-free opium
poppy may have been wiped out even though man of that time
did not understand the chemical differences involved. Our
own work was at first directed towards increasing the
morphine yield in the cultivation of opium poppy[5]. However
the presence of morphine was only of secondary importance
in poppy cultivation. The primary concern was the important
production of oil from the seeds. Since this seed production is strongly influenced by climatic factors and by
parasites (e.g., *Helminthosporium*), the cultivation of
poppy for oil became economically uninteresting. Simultaneously morphine lost its immediate importance for medicine as
time analgesics were developed. Furthermore, morphine as a
drug of abuse became a great problem for all the civilized
world. However morphine still remained important as the
starting material for the manufacture of the majority of
the medicinally used codeine. As long as any interest in
morphine existed, its abuse could not be excluded. There
were two possibilities to obtaining codeine while avoiding

morphine. One could hope to find or to breed a codeine
poppy. But among the races of P. *somniferum* we did not
find a plant that forms codeine but which does not transform it to morphine. It could, however, not be excluded
that such a poppy may exist.

Based on our experiences with *Claviceps purpurea* we
hoped to find a codeine or at least a thebaine poppy. In
1958 we wrote (Kleinschmidt and Mothes) "Thus one will be
able to have a thebaine poppy or a codeine poppy as one has
ergot races which contain exclusively or almost exclusively
the one or the other alkaloid." The progress of our work
was decisively influenced by a publication of Gadamer (1913)
who thought that the red poppy, declared to be *Papaver
orientale* contained thebaine during the main time of its
vegetative development, but, in the autumn contained
isothebaine (Figure 2). Gadamer considered a transformation
from thebaine to isothebaine to be possible. But at that
time a detailed knowledge of the molecular structures of
these alkaloids was still lacking. Dawson and James (1956)
showed that Gadamer's results were probably caused by
harvests of different *Papaver orientale*. More recently
Stermitz et al. (1961), Battersby (1965), and Battersby
et al. (1965) established that a transformation from
thebaine to isothebaine and vice versa does not take place
(Figure 2). The synthesis of thebaine starts from
reticuline, the synthesis of isothebaine from orientaline.
But the fact remained that Gadamer must have had a thebaine
poppy in his hands. This poppy was to be found. Among
thousands of specimens we found only one poppy with big
red flowers, and this essentially contains only thebaine[49].
Bohm (1967) then demonstrated that the juvenile stages of
our variety contained several more alkaloids, among them
alpinigenine. These alkaloids disappear during flower
formation so that, when seeds are ripe, practically only
thebaine is found in the plant. Further cytological,
morphological, chemical and biochemical investigations
and the continuation of quality breeding and the breeding
for alkaloid yield per area unit were done[28]. It could
be shown beyond doubt that the new poppy with big dark
red flowers does not belong to P. *orientale* but to the
species P. *bracteatum* which had already been described
Lindley.

Our variety received the name P. *bracteatum* Halle III.

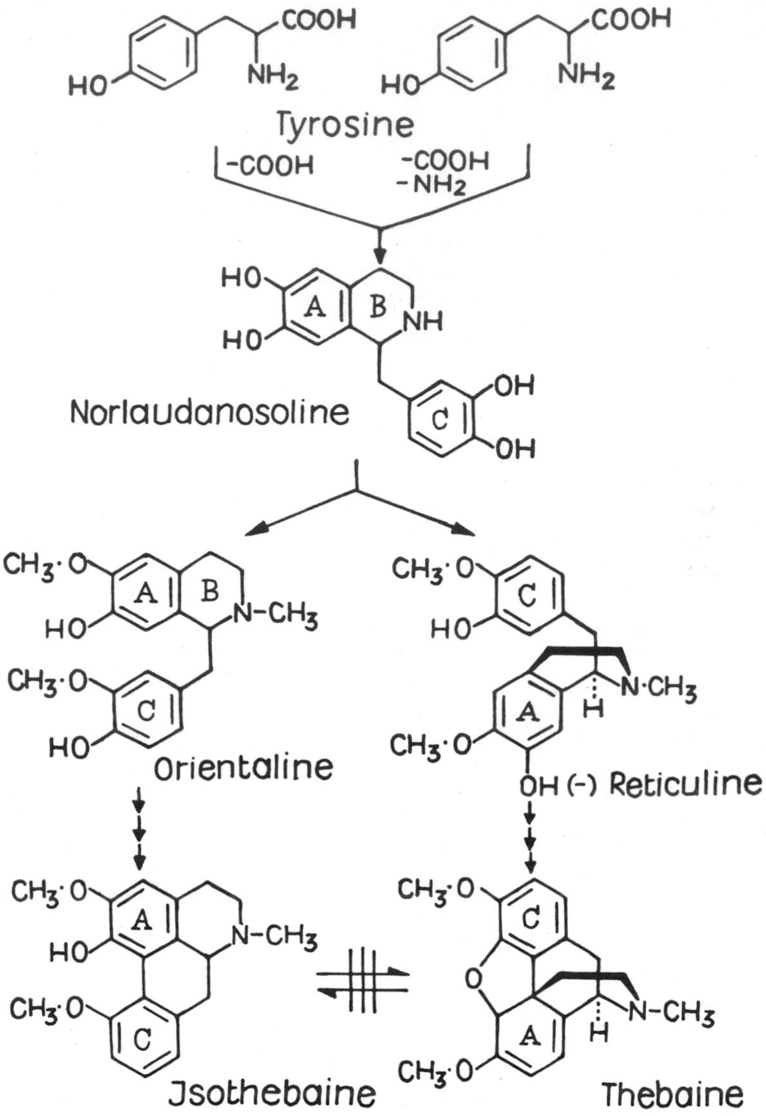

Figure 2.   Aspects of the biosynthesis of thebaine
            and isothebaine.

It is characterized by an almost 100% thebaine content.
It has perennial character and thick storage roots which
are rich in alkaloids. This allowed a new technology for
alkaloid production. The harvest of these turnip-like roots
and also their storage is quite independent of climatic
conditions. In the meantime, investigations on *P. bracteatum*,
have been started in various places. Confusions with
P. orientale, or *P. pseudoorientale* were, however, not
always avoided (concerning taxonomy cf.Gunther and Bohm
1968; Goldblatt 1974; Gunther 1975). Our assumption that
this thebaine poppy originates from the mountains of Iran
and the northeastern Caucasus has been confirmed by
Lalezari (1974).

Since the problem of a non-narcotic source of codeine
is now a world problem, the division of Narcotic Drugs of
the United Nations has established a special working group
for *Papaver bracteatum* with its residence in Geneva (head:
Dr. Braenden). The occurrence of thebaine as well as of
codeine and morphine has been described for quite a number
of Papaveraceae. According to our experiences it would be
important to verify these statements.

The occurrence of morphine and codeine associated with
thebaine seems to be certain only for *P. somniferum* and
P. setigerum. That this synthetic chain can be interrupted
at the thebaine stage is clearly shown by our example. That
codeine as an intermediate of morphine formation is not
completely transformed even in ripe *P. somniferum* is seen
from numerous analyses of *P. somniferum* of different origins
and different developmental stages. The accessory alkaloids
of the young *P. bracteatum* plants are neither codeine nor
morphine. Whether they are intermediates of thebaine formation remains to be determined.

A special position is occupied by alpinigenine which
in the variants of our Halle III can be present even in the
ripe plant. Bohm (1971) and Ronsch (1972) have reported
on its biosynthesis. So far *P. bracteatum* remains an
exceptional case which, however, finds a parallel in the
behavior of some strains of *Claviceps purpurea*. This
thebaine poppy can only be fully appreciated if one
considers that the genus *Papaver* possesses about 130
alkaloids of the benzylisoquinoline type and that a few
varieties of *P. somniferum* contain more than 30 alkaloids;

the biosynthetic pathways of some of which separate very
early, after norlaudanosoline.

Characteristic for *Papaver* and the Papaveraceae as a
whole is the occurrence of benzisoquinolines. An enormous
qualitative and quantitative variability is attained in
this way, which is typical for the secondary plant substances. In animals these substances produce very
different and sometimes opposing pharmacological effects.
Therefore, it is unlikely that they take part in comparable
physiological reactions in the plant. Furthermore they are
absent in various chemical races and occur in others in
high concentrations. It can therefore be excluded that
these substances have a physiological importance. The
breeder can use this variability, as it occurs in nature,
by selection and hybridization to reach certain goals.
But he will certainly also obtain the same variations by
induced mutations.

To allow for regulation of the formation of secondary
substances not only are specific enzymes, precursors and
other substances required, but in many cases special cellular
compartments are needed. This was shown for the synthesis
and accumulation of dihydroxyphenylalanine (Dopa) in
Euphorbia lathyrus [34]. Neumann and Muller (1974) observed
with *Macleaya* (Papaveraceae) that the appearance of alkaloid
coincides with the formation of alkaloid cells (latex tubes).
This will be true in general also for the ontogeny of the
intact plant. Phylogenetically this means a co-evolution
of alkaloid formation and of the development of spacies of
proper chemical and anatomical qualities for their
deposition.

I cannot discuss here the indoubtedly great importance
of secondary substances in the social structure of living
nature and their ecological importance. Their role in
plant life must be considered to be very complicated
(Fraenkel 1959; Swain 1974). It covers not only the
attraction of pollinating insects and other animals by
colored and odoriferous substances, but also the attraction
of animals which function in fruit and seed dispersal. It
also includes the function of repellents against animal
enemies. On the other hand the effect of excluding most
parasites produces ecological niches into which other
animals can enter, to which those repellents are attractants.

The interaction of higher plants and insects is magnificent
in those cases where insects and also other animals take up
specific plant substances and excrete them sometimes unchanged through special glands. They use them as repellents
against other animals or as intra-specific sex attractants
(pheromones)[15,21,37,52]. These ecological relations are
very numerous and they caused some authors to consider
secondary metabolites to be an essential attribute for
the existence of plants and for the existence of social
communities. I do not want to enlarge the great number
of speculations uttered to understand the "sense" and the
evolution of the secondary substances. There are many
examples where the loss of these substances does not appear
to reduce the vitality of plants living under present day
natural conditions.

   The secondary substances are a reflection of the enormous chemical capability of plants, especially of the higher
plants. One belittles the greatness of living nature
one considers this capability under the aspect of simple
usefulness.

   I already mentioned in the beginning that the lower
plants and procaryotes have a much greater chemical potency
than we had anticipated until recently. The rapidly
developing field of chemical ecology has demonstrated a greater
than expected synthetic capacity even for the animals. The
poisons in fish and molluscs and the repellents of insects
are a vast field[18,56].

   An interesting finding of Eisner et al. (1971) increases
the understanding of the phenomenon of "storage excretion".
In the defense secretion of the grasshopper *Romalea
microptera* in addition to phenols, diphenols, quinones,
terpenes and hydroaromatic compounds they also identified
2,5-dichlorophenol. They believe that this substance
originates from a herbicide which has been taken up with
the food. Indeed we do not know any analogous substance
from the animal or plant kingdom. The inner excretion into
the defense system shows that substances with certain
physical and chemical properties can be subjected to this
special excretion which, on the other hand, underlines the
excretional character of secondary substances.

   Very interesting also are such findings as the

synthesis of norlaudanosoline from dopamine by rat brain-
stem homogenates[11] and the synthesis of 1,2,3,4-tetraiso-
quinolines from epinephrine by adrenal tissue[9]. These
reactions are the same in principle as the ones which take
place in plants and which in plants lead to the synthesis of whole classes of alkaloids. It is not impossible
that ubiquitous enzymes are involved here and that perhaps
spontaneous reactions take place in the formation of the
base.

Some scientists are inclined to think that the genetic
prerequisites for the synthesis of all the different secondary substances of plants exist in all plants. The formation of specific secondary substances in closely defined
taxa would then be only a consequence of a specific gene
expression. Undoubtedly the DNA content of the nuclei of
higher plants would be sufficient to represent the necessary
number of genes for this. However, I do not want you to
think that I advocate such an extreme proposition.

In general the   secondary substances are a play of
nature which only secondarily have acquired a particular
significance. Since they are often toxic even for the
organism that forms them, either a rapid excretion (as in
animals) or a good internal protection in special compartments (storage excretion) is necessary[57,66].

Whenever a molecular structure, by substitutions or
cleavage of ring systems and renewed ring closures, allows
for considerable modification and variation of a secondary
substance,nature will use this possibility. The great
number of benzylisoquinoline alkaloids, whose carbon
skeletons are built up from only two molecules of tyrosine
and some methyl groups, shows this play of nature with
only a few building blocks. In the tryptamine-monoterpene
alkaloids, of which 800 may exist, this use of structural
possibilities is carried to the extreme selection and
induced mutations are the means given to the investigator
to repeat what in the course of evolution may often have
happened. Not all such aimless creations remained without
negative consequences for the vitality of the wild form.
Thus, in most varieties of his cultivated plants man
possessed variants of a fundamental type which under natural
conditions and without cultivation by man could not exist.
We should not only, with the modern knowledge of genetics,

evaluate the secondary plant substances, but also, in the
interest of medicine, we should be occupied with problems
which nature may have solved several times before, but
without leaving any evidence of the solution. Just as a
child with only a few building blocks builds up bold and
always new structures,and as much of it breaks down before
the marvellous architecture has been shown to the parents,
so nature is building. Nature has greater imagination than
we have. But when we have understood how she erects her
buildings, we will be able to play even more extensively
than she does.

Chemical activities of microbes and of tissue or
callus cultures open a wide field for the construction of
new molecules and for the easier accomplishment of known
syntheses.

But all of this is only valid for the secondary substances.
They evidently show us the chemical capacity of living
matter. The ability to undergo variation, of course, is
also an attribute of the primary substances. However,
a mutated proteinogenic amino acid will be useless for
protein synthesis. There may be a continuous change in the
sequence of the amino acids in an essential protein, but
only in rare cases will such a variant be able to live and
in even rarer cases will it survive under ecological conditions. Sequential analysis of defined plant proteins (e.g.,
(cytochrome C, ferredoxin etc.) point out the long road of
protein changes[7]. For phylogenetic discussions they are
likely to be more useful than the secondary substances,
although the latter by virtue of their micromolecular
character and their mostly harmless diversity present good
possibilities for chemotaxonomical speculation.

I want to conclude with a remark on the breeding of a
cultured plant which is of highest importance in nutritional
physiology and choose the grain as an example. Proteins
of different quality are deposited in different sites
in the embryo and in the endosperm. Judged from
the need of man, the proteins of the embryo are more
valuable. To improve them qualitatively by mutation seems
almost impossible. One may, however, be able to increase
their amount, if one succeeds in shifting the ratio of
embryo to endosperm. To change the chemical quality of
the reserve protein of the endosperm by selection or

mutation seems quite possible. These proteins are similar
to secondary substances. They have no physiological
function, but represent reserves for amino acids. Thus I
am closing a circle which I started to draw with the
discussion of the fatty acids in rape seed.

ACKNOWLEDGEMENTS

    I thank Dr. Lisabeth Engelbrecht for translating
this manuscript.

REFERENCES

1. Battersby, A. R. 1965. 3. Intern. Alkaloidsymp. Abh.
      D. Akad. Wiss., Kl. Chemie, Nr. 3 (1966) 295-307.
2. Battersby, A.R. 1965. *In* "Beitr. zur Biochemie und
      Physiologie von Naturstoffen." Verlag Fischer,
      Jena, 81-91.
3. Battersby, A.R., R.T. Brown, J.H. Clementz and G.G.
      Iverach. 1965. *Chem. Commun*. No. 11; 230-232.
4. Bohm, H. 1967. *Planta Medica 15*:215-220.
5. Bohm, H. 1970. *Planta Medica 18*:93-109.
6. Bohm, H. 1971. *Bioch. Physiol. Pflanz. 162*:474-477.
7. Boulter, D. 1972. *Progr. in Phytochem. 3*:199-229.
8. Charney, W. and H. Herzog. 1967. *Microbiol Transform.
      of Steroids* (Academic Press) p. 728.
9. Cohen, G. and M. Collins. 1970. *Science 167*:1749.
10. Czapek, F. 1922-1925. *Biochemie d. Pflanzen,* 3.Bde.,
      III Bd. p. 852 (Jena, Fischer-Verlag).
11. Davis, V. E. and M. J. Walsh. 1970. *Science 167*:
      1005.
12. Dawson, R. F. and C. James. 1956. *Lloydia 19*:59-64.
13. Dopp, H. 1974. Umschau *34*:517.
14. Dopp, H. and H. Musso. 1973. *Chem. Ber. 106*:3473-3482.
15. Ehrlich, P. R. and P. H. Raven. 1964. *Evolution 18*:
      586-608.
16. Ehrlich, P. R. 1967. *Scientif. Amer. 216*:105-113.
17. Eisner, Th., L. B. Hendry, D. B. Peakall and J. Meinwald.
      1974. *Science 172*:277-278.
18. Eisner, Th., and J. Meinwald. 1966. *Science 153*:
      1341-1350.
19. Floss, H. G. 1976. *Tetrahedron Reports* (in press).
20. Fraenkel, G. 1959. *Science 125*:1466.
21. Fraenkel, G. 1969. *Entom. exp. a. appl. 12*:473-486.
22. Gadamer, J. 1973. *Zeitschr. angew. Chem. 26*:625.
23. Goldblatt, P. 1974. *Ann. Miss. Bot. Gard. 61*:264-296.
24. Groger, D. 1975. *Planta Medica 28*:269-288.
25. Groger, D. and H. P. Schmauder. 1969. *Experientia 25*:
      95-96.
26. Grutzmann, K. D. and H. B. Schroter. 1966. *Abh. D.
      Akad. Wiss., Kl. Chemie 3*:347.
27. Gunther, K. F. 1975. *Flora* (Jena) *164*:393-436.
28. Gunther, K. F. and H. Bohm. 1968. *Osterr. Bot. Z.
      115*:1-5.

29. Hegnauer, R. 1975. *In* "Crop Genetic Resources for Today and Tomorrow". Cambridge Univ. Press, 249-265.

30. Hofmann, A. and H. Tscherter. 1960. *Experientia 16:* 414.

31. Kleinschmidt, G. and K. Mothes. 1958. *Pharmazie 13:* 357-360.

32. Kobel, H. and J. J. Sanglier. 1973. *In* "Genetics of Industrial Microorganisms". **Akad.** Verlag. Prague, p. 421.

33. Kuhn, L. and S. Pfeifer. 1963. *Pharmazie 18:*819.

34. Liss, I. 1961. *Flora* (Jena) *151:*351-367.

35. Luckner, M. 1969. "D. Sekundarstoffwechsel in Pflanze und Tier". Jena, Fischer-Verlag, p. 360.

36. Luckner, M. 1972. "Secondary Metabolism in Plants and Animals". Chapmann and Hall, London, p. 404.

37. Karlson, P. and D. Schneider. 1973. D. Naturwiss. *60:*113-121.

38. Marnati, M. P., A. Minghetti and C. Spalla. 1975. *In* "Conf. on Med. Plants" (Marianske Lazne) p. 68.

39. Meinicke, R. 1956. *Flora* (Jena) *151:*351-367.

40. Miller, R.J., C. Jolles and H. Rapoport. 1973. *Phytochem. 12:*597-603.

41. Mothes, K. 1965. *D. Naturwiss. 52:*571-585.

42. Mothes, K. 1966. *D. Naturwiss. 53:*307-323.

43. Mothes, K. 1972. *Abh. Sachs. Akad. Wiss. 52(1):*3-29.

44. Mothes, K. 1973. *Ber. Osterr. Akad. Wiss. Abt. I 181:* 1-37.

45. Mothes, K. 1975. *Ber. Akad. Wiss. DDR* 9/2 IV, 17-24.

46. Mothes, K. 1975. *Wissenschaft u. Fortschr. 25:*300-304.

47. Mothes, K. and L. Engelbrecht. 1956. *Flora* (Jena) *143:* 428-472.

48. Mothes, K. and A. Romeike. 1954. *Flora* (Jena) *142:* 109-131.

49. Neubauer, D. and K. Mothes. 1963. *Planta Medica 11:* 387-391.

50. Neumann, D. and E. Muller. 1974. *Biochem. Phys. Planzen 165:*271-282.

51. Paech, K. 1950. "Biochemie u. Physiol. d. sekund. Pflanzenstoffe. Springer-Verlag, Berlin. p. 268.

52. Reichstein, T. 1967. *Naturw. Rundschau. 20:*499.

53. Robinson, Tr. 1974. *Science 184:*430-435.

54. Robbelen, G. 1974. *Angew. Bot.* (D. Botaniker Tagung, Wurzburg).

55. Ronsch, H. 1972. *Eur. J. Bioch. 28:*123-126.

56. Schildknecht, H. 1970. *Angew. Chemie 82*:17-25.
57. Schnepf, E. 1973. *In* "Grundl. d. Cytologie". Fischer-
    Verlag, Jena 461-478.
58. Silber, A. and W. Bischoff. 1955. *Arch. d. Pharmaz.*
    258/260:124-195.
59. Silber, A.,K. Mothes and D. Groger. 1955. *D.*
    Kulturpflanze 3:90-104.
60. Stermitz, F. H. and H. Rapoport. 1961. *J. Amer. Chem.*
    Soc. 83: 4045-50.
61. Swain, T. 1974. *In* "Comprehensive Biochem.". *29A:*
    125-298.
62. Thomas, R. and R. A. Basset. 1972. *Progr. in Phytochem.*
    3:47-111.
63. Tsuda, K. 1964. *Int. Ass. Microbiol. Sympos. on*
    Microbiol. Nr. 6, Tokyo.
64. Vanek, Z., Z. Hostalek and J. Cudlin (edit.). 1973.
    "Genetics of Industrial Microorganisms", Praha,
    Akademia, p. 511.
65. Veliky, I. A. 1972. *Phytochem. 11*:1405.
66. Wigglesworth, V. B. 1972. "Principles of Insect
    Physiology" (2. ed.), Chapman, London, p. 579.
67. Witkop, B. 1971. *Experientia 27*:1121-1148.

414

DATE DUE